out

h

COMPARATIVE AVIAN NUTRITION

Comparative Avian Nutrition

Kirk C. Klasing

Department of Avian Sciences
College of Agricultural and Environmental Sciences
University of California
Davis
California, USA

CAB INTERNATIONAL

CABI *Publishing* is a division of CAB *International*

CABI Publishing
CAB International
Wallingford
Oxon OX10 8DE
UK

Tel: +44 (0)1491 832111
Fax: +44 (0)1491 833508
Email: cabi@cabi.org
Web site: http://www.cabi.org

CABI Publishing
10 E. 40th Street
Suite 3203
New York, NY 10016
USA

Tel: +1 212 481 7018
Fax: +1 212 686 7993
Email: cabi-nao@cabi.org

A catalogue record for this book is available from the British Library, London, UK.

Library of Congress Cataloging-in-Publication Data
Klasing, Kirk C.
    Comparative avian nutrition / Kirk C. Klasing.
        p. cm.
    Includes index.
    ISBN 0-85199-219-6 (alk. paper)
        1. Birds–Nutrition.   2. Captive wild birds–Nutrition.
    3. Poultry–Nutrition. I. Title.
    QL698.K58   1998
    573.3'18-dc21

97-33666
CIP

ISBN 0 85199 219 6

First printed 1998
Reprinted 2000

Typeset in Souvenir by Solidus (Bristol) Limited
Printed and bound in the UK at the University Press, Cambridge

CONTENTS

PREFACE

Birds are fascinating because of their conspicuous and natural beauty, exceptional variety, and diverse social behaviors. In captivity, birds have utility as a food source, as pets, and as a genetic reservoir for diminished wild populations. Wild birds have evolved diverse feeding, migratory, and reproductive behaviors that have enabled them to occupy an exceptional number of ecological niches. These adaptations make Aves one of the most exciting targets in many areas of investigation (Konishi *et al.*, 1989). Observations on the adaptations of the beak of the finch for nutritional purposes were a primary contributor to Charles Darwin's seminal theories on evolution. More recently, chickens and pigeons have been the target of research that led to an impressive number of Nobel prizes for discoveries in developmental biology, physiology, biochemistry, behavior, and nutrition.

Nutrition is the science that describes the process of providing the cells of the body with simple and complex molecules obtained from the external environment at the appropriate rates and ratios to optimize health, growth, and reproduction. Avian nutrition integrates the disciplines of anatomy, biochemistry, physiology, behavior, and ecology into a unifying concept of the interactions between the bird and its food supply. Most importantly, nutrition is a quantitative science, requiring not only an accurate description of the molecular details of digestion, metabolism, and excretion but also an accurate estimation of their rates. Accurate estimations of the daily nutrient budgets for purposes as diverse as growth, migration, and reproduction are a difficult but important area of investigation.

A thorough understanding of avian nutrition is central to securing the survival and productivity of all birds, whether free-ranging or captive. There are more than 9000 species of birds (more than twice as many as mammals), representing an incredible diversity and filling a large number of the possible niches within the food web. In the wild, birds are responsible for securing their own nutrition, but it is rare that this process is not touched or shaped by the activities and interest of humans. In the captive situation, providing an appropriate diet is critical for successful management and reproduction. Most of our scientific understanding of avian nutrition comes from four somewhat independent disciplines: nutritional ecology, zoo biology, poultry production, and basic biomedical sciences. This book is an attempt to integrate these diverse areas of study.

Information on the nutritional ecology of a wide variety of bird species gives us an appreciation for the behavioral strategies of feeding and for the primary foods consumed. In the USA, 'economic ornithology' was initiated by the Department of Agriculture in 1885 to study the interrelation of birds and agriculture by investigating food, habits, and migration of birds in relation to economically important insects and plants. More recently, surveys on food consumption of birds have been augmented by research in physiological ecology, giving us an appreciation for the energetic costs of a variety of processes and the crucial role of nutrition in survival and evolution.

Modern zoos and private institutions have adopted the propagation and conservation of rare species as part of their mission. Emphasis on the health and reproduction of captive birds has prompted many zoos to include nutritionists on their staff to conduct research toward the understanding of nutritional needs of each species so that palatable diets with optimal nutrient composition can be provided. Although the small number of animals available at many zoos precludes elegant short-term experiments, an invaluable life-cycle perspective has been gained for an extensive number of avian species (Dierenfeld, 1996).

The chicken, turkey, and domestic duck are economically important as a source of human food and consequently have been the subject of thousands of studies aimed at low-cost production of wholesome meat and eggs. This research has generated quantitatively precise information on the requirements of poultry and the nutritional value of foodstuffs. Estimates of the nutrient needs of all other avian species have been based on knowledge generated with poultry, but these inferences generally await confirmation by investigative research.

Detailed physiological, metabolic, biochemical, and biomedical studies have been done on a relatively small number of avian species. In particular, two Galliformes, the domestic chicken and the Japanese Quail, have been the subject of intensive basic research and are the foundation of our understanding of many of the fundamental underpinnings of avian nutrition.

Dynamics of Avian Nutrition

Nutrition encompasses the procurement, digestion, absorption, and metabolism of food items. The quantity of each specific nutrient required by birds is exceptionally diverse. For example, a young Japanese Quail chick requires a diet with 24% protein, but only 0.0000003% of vitamin B_{12}. Furthermore, the nutritional needs of birds change during their life cycle, being highest at hatching, decreasing continuously throughout the growth period, and then increasing again in females during the breeding season. Within any given life stage there are also profound differences due to the environment in which the bird is located. This is well illustrated by the large difference in energy requirements between an active bird in the wild and a more sedentary bird kept under thermoneutral conditions in captivity. This profound difference is reflected in both the quantity and the composition of food consumed in an optimal diet.

Proper nutrition comes at a substantial cost of time and energy to the individual bird in the wild and at considerable monetary cost to humans keeping or attracting birds. In the wild, food procurement is often the predominant activity in which a bird engages. For example, in the winter many nonmigrating birds spend in excess of 80% of their daylight time in search of food. Interest in attracting birds to backyards and other habitats for the purpose of observation or to increase the range of rare birds has grown rapidly and is commercially important. According to a 1991 survey conducted and analyzed by the US Fish and Wildlife Service, more than 63 million people spent more than $1.5 billion on commercially packaged wild-bird feed in the USA. In captivity, feed is the predominant cost of keeping a bird during its lifetime, almost always exceeding housing costs or veterinary care. In commercial poultry industries, feed accounts for more than two-thirds of the cost of meat and egg production. In the USA alone, the poultry industry spent about $15 billion dollars on feed in 1996.

Use of This Text

It is hoped that this text will be useful to advanced undergraduates and graduate students, and as a general reference for the ornithologist, veterinarian, field biologist, practicing animal nutritionist, and aviculturalist. Several texts are available that cover specific aspects of avian nutrition, such as those relevant to poultry (Scott *et al.*, 1982; Leeson and Summers, 1991; Scott and Dean, 1991; Larbier and Leclercq, 1992) or birds in context with other wildlife (Robbins, 1993; Chivers and Langer, 1994; Stevens and Hume, 1996). This text is intended to cover the principles of nutrition as applied to birds in general, whether free-living or captive. More importantly, it is hoped that an appreciation is given to the diverse adaptations that birds have adopted to accommodate very different foodstuffs. The first four chapters cover dietary patterns and the digestive anatomy and physiology needed to process diverse foods in order to assimilate their nutrients. The next seven chapters consider metabolism and requirement of each nutrient.

In keeping with the broad audience for whom this book is intended, specific species are referred to by their common English names, as listed by Clements (1991). Taxonomic names of particular species (those capitalized) referred to in the text may be found in the Appendix. References included in this text are intended to provide additional background and depth and are grouped to improve the readability of the text.

Acknowledgments

Students are the real teachers and I would like to thank my graduate and undergraduate students for their help. Sometimes the help comes from reviewing chapters and surveying the literature. More important are those 'dumb' questions that I can never seem to answer, but sure get me thinking. I appreciate the invaluable advice of a number of colleagues in reviewing chapters or listening to ideas, especially Carlos Basque, Chris Calvert, Doug Conklin, Bob Elkin, Howard Kratzer, Matthias Stark, Pran Vohra, Rosemary Walzem, and Wes Weathers.

This book is dedicated to my wife, Susan, and daughters, Samantha and Jillian. Their love and patience are the most important nutrients of all.

References

Chivers, D.J. and Langer, P. (1994) *The Digestive System in Mammals: Food, Form and Function.* Cambridge University Press, Cambridge.

Clements, J.F. (1991) *Birds of the World: A Check List.* Ibis Publishing, Vista, California.

Dierenfeld, E.S. (1996) Nutritional wisdom: adding the science to the art. *Zoo Biology* 15, 447–448.

Konishi, M., Emlen, S.T., Ricklefs, R.E. and Wingfield, J.C. (1989) Contributions of bird studies to biology. *Science* 246, 465–472.

Larbier, M. and Leclercq, B. (1992) *Nutrition and Feeding of Poultry.* Nottingham University Press, Loughborough.

Leeson, S. and Summers, J.D. (1991) *Commercial Poultry Nutrition.* University Books, Guelph.

Robbins, C.T. (1993) *Wildlife Feeding and Nutrition,* 2nd edn. Academic Press, San Diego.

Scott, M.L. and Dean, W.F. (1991) *Nutrition and Management of Ducks.* M.L. Scott of Ithaca, Ithaca, New York.

Scott, M.L., Nesheim, M.C. and Young, R.J. (1982) *Nutrition of the Chicken,* 3rd edn. M.L. Scott & Associates, Ithaca, New York.

Stevens, C.E. and Hume, I.D. (1995) *Comparative Physiology of the Vertebrate Digestive System.* Cambridge University Press, Cambridge.

CHAPTER 1
Dietary Patterns

The dietary preferences, gastrointestinal tract morphology, and metabolic capabilities of birds have been intimately intertwined during their evolution. The capacity for flight and migration has given birds a competitive advantage over terrestrial animals in foraging for high quantities of nutritious foods, but flight has also placed physical limitations on digestive anatomy. In general, few avian species have adapted to poorly nutritious foods, such as leaves and grasses, which are commonly consumed by mammalian herbivores. Many factors influence the exact composition of a bird's diet, but certainly the morphological features for food procurement, the anatomy and physiology of the gastrointestinal tract, the metabolic capacity, and the life-cycle nutrient requirements set the broad limits of potential foods for consumption.

Dietary patterns can be classified according to a variety of different schemes. Nomenclature for dietary patterns is extremely variable throughout the literature, with ecologists, evolutionary biologists, and nutritionists frequently using the same terms differently. Some standardization is much needed and Langer and Chivers (1994) have proposed a unifying scheme, which is based on feeding strategies of mammals and has sufficient resolution to be useful to nutritionists. These food-consumption categories are applicable to birds, with a few simplifications to account for the absence of birds in niches occupied by ruminants and other large mammals that nonselectively consume high-roughage foods (Table 1.1).

Classification of birds by the trophic level in which they primarily feed (e.g., herbivore, insectivore) is a useful way to identify morphological convergence due to similar nutritional and ecological selection pressures (Lein, 1972; De Graaf *et al.*, 1985). For this reason, trophic-level classification permits simplifications and generalizations needed for prediction of nutritional needs of diverse and unstudied species. Classification is easy when essentially all of the food consumed by a bird is of the same general type and the bird can easily be considered an oligivore. More often, birds are omnivores (polyvore) and select among several food categories, or trophic levels, at any given time. The term omnivore is most appropriate for species that consume both animal and plant foods. In the wild, their exact choice is determined by a combination of factors, including: seasonal availability of food, foraging efficiency, changing nutrient requirements, palatability, and predator patterns. Many omnivorous species fit into one of the oligivore categories seasonally or during some phase of their life history.

Birds consuming foods almost exclusively of animal origin are known as faunivores. Birds that prey upon vertebrates are divided into two categories, the

Table 1.1. Classification of food consumption patterns of birds.[*]

Category	Examples
Generalist feeder: omnivore (polyphage)	Tinamous, bustards, quail, pheasants, cranes, crows
Specialist feeder: oligivore (oligophage)	
Animal matter: faunivore (zoophage)	Penguins, grebes, petrels, auks, herons, albatrosses, terns
Invertebrate: microfaunivore	Plovers, sandpipers, some ducks
Arthropod	
Insect: insectivore	Cuckoos, nightjars, swifts, woodpeckers, swallows, wrens, thrushes
Crustacean: crustacivore	Crab plovers, some rails, penguins, and auks
Mollusk: molluscivore	Limpkins, Snail Kite, oystercatchers, Kiwi
Zooplankton: planktonivore	Flamingos
Vertebrate: macrofaunivore	
Fish: piscivore	Loons, pelicans, storks, cormorants, mergansers, Osprey
Terrestrial vertebrates: carnivore	Hawks, owls, eagles, falcons, vultures
Plant matter: florivore (phytophage)	
Bulk and roughage consumer[†]	
Browser	
Leaves, buds, shoots, grasses: herbivore	Ostrich, grouse, some ducks
Grasses: graminivore (phoëphage)	Geese, swans
Leaves: folivore	Hoatzin
Concentrate selectors	
Grains and hard seeds: granivore	Sparrows, finches, waxbills, some ducks, pigeons, and parrots
Fruits: frugivore	Toucans, manakins, birds of paradise, tanagers, some pigeons and bulbuls
Nectar: nectarivore	Hummingbirds, lorikeets, sunbirds, honeyeaters
Fungus: fungivore	Pygmy Parrot
Lichens: lichenivore	
Moss: bryophytivore	
Exudates (saps, gums, resins): exudativore	Sapsuckers

[*]Greek (-phage) terms are shown in parentheses and are considered synonymous with the more commonly used Latin (-ivore) terms. Some individual species of the example taxa listed may have other primary consumption patterns.
[†]This category occurs in mammals, but not birds.

piscivores, which consume fish, and the carnivores, which consume terrestrial vertebrates, such as reptiles, amphibians, mammals, and other birds. Insectivores, crustacivores, and molluscivores concentrate their feeding effort on insects, crustaceans, and mollusks, respectively. Birds that filter or select zooplankton from the water are known as planktonivores. The Sharp-beaked Ground Finch found on one of the islands of the Galapagos is described as a sanguinivore because it consumes blood from other birds as a primary food item; this dietary specialization is rare among Aves.

Those birds that consume foods of plant origin are florivores. Usually, plant material is carefully chosen for consumption due to selective browsing, which is greatly facilitated by flight. The advantage afforded by flight also comes with evolutionary pressure to keep the size of the digestive tract and the weight of its contents minimal. Presumably these two factors make it impractical for flying birds to fill the niches occupied by mammals that consume high-roughage plant material indiscriminately (bulk consumers; see Chapter 4). Birds that select the most nutritious leaves, buds, rhizomes, shoots, or grasses are herbivores. Those that concentrate on grasses and leaves are graminivores and folivores, respectively. Many birds select the components of the plant that have the highest concentration of digestible nutrients, such as that found in grain, fruit, or nectar, and are granivores, frugivores, and nectarivores, respectively. Birds that probe or drill for the saps, gums, or resins in plants are known as exudativores.

Many of the terms that describe the consumption categories are not botanically correct in usage by nutritionists. The best example of this is the term 'frugivore'. To a botanist, frugivores should include all birds that eat the fertile ovaries of plants and their associated structures. Many ecologists consider this term too vague and consider those birds that disperse seeds, and thereby aid in plant reproduction, as 'true frugivores'; those birds that destroy seeds by digesting them are considered as granivores. Taxonomically, only the subset of birds that eat grains produced by *Gramineae*, the grasses, should be considered granivores. Nonetheless, the nutritional definition of frugivore and granivore mostly considers the nutritional aspects of the food, which are the aspects that affect digestive morphology and metabolic diversity. Nutritionally, frugivores eat soft, moist fruits, which are relatively nutrient-dilute, whereas granivores eat hard, dry, nutrient-dense fruits (e.g., beans, nuts), regardless of the plant family of origin. Multiple definitions have resulted in similar confusion for the terms carnivore, florivore, herbivore, and graminivore (see Chapter 4).

A variety of terms are often used to describe how strictly a species adheres to a pattern of consumption, in addition to the terms 'oligivore' and 'omnivore'. Species of birds that consume a particular category of food during all of their adult life may be referred to as obligate consumers, e.g., obligate herbivore or obligate planktonivore. This situation is somewhat rare and the categories of food preferences are often best applied cautiously and with qualifications. The term 'primary' better describes the consumption pattern for most birds where consumption of one food category predominates, but modest amounts of other food categories are also eaten (Table 1.1). For example, almost all nectarivores and frugivores supplement their primary food with insects. Even the term 'primary' may be difficult to apply to a species because there may be marked variations in the dietary preferences within a single species, depending upon such things as their location or sex. For example, the Northern Shrike feeds on small birds in Sweden, on arthropods and lizards in Spain, and on insects in North Africa.

Many birds in temperate climates switch between an insect-dominated diet in the spring to a fruit-dominated diet in the fall. This is referred to as facultative frugivory and facultative insectivory. The degree of dietary specialization is a central topic in community ecology, where the terms 'specialist' and 'generalist' are used. 'Specialist' indicates a species that feeds within one trophic level (e.g., insects, fruits, or seeds) or on a restricted subgroup within a trophic level (e.g., only aquatic insects, high-lipid fruits, or pine seeds). Conversely, generalists switch between trophic levels, either seasonally or within a single day. The exact meaning of specialist and generalist often varies depending on the context of their use (Sherry, 1990).

Some birds are highly adapted to consuming a very restricted number of food items within their consumption category and are called stenophagous; if only a single food item is consumed, the species is monophagous. Monophagy is the highest degree of specialization but probably is never absolute in the wild. For example, the Snail Kite is close to being monophagous because it primarily eats a single species of freshwater snail, *Pomacea*, which it extracts with its peculiarly

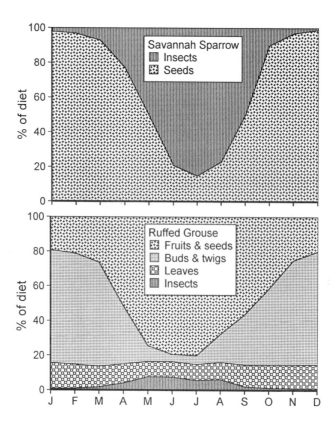

Fig. 1.1. Seasonal changes in food consumption by Ruffed Grouse, which are omnivorous, and Savannah Sparrows, which are facultative insectivorous/granivorous.

Table 1.2. Numbers of families feeding on different food categories.*

Food category	Primarily	Regularly	Sparingly	Total columns 2–4	Infrequently if at all
Green plants, buds	2	13	13	28	140
Seeds	4	31	24	59	109
Fruit	19	34	21	74	94
Nectar	3	5	7	16	152
Insects	50	58	32	140	28
Other terrestrial invertebrates	2	10	23	35	133
Littoral and benthic invertebrates	6	10	13	29	139
Small vertebrates (< 5 kg)	4	14	29	47	121
Large vertebrates (> 5 kg)	0	0	3	3	165
Fish, crustacea, squid, etc.	20	10	11	41	127
Carrion	1	4	6	11	157
General	–	–	–	56	112

*Data are derived from a detailed family-by-family analysis of the feeding habits of birds by Morse (1975). These categories reflect common divisions used by ecologists and are less specific than those used by nutritionists (Table 1.1). The 'General' category refers to those families that do not primarily consume foods in any one category.

hooked beak; but when this snail is not available it will consume other species. Obviously the population dynamics of a highly stenophagous species is tied closely to its sole food source. For example, the range and population dynamics of the Lesser Flamingo corresponds with the abundance of the filamentous blue-green alga, *Spirulina*, on which it is almost totally dependent. Many birds, such as some eagles, hawks, penguins, woodpeckers, kingfishers, parrots, and geese, tend to be stenophagous during specific seasons of the year. At the furthest extreme, the Ruffed Grouse has been observed consuming in excess of 300 different species of plants and 100 different species of small animals in a single year and is referred to as highly euryphagous (Fig. 1.1).

The number of species supported on a given amount of food resource is determined by the degree of dietary specialization. A larger number of species can be supported if each specializes in consuming a specific type of food. In contrast, a smaller number of generalist species, or 'jacks-of-all-trades,' can be supported. From the bird's point of view, specialization is favored when a variety of unique foods are stable in supply, and generalization is favored when the food resources vary in abundance over time or location. Species that utilize few food types can exploit each of them more efficiently than species that utilize a wide range of food resources. The nutritional niche refers to the exact foods consumed within a continuum of foods available. The term 'guild' is used in community ecology to refer to the collection of different species utilizing the same food

Box 1.1. Substrates or places where a food item is found or taken (from De Graff *et al.*, 1985).

Air: caught in the air
Bark: on, in, or under bark of trees
Coastal: waters along coast (can include brackish as well as salt water)
Coastal beach: beaches and/or tidal flats along coast
Coastal bottom: floor of continental shelf along coast
Coastal rock: rocks along coast
Coastal surface: surface of coastal waters
Floral: on or in flowers
Fresh marsh: freshwater marshes (on mud, in shallow water, or on marsh plants)
Fresh water: freshwater habitats (ponds, lakes, rivers, streams)
Freshwater bottom: bottoms of freshwater ponds and lakes
Freshwater shoreline: shores of freshwater ponds, lakes, rivers, or streams
Freshwater surface: surface of freshwater habitats
Ground: on the ground or on very low, weedy vegetation
Lower canopy/shrub: on leaves, twigs, and branches of shrubs, saplings, and lower
    crowns of trees
Marsh: fresh, brackish, or saltwater marshes (on mud, in shallow water, or on marsh
    plants)
Mud: inland on mud flats (wet fields, meadows, tundra, or associated with freshwater
    habitats)
Pelagic: ocean waters away from coastlines
Pelagic surface: surface of ocean waters
Riparian bottom: bottoms of rivers and streams
Salt marsh: brackish or salt marshes (on mud, in shallow water, or on marsh plants)
Shoreline: along shoreline of both freshwater and saltwater (coastal) habitats
Upper canopy: on leaves, twigs, and branches of trees in main canopy
Water: brackish, fresh-, and saltwater habitats
Water bottom: on bottoms of fresh-, brackish, or saltwater habitats (benthic)
Water surface: on surface of fresh-, brackish, or saltwater habitats

resources. Guilds can be classified by a combination of the food type (Tables 1.1 and 1.2), the substrate where the food is procured (Box 1.1), and the method for procuring the food (Box 1.2).

In general, the nutrient balance, density, and digestibility of plant-based diets are inferior to those of animal-based diets. For this reason, coupled with the extremely high nutrient needs of fast-growing chicks, most florivorous adult birds feed their young a more faunivorous diet. However, a species is usually classified according to the consumption category of the adult. Morse (1975) summarized the primary consumption patterns of 168 bird families (Table 1.2). Clearly arthropods are the predominant avian food, with fruits and seeds also commonly consumed. Foods that require extensive physical adaptations for procurement (e.g., nectar, vertebrates, benthic invertebrates) or for digestion (green plants) are not consumed by most families. The dietary pattern of birds reflects their

---

Box 1.2. Techniques or manner in which food is obtained.[*]

Ambusher: slowly stalks or waits for prey to come within reach

Chaser: pursues prey on ground

Dabbler: submerges head and neck or tips up (various water substrates)

Diver: dives from surface for underwater food

Excavator: locates food in bark by drilling holes

Food pirate: steals food from other species, usually other birds

Foot plunger: catches prey by plunging from air to water surface (or ground) and seizing prey in talons

Forager: takes almost any food items encountered upon the substrate (includes all herbivores and omnivores feeding on terrestrial habitats or vegetation, except grazers and grubbers)

Gleaner: selects particular food items from the substrate

Grazer: feeds on grasses, sedges, or grains in fields or meadows

Grubber: digs up roots and tubers of either terrestrial or aquatic plants

Hawker: flies after prey and captures it either in air or on ground

Hover-gleaner: hovers in air while selecting prey (from vegetation or ground)

Plunger: dives from air into water to capture prey in bill or gular pouch

Prober: inserts bill into substrate (beach, mud, ground) and locates prey by touch

Sallier: perches on exposed branch or twig, waits for insect to fly by, and then pursues and catches insect in air.

Scaler: exposes prey under bark by scaling off loose bark

Scavenger: takes a variety of items, including refuse or carrion

Screener: flies with bill open and screens prey from air

Skimmer: flies low over water and skims food from water surface with lower mandible in water

Strainer: strains food items from water or mud through lamellae along edge of bill

[*]Some techniques are associated with particular food types and/or substrates (De Graff *et al.*, 1985).

---

evolutionary history. Primitive birds are thought to have fed on arthropods, especially insects. Many of these insects were caught while birds were foraging on vegetation. This close association between birds and plants presumably gave rise to the consumption of seeds, berries, and nectar. Herbivory developed late in the evolution of birds.

The nutrient composition of foods consumed by birds is extremely variable (Fig. 1.2). In general, faunivores consume a diet that is rich in easily digestible protein and fat. Florivores consume diets that are lower in protein and high in carbohydrate. At the extremes are piscivores and nectarivores. Fish are composed of mostly protein and fat with less than 1% carbohydrate, whereas nectar is 95% carbohydrate and less than 5% protein and fat. Essentially all important anatomical, physiological, and metabolic systems of the bird's body are adapted to accommodate specialization on these vastly different diets. These adaptations are considered in depth in the next three chapters.

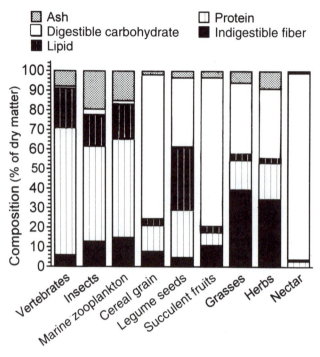

Fig. 1.2. Composition of typical foods from various consumption categories, illustrating the large variability in the principal nutrients in these diverse foods.

References

De Graff, R.M., Tilghman, N.G., and Anderson, S.H. (1985) Foraging guilds of North American birds. *Environmental Management* 9, 492–536.

Langer, P. and Chivers, D.J. (1994) Classification of foods for comparative analysis of the gastro-intestinal tract. In: Chivers, D.J. and Langer, P. (eds) *The Digestive System in Mammals: Food, Form and Function.* Cambridge University Press, Cambridge.

Lein, M.R. (1972) A trophic comparison of avifauna. *Systematic Zoology* 21, 135–150.

Morse, D.H. (1975) Ecological aspects of adaptive radiation in birds. *Biological Reviews* 50, 167–214.

Sherry, T.W. (1990) When are birds dietarily specialized? Distinguishing ecological from evolutionary approaches. *Studies in Avian Biology* 13, 337–352.

CHAPTER 2
Anatomy and Physiology of the Digestive System

The anatomical plan of the digestive tract of birds is extremely variable (Figs 2.1 and 2.2), but many generalizations are possible. Distantly related species of birds that have similar dietary preferences often have digestive tracts of similar structure and function. Species consuming easily digested foods (nectar, fruits) have short, simple digestive tracts. Species consuming foods that require more enzymatic attack (animal matter and seeds) have large stomachs and relatively small lower intestines. Some species have digestive tracts dominated by large ceca for the fermentation of plant cell walls, which are plentiful in the environment but very difficult to digest. Across the > 9000 species of birds, there is an almost continuous distribution of digestive-tract morphologies, from the simplest (nectarivores) to the most elaborate (herbivores), with the largest number of species being toward the center of this distribution (omnivores, insectivores).

Clearly the dimensions of the digestive tract are highly variable and influenced by phylogenetic relationships and nutritional strategy, but some rough generalizations are possible (Ricklefs, 1996). Across species, the diameter, length, volume, and surface area of the intestine increase in size with increasing bird mass. When comparing a wide variety of species of Passeriformes, the volume of digesta that can be held by the intestines is directly proportional to the mass of the bird (Fig. 2.3). In other words, a plot of the volumes of the digestive tract versus bird mass gives a slope of about 1. This value is referred to as the allometric constant and indicates that digesta volume increases at a rate proportional to the body mass to the first power. To accommodate this proportional increase in volume, the length and diameter of the gastrointestinal tract increase with increasing bird size at an allometric constant of about $\frac{1}{3}$. The total surface area of the intestines scales at about 0.72 body mass. It is interesting that the rate of increase in intestinal surface area with increasing body mass is very similar to the rate of increase in the metabolic rate (allometric constant = 0.73). Apparently a specific area of intestinal epithelium is required to obtain the nutrients necessary for a particular metabolic rate. The time that digesta are retained within the tract, known as the retention time, increases only slightly with increasing bird size (allometric constant of 0.21). Thus, as birds get larger they maintain digestive efficiency predominantly by increasing the surface area of their intestines, with only a slight increase in the time that they subject food to digestive processes in the tract.

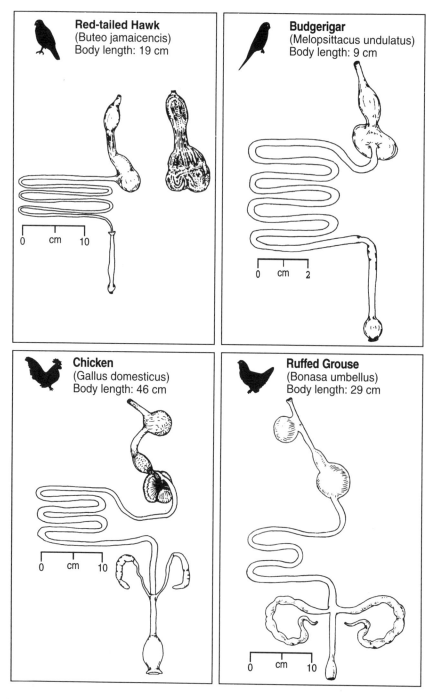

Fig. 2.1. Gastrointestinal tracts of the Red-tailed Hawk (with sagittal section of foregut), Budgerigar, chicken, and Ruffed Grouse (used with permission from Stevens and Hume, 1995).

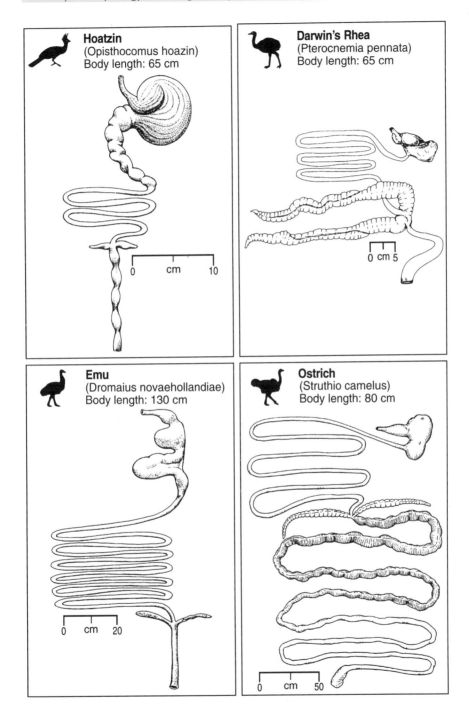

Fig. 2.2. Gastrointestinal tracts of the Hoatzin, Darwin's Rhea, Emu, and Ostrich (used with permission from Stevens and Hume, 1995).

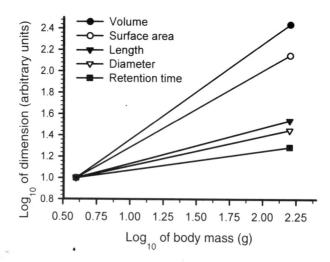

Fig. 2.3. Morphometry of the digestive tracts of passerines. Regression lines are from a sample of approximately 30 representative species of passerines taken from North and Central America and weighing between 6.8 and 141.1 g (Ricklefs, 1996). The length and the diameter of the intestine were measured in cm. The surface area of the intestinal epithelium is expressed as cm^2. The retention time is calculated in $time^{-1}$.

Anatomy of the Digestive System

The distinctive anatomy and physiology of the avian gastrointestinal tract reflect the constraints of flight, including minimizing weight through the reduction in length and volume and centralizing the tract and its associated weight within the body cavity. The avian gastrointestinal tract has a greater number of organs, which have greater interorgan cooperation, than their mammalian counterparts. The avian digestive tract starts with the beak, followed by a toothless mouth, tongue, pharynx, esophagus, crop, proventriculus, ventriculus or gizzard, intestine, ceca, rectum, cloaca and vent (Fig. 2.4). Accessory organs include the salivary glands, biliary system, pancreas, Peyer's patches, and bursa. In the following description, anatomical terms describe positions or movements in a digestive tract dissected out of the bird, uncoiled and stretched linearly from beak to vent. The term 'anterior' designates towards the tip of the beak and 'posterior' designates towards the vent. In this context, anterior is generally synonymous with cephalad, or rostral, and posterior is synonymous with caudal. Distal and proximal are used when an organ (e.g. ceca) diverges from the main line of the gastrointestinal tract, with distal referring to the direction away, and proximal towards, the main tract. In this context, distal and proximal are synonymous with lateral and medial, respectively.

Histologically, the digestive tract is lined with a continuous mucous membrane from the mouth to the vent. This mucous membrane consists of an innermost epithelial layer, beneath which are the lamina propria and lamina muscularis, and in many areas of the tract there is a submucosa. These layers

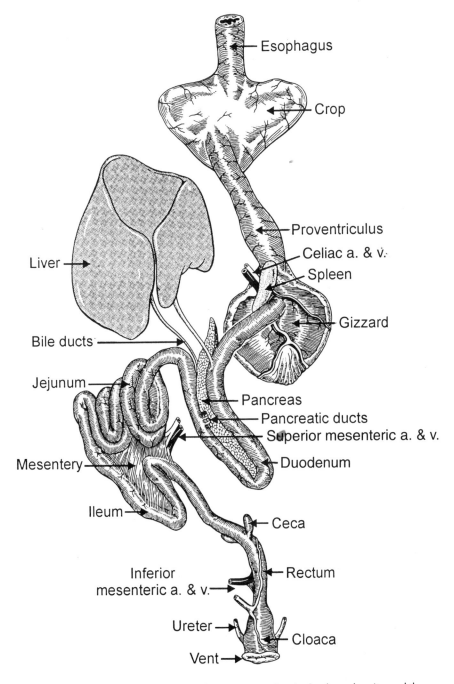

Fig. 2.4. The digestive organs of the Rock Dove excepting the beak, oral cavity and the anterior esophagus. The length of the bile ducts has been greatly exaggerated for diagrammatic purposes. a, Artery; v, vein. (Used with permission from Proctor and Lynch, 1993.)

provide protection from abrasion by the food as it passes through and prevent the entrance of microorganisms. The epithelial surface is also specialized for the selective absorption of nutrients and the exclusion of nonnutrients. The submucosa is surrounded by a muscle coat, composed of an inner circular layer and an outer longitudinal layer of smooth muscle, which are important in the movement and mixing of the digesta. The outermost layer is the serosa, which provides structural integrity to the digestive organs and protects them from abrasion and trauma. The intestines are attached to a mesenteric membrane, which contains the blood vessels that perfuse this region of the tract.

Beak

The avian jaw lacks teeth and lips and birds swallow their food in gulps without chewing. The grasping and particle-size-reduction functions of these structures are taken on by the avian beak, tongue, and gizzard. Presumably this adaptation permits the weight associated with heavy jaws, teeth, and muscles to be moved more centrally to accommodate flight. Certainly, the proportion of the weight within the head that is associated with food acquisition is considerably less in birds than in mammals.

In the past, 'beak' was reserved for describing the sharply curved beaks of birds of prey and 'bill' described the broad flat bills of ducks and geese. The terms 'beak' and 'bill' are often used synonymously in the modern literature. Birds depend on their beaks for acquiring food, but also for preening their feathers, climbing, building nests, defense, and courtship displays. The anterior upper and lower mandible are covered with a horny sheath, the rhamphotheca, composed of keratin, which forms the visible shape of the beak. The keratin is continually lost by wear and replaced by new growth. The location and rate of growth and wear influence the exact shape of the beak, and subtle changes may occur as food types change. The type of pigmentation incorporated into the keratin of the beak influences its hardness and rate of wear (Bonser and Witter, 1993). The beak is highly adapted in shape and size according to the type of food that is consumed and how it is processed. It is used to crack and hull seeds, grasp, skim, spear, tear, probe, sieve, etc. (Fig. 2.5). Even in species where the beak is very large, such as toucans and hornbills, it is quite light because of vast air-filled cavities strutted with trabeculae. The edges of the beak, or tomia, are often sharp to facilitate cutting and may be serrated, notched, or dentate, or possess lamellae for filtration. The relationship between the beak and diet has been the subject of an extraordinary number of studies and will be referred to in more detail in Chapter 4. The morphology of the beak is the most constant component of the digestive tract within a bird's lifetime and represents the results of generations of selection driven by the physical requirements of obtaining food. For this reason, it is easily studied and represents a window on the nutritional adaptations of a bird species, unbiased by acute changes in diet.

Several structural attributes of the beak, jaws, and skull provide unique functional characteristics that are important in obtaining food (Bock, 1964, 1966; Burton, 1974). The articulation of the lower jaw is similar to that of reptiles, allowing great mobility so that birds can open their beaks widely. The diameter

Fig. 2.5. Examples of specialized beak shapes include: White-throated Sparrow, light beak adapted for small seeds and plant material; Evening Grosbeak, heavy beak for cracking seeds with tough coats; Atlantic Puffin, tall but thin beak for catching and holding small fish; Golden-winged Warbler, thin beak for gleaning small insects; Whimbrel, long beak for probing mud flats for invertebrates; Pileated Woodpecker, heavy chiseling beak for splintering wood to expose insects; Keel-billed Toucan, long beak for grasping fruits; Great Blue Heron, long sharp beak for stabbing and spearing small vertebrates; Greater Flamingo, sharply downturned beak modified into a water filtration system for obtaining small invertebrates; Sword-billed Hummingbird, extremely long beak for probing deep into tubular-shaped flowers; Red-tailed Hawk, beak adapted to tearing flesh from small mammals. (Used with permission from Proctor and Lynch, 1993.)

of the opening into the oral cavity at the most restrictive point is referred to as the gape. In many birds, such as flamingos, parrots, woodpeckers and hornbills, the upper mandible articulates in the cranium at the nasofrontal hinge, permitting an increased gape of the beak. This flexion, called prokinesis, also absorbs some of the shock associated with pecking, drilling, and seed cracking. In Charadriiformes, such as plovers, sandpipers, avocets, and oystercatchers, the articulation is not at a specific hinge, but spread over a wider zone along the upper mandible (rhynchokinesis), and increases the manipulative capabilities of a long beak. Specific adaptations of the beak and the interplay between the beak and the tongue for food prehension, sorting, and processing are covered in Chapter 4. The beak is also the entry point of the respiratory system and the upper beak is perforated by the nostrils or nares.

Oral cavity and pharynx

The oral cavity, or mouth, is between the upper and lower mandibles and is illustrated in Fig. 2.6 (Fitzgerald, 1969; Ziswiler and Farner, 1972; King and McLelland, 1984). The functions of the oral cavity include grasping, testing, mechanical processing, such as crushing or shelling, lubricating, and propelling food to the esophagus. The pharynx is posterior to the oral cavity and connects it with the esophagus. Together, the oral cavity and the pharynx are referred to as the oropharynx. The anterior part of the oral cavity is roofed with a hard palate, but the mammalian equivalent of a soft palate is lacking in avian species. The hard palate is formed by the upper mandibular bones, is covered by a highly keratinized surface and becomes softer posteriorly. The choanal opening, a median slit in the hard palate, connects the oral cavity to the nostrils. Several backward-pointing rows of papillae run transversely across the roof and floor of the oropharynx and presumably act to direct food posteriorly during swallowing. In chickens and Japanese Quail, a large transverse row of papillae demarcates the posterior extremity of the hard palate. On the roof of the oral cavity of most bird species are two ridges, one distal and one anterior to the choanal opening. These ridges are especially well developed in seed-eating passerines, which use them in cooperation with their tongue to remove the seed coat. The epithelium of the roof and bottom of the oral cavity has a high degree of keratinization at those areas which are subject to excessive wear during food prehension. In pelicans and in some passerines, the floor of the oral cavity and pharynx is expandable and serves as an area to store food. In many finches, this storage area expands during the breeding season and is used to store food to provision the young. The oral cavity of chicks in some species is colored with distinctive markings and patterns. This coloration combined with an energetic display of begging with a gaping mouth stimulates the parents to feed the chick.

TONGUE

The avian tongue is not composed of overlapping muscle layers, as in mammals, but is mobilized by the hyoid apparatus, consisting of multiple articulating bones and their musculature. The parrots (Psittacidae) are an exception in that they have muscles in the tongue that are independent of the hyoid apparatus and

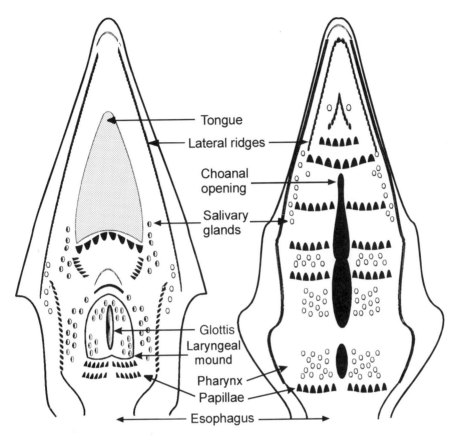

Fig. 2.6. The oropharynx of the chicken, with the floor containing the tongue and glottis shown on the left and the roof containing the choanal opening shown on the right (after King and McLelland, 1984).

permit the flexibility required for manipulating seeds. The avian tongue is variable in length and form in a manner similar to the beak. It is usually sharply pointed in the front and has papillae to the rear. King and McLelland (1984) have characterized three forms of adaptations for food collection, food manipulation, and food swallowing.

1. Tongues adapted for food collection are long, narrow, and protrusile and function as probes, brushes, or spears. For example, the tongue of many woodpeckers can be thrust forward several times the length of the beak and its most anterior edges have barbs and spiny papillae covered with sticky saliva that entraps insects. The tongues of sap-sucking woodpeckers, lories, and lorikeets are relatively shorter but end in fine hair-like processes for the collection of sap or nectar through capillary action.

2. Tongues adapted for the manipulation of food are usually not very protrusile. Parrots, finches, and crossbills have thick, muscular tongues, which act as fingers for the manipulation and extraction of seeds from their husks or cones. The tongues of many fish-eating birds, such as penguins, mergansers, and shear-waters, have extensive papillae that are sharp, stiff, and useful for holding slippery prey. In raptors, the stratified squamous epithelium is highly keratinized toward the tip to provide a tough raspy surface. Many dabbling waterfowl have well-developed, thread-like papillae on the lateral edges of the tongue. These papillae interdigitate with the lamellae of the beak to form a filtering apparatus that retains solid food particles while permitting water to leave through the sides of the bill (Fowler, 1991).

3. Tongues adapted for swallowing are short, nonprotrusile and usually have papillae facing posteriorly, which aids in moving the food bolus to the esophagus. Examples include chickens, game birds, pelicans, cormorants, storks, spoonbills, ibises, Ostrich and cassowaries (Goodman and Fisher, 1962; Fowler, 1991).

SALIVARY GLANDS

Salivary glands are best developed in birds that consume dry diets, such as grasses, seeds, or insects (Belman and Kare, 1961; Pritchard, 1972). Simple, branched, or compound tubular glands are scattered in groups around the lower oral cavity, tongue, and pharynx. In chickens they secrete 7–30 ml day^{-1} of a mucinous saliva. The saliva is adequate to lubricate a bolus of food but usually is insufficient to moisten the food enough for extensive enzymatic digestion. Amylase and lipase have been reported in chicken saliva but their significance is minor. In species that consume slippery food (e.g. piscivores), the salivary glands are greatly diminished or missing. Salivary glands are large and numerous in woodpeckers, swifts and swiftlets, providing a sticky saliva, which aids in adhering insects.

TASTE BUDS

Birds generally have poorer taste acuity than mammals (Stresemann, 1927–1934; Berkhoudt, 1985). This is probably due to the rapid transit of food through the mouth, lack of mastication, and relatively low saliva addition to the food. Low taste acuity is reflected in very low numbers of taste receptors: 62 in Japanese Quail, 24 in Blue Tits, 375 in Mallards, 350 in parrots, compared with 9000 in humans and 17,000 in rabbits. The location of taste buds is variable and generally reflects the relative role of the tongue versus other oral structures in mediating food acceptance or rejection. In many species, taste buds are located on the palate and on the posterior tongue (e.g. chickens, pigeons, swifts, Falconiformes, and some passerines). The palatal taste buds are usually located in regions where the epithelium is soft and glandular, typically near the salivary glands. Unlike mammals, few taste buds are found on the anterior surface of the avian tongue, nor are taste buds usually found on special papillae.

The tongue, oral cavity, and beak of birds have a rich supply of touch receptors, and they augment the bird's relatively poor gustatory capacity with a strong tactile sense. For example, in Mallards the concentrations of touch receptors found on the internal aspect of the tip of the bill and on the roof of the oral cavity are even greater than the density found on the human index finger.

Taste buds are also scattered throughout this region. This positioning of touch and taste receptors corresponds with the pathway food takes in its journey through the oral cavity during filtration and permits the immediate rejection of objectionable items. Sandpipers and woodcocks have taste buds and touch receptors concentrated near the tip of the lower beak and throughout the length of their tongue so that food items may be discriminated during probing into sand or mud (Dorst, 1974).

From scattered evidence, it can be generally concluded that birds can taste the same four primary flavors (sour, sweet, bitter, salty) as humans but with considerably less acuity (Navarro and Bucher, 1992). Birds' perception of taste may also vary considerably from that of mammals. For example, many birds will consume red peppers containing levels of capsaicin that completely repel rodents and humans. Additionally, saccharin is not perceived as sweet in birds.

The sense of smell is not well developed in birds, with a few possible exceptions, such as the Kiwi, and some vultures and seabirds that use smell for the location of food (Healy and Guilford, 1990). Nocturnally active birds have large olfactory bulbs, but sight and touch are clearly the predominant sense used by most birds for discriminating food and nonfood items. Flowers and fruits that rely on birds for pollination and seed dispersal are not scented, unlike those that attract mammals.

Esophagus and crop

Posteriorly, the pharynx joins the esophagus, which extends along the neck, into the thoracic cavity, and terminates in the proventriculus. It has a relatively greater diameter than that found in mammals in order to accommodate food that has not been reduced in size by chewing. To aid in swallowing large food items, the esophagus is expandable, due to a series of longitudinal folds, which are enriched in mucous glands to provide lubrication. The diameter is especially great in birds (e.g., carnivores, piscivores, and frugivores) that feed on large items and is often smallest in granivores, insectivores, and birds, such as many parrots, that crush their food. The epithelial lining is thick and cornified for protection against mechanical damage as a result of swallowing foods whole. The thickness of this cornified layer and the size of mucous glands is greatest in granivores and herbivores (Ziswiler, 1985).

The main function of the esophagus is to pass food from the mouth to the proventriculus by the peristaltic contraction of inner circular and outer longitudinal muscles. The esophagus of most species also serves a storage function. This is accomplished by the extreme elasticity of the esophagus in many species, such as grebes, cormorants, penguins, geese, ducks, owls, woodpeckers, gulls, and petrels. Birds are apparently not discomforted by the presence of food occluding their esophagus. The presence of a fish extending from the proventriculus to the beak is a common sight among piscivorous seabirds and may persist for an hour or more. In some species, the storage function of the esophagus is enhanced by its widening just prior to entering the thoracic cavity to give one or more clearly partitioned diverticula, known as a crop. Generally, a defining characteristic of a true crop is the presence of a controllable sphincter that regulates the entrance

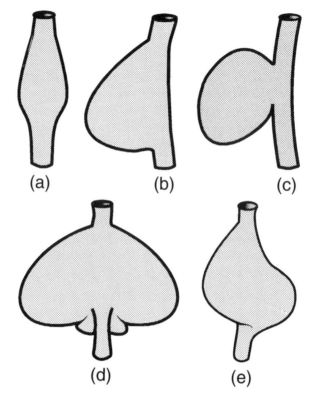

(a)   (b)   (c)

(d)   (e)

Fig. 2.7. Crops may be classified into four general shapes: (a) an enlargement of the posterior esophagus, as in cormorants; (b) an enlarged diverticulum, as in vultures; (c) a true crop with musculature to control the diameter of the entrance, as in chickens and other Galliformes; and (d) two enlarged diverticula on either side of the esophagus as in pigeons. The crops of some birds, such as Budgerigars (e), do not fit neatly into one of these categories. (Redrawn from Pernkopf and Lehner, 1937.)

and exit of food. An enlarged diverticulum without specific musculature to control food passage is referred to as a false crop, as in the redpoll. Ziswiler and Farner (1972) have divided crops into four types according to their size and morphology (Fig. 2.7). In most types, the presence of large folds of mucosa allows considerable expansion and shrinkage depending on the amount of contents. Crops are particularly well developed in granivores and in scavengers, such as vultures (Fisher and Dater, 1961).

An enlarged esophagus or a crop provides a temporary storage area that permits a bird to forage for large amounts of food rapidly and then fly off to digest the meal in safe cover (Hainsworth and Wolf, 1972; Buyse *et al.*, 1993). A crop also permits 'tanking up' in the evening so that food can be slowly released to supply nutrients during the nighttime. In hummingbirds, the filling of a large crop by extensive foraging at dusk fuels their high metabolic rate for several hours into the night. In chickens, food stored during an evening feeding bout supplies 75%

of the nocturnal energy needs. An additional function of the crop in granivores and herbivores is the provision of a moist environment where food begins to soften, permitting more efficient digestion. However, in chickens and quail, mucous glands are only present near the entrance of the crop, so most of the water needed for softening must be consumed with the meal. There are no enzymes secreted into the crop but those present in the food may provide some digestion while residing in the crop. In the Hoatzin, the crop serves as an organ for microbial fermentation, permitting the efficient use of plant leaves containing high dietary fiber (Grajal *et al.*, 1989).

The crop and the esophagus play an important role in nourishing the young of many species. In hawks, penguins, storks, pelicans, darters, gannets, cormorants, pigeons, parrots, and some waxbills and finches the crop and/or esophagus serves to store food that is later regurgitated to feed the young. In pigeons and doves (Columbidae), the crop of both sexes produces a 'crop milk' for feeding nestlings. The production of crop milk has many similarities with lactation in mammals. Brooding stimulates prolactin secretion, which triggers differentiation of the crop epithelium around the sixth day of incubation. By the thirteenth day, the epithelium thickens considerably and becomes rich in blood vessels, making the crop lining reddish in color. The epithelial cells accumulate fat and protein, and are shed into the lumen of the crop to give the milk. This holocrine secretion has a cheese-like texture, is very low in carbohydrate, and contains about 60% protein, 5% ash, and 35% fat. As in mammalian milks, the fat in crop milk is predominantly medium-chain triglycerides. Crop milk is fed to the squab for about 2 weeks following hatching. Milk-laden cells are initially sloughed off only when the crop is empty, ensuring that the milk is not diluted by adult foods. As the chick matures, crop milk is fed in combination with other foods and milk-laden epithelial cells are shed only during the times of the day when parents are provisioning their young. Crop milk assures a steady supply and balance of nutrients to the rapidly growing altricial chicks. The nutritional advantage afforded by the production of crop milk has been suggested as a major factor in the pandemic success of the Columbidae (Dumont, 1965; Horseman and Buntin, 1995).

Male Emperor Penguins provide their newly hatched chick with a secretion produced by the desquamation of esophageal epithelial cells (Prevost and Vilter, 1963). The esophageal mucous glands of the Greater Flamingo produce a nutritive merocrine secretion that is regurgitated and fed to the young. This secretion also contains red pigments necessary for coloration of developing feathers (Wackernagel, 1964).

Birds have the capacity to regurgitate food from their esophagus or crop. Regurgitation may function to reduce preflight weight to permit takeoff in heavy birds (e.g., seabirds and vultures) that eat large meals. Physical stresses, emotional stresses (capture), and noxious chemicals (crude oil) may also initiate regurgitation in some species. Emetics, such as antimony potassium tartrate, are sometimes used experimentally to induce regurgitation so that food consumption can be itemized (Poulin and Lefebvre, 1995).

Stomach

The stomach of most birds consists of two parts: the proventriculus (glandular stomach) and the gizzard (muscular stomach), which is sometimes called the ventriculus (Fig. 2.8). The proventriculus secretes hydrochloric acid (HCl) and pepsin, which have much of their action in the gizzard, where the food is 'chewed' by muscular contractions. Often one of the two organs predominates in size, depending on the diet. In many carnivores and piscivores, the proventriculus is large and the gizzard is thin-walled and weak; this is taken to the extreme in pelicans, where the gizzard is essentially absent. In granivores and herbivores, the gizzard is very muscular and dominates the proventriculus in size.

PROVENTRICULUS

The junction between the esophagus and proventriculus is gradual and demarcated by the presence of gastric glands in a thickened mucosa. The lamina propria near the junction is richly supplied with lymphatic nodules, which are often called the esophageal tonsils. The mucosa of the proventriculus has an abundance of glands of two principal types: the tubular glands, which secrete mucus, and the gastric glands, which secrete HCl and pepsin. Mucus is discharged from tubular glands immediately after the initiation of feeding, and HCl and pepsin are secreted when food arrives in the lumen of the proventriculus. A unique gastric gland cell, which combines the functions of mammalian chief and parietal cells, secretes both acid and pepsin. Compound gastric glands are particularly well developed in granivores, but are less developed in omnivores. In some species, including chickens and quail, the proventriculus has a series of conspicuous papillae projecting into the lumen and the gastric glands open at their apex. The proventriculus is highly distensible in species such as albatrosses,

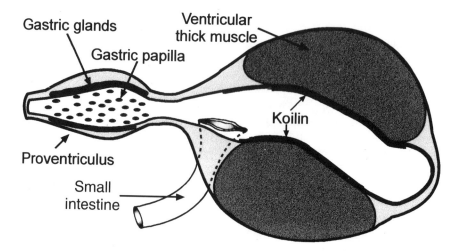

Fig. 2.8. A cross-section of the stomach of a chicken, which consists of the proventriculus and the ventriculus, or gizzard. The koilin lining, or cuticle, is especially thick directly over the two thick lateral muscles.

petrels, fulmars, gulls, cormorants, owls, ratites, and storks, permitting large food items to be swallowed and digested (Fitzgerald, 1969; Turk, 1982; Ziswiler, 1990).

The proventriculus can serve as a storage area as well as an area for gastric digestion in some woodpeckers, hawks, cormorants, petrels, herons, gulls, terns, Ostriches and Emus (Degen *et al.*, 1994). The proventriculus of the Ostrich stores water in addition to food. The water is slowly released until the next drinking opportunity, permitting foraging over a large radius from a water source. Procellariiformes, such as albatrosses, shearwaters, petrels, and storm petrels, concentrate dietary nonpolar lipids (stomach oils) in their proventriculus. The accumulation is the result of unique proventricular anatomy and associated motility, as well as high dietary lipid load. These pelagic birds travel long distances between breeding sites and feeding areas. The ability to concentrate and retain high-energy oils, which occupy small volumes and have low osmotic loads, facilitates the provision of chicks with food harvested at distant locations. Nearly 80% of the energy delivered to chicks of Wilson's Storm-Petrel resides in stomach oils. Some Procellariiformes can eject the pungent oil with an explosive force to incapacitate their predators (Duke *et al.*, 1989; Place *et al.*, 1989; Obst and Nagy, 1993).

GIZZARD

The posterior end of the proventriculus is constricted, forming an isthmus that connects to the gizzard. The function of the gizzard is to mechanically massage and grind food to reduce its size and increase its surface area. It also serves as a location for the action of HCl and pepsin added to the food during passage through the proventriculus. The gizzard of birds (e.g., carnivores, piscivores, nectarivores, and frugivores) feeding on foods that are soft and easily digested is a relatively round organ, similar in thickness and muscularity to the proventriculus. In some nectarivores and frugivores, the gizzard is only a very small diverticulum at the junction of the proventriculus and the small intestine, where soft-bodied insects that are occasionally consumed may be diverted for mechanical and peptic digestion. In many insectivores, herbivores, and omnivores that feed on coarse food items, the gizzard is a larger and more muscular organ. The amount of muscularity is greatest in birds that consume grain and other hard seeds. The gizzard is also well developed in molluscivores and crustacivores, in order to crush shells and exoskeletons. It is less muscular in parrots (Psittacidae), which remove the hull from seeds and fragment them with their bill prior to swallowing.

The gizzards of chickens, turkeys, and Japanese Quail are particularly well studied (Hill, 1971; Ziswiler and Farner, 1972; Akester, 1986). Two pairs of smooth muscles are arranged in distinct bands, which both originate and terminate on a circular tendon. This unique muscle arrangement and the differing size of the two muscle pairs forms a disk-shaped organ. The muscles are asymmetrically arranged relative to the long axis, giving both rotary (mixing) and crushing movements during contraction. The distinctive red color of the gizzard muscle arises from its high myoglobin content. The mucosa of the lumen contains numerous deep tubular glands, which secrete a protein-rich fluid that forms horny plates, known as the cuticle (or sometimes called the koilin membrane).

The cuticle acts as a grinding surface and protects the underlying mucosa from acid and pepsin. It is constructed in a similar way to reinforced concrete. Most of the cuticle is a combination of protein, produced by the underlying epithelium, mixed with dead cells (i.e., the 'concrete'). The reinforcement is provided by an abundance of hard rods, which project from the underlying tubular glands and protrude through the surface of the cuticle. These rods act to increase the abrasiveness of the cuticle and are sometimes referred to as dentate processes, or sometimes 'gizzard teeth,' but the small size of the rods ($\approx$20 µm in diameter) makes this arrangement more like sandpaper than teeth. Sand and small pebbles (grit) may also lodge in the gizzard to add additional abrasive action and their function is more analogous to that of mammalian teeth. The gizzards of domestic geese, turkeys, and ducks contain on average 30, 10, and 45 g of grit, respectively. When compared across a wide variety of species, it is clear that the amount of grit in the gizzard varies directly with the hardness of the dietary items consumed. Among florivores, the amount of grit is highest in herbivores and granivores. Insectivorous birds commonly have grit in their gizzards, and it tends to be highest in those species that eat hard-bodied insects. The small amounts of grit found in frugivores and nectarivores may function primarily as a source of calcium, i.e., calciferous grit. Birds that shift seasonally between soft and hard food items (e.g., facultative frugivore–granivore) have corresponding changes in the grit content of their gizzards (Soler *et al.*, 1993; Gionfriddo and Best, 1996).

The thickness and physical properties of the cuticle are highly correlated with food consumed, being especially thick in granivores, those molluscivores that consume shells, and insectivores consuming hard exoskeletons, but thin and soft in frugivores. It is thickest directly under the thick muscles, which provide much of the grinding within the gizzard. The cuticle lining wears down steadily and undergoes continual renewal. It is often green, brown, or yellow in color due to the reflux of bile pigments from the small intestine. The cuticle's loose attachment to the mucosa permits it to be peeled off for inspection during autopsy. In some species, the cuticle is periodically shed and excreted, but in male hornbills it is filled with seeds, regurgitated, and fed to its nesting mate.

The gizzard sometimes serves to sort food components, so that indigestible materials are egested and not permitted to disrupt digestion in the small intestine. The gizzard of carnivores, piscivores, and insectivores serves as a trap for indigestible fragments of bones, cartilage, feathers, fur, and chitin, which are formed into a pellet and egested. Large hard items are usually wrapped with softer items, such as hair or feathers. Some species (e.g., grouse) egest a pellet containing large amounts of plant fiber, and many frugivorous birds egest seeds that were stripped of their pulp by the gizzard. Some birds also egest the hard coats of seeds or fruits.

The size of the gizzard can change with diet (Spitzer, 1972; Piersma *et al.*, 1993). Herbivores, such as geese or ratites, fed a diet high in grass or leaves have greatly enlarged gizzard musculature compared with those fed a diet high in ground grain. Similarly, shore birds have larger, more muscular gizzards when consuming mollusks with shells than when adapted to a diet of soft prey. In many birds, gizzard size follows a seasonal rhythm. For example, facultative granivores–insectivores have a large, muscular, and grit-filled gizzard with a hard cuticle in the winter, when they eat mostly seeds, but, in the summer, when the birds eat mostly soft insects, the gizzard weighs half as much and has a softened cuticle with little grit.

The posterior end of the gizzard of most species tapers into a pyloric region, which joins the small intestine (Ziswiler and Farner, 1972; King and McLelland, 1984). In chickens this region is very short (5 mm) and has an epithelium that is intermediate between that of the gizzard and that of the intestine. Many species (e.g., herons, storks, cormorants, penguins, pelicans, and some ducks, geese and rails) that consume fish, aquatic invertebrates, or aquatic plants have a chamber-like pyloric region, called the pyloric stomach. The mucosa of the pyloric stomach may contain long papillae (darters) or may trap ingested feathers (grebes), which apparently act to retain undigested food components, but permit the passage of liquefied digesta into the small intestine. The pyloric region contains tubular glands, which produce mucus for protection of the epithelium from proventricular secretions.

Small intestine

The small intestine functions in enzymatic digestion and absorption of the end products of digestion. Across species, the small intestine is considerably less variable in form and function compared with the more anterior organs. This is because the diverse physical constitution of different foods is reduced to a relatively uniform and fluid suspension, or chyme, by the action of the proventriculus and gizzard. By convention, the small intestine consists of a duodenum, jejunum, and ileum, although these segments are not clearly demarcated in most birds. The duodenum originates from the gizzard and forms a loop around the pancreas. The bile and pancreatic ducts enter the duodenum. Posterior to the duodenal loop is the jejunum, followed by the ileum. Histological differentiation of the three sections is unreliable and gross anatomical landmarks are used for identification. By definition, the duodenum ends at the point where the small intestine leaves its association with the pancreas; the jejunum extends from the end of the duodenum to the vitelline diverticulum (Meckel's diverticulum), which is the remnant of the yolk sac, located approximately midway along the length of the small intestine; and the ileum extends from the vitelline diverticulum to the cecal junction. The vitelline diverticulum of adults of different species is highly variable in size. In the chicken it is easily visible without magnification, but in passerines, woodpeckers, parrots, doves, and pigeons it is usually only seen histologically, as a collection of lymphatic follicles in the intestinal wall. In cormorants (Phalacrocoracidae) the vitelline diverticulum matures into a conspicuous lymphoid organ. In Japanese Quail, the duodenum is about 20% of the length of the entire intestine and is wider in diameter than the jejunum or the ileum, each of which makes up about 40% of the length. The intestine is usually loosely coiled within the abdominal cavity; however, in some petrels the small intestine has very tight helical coils (Mitchell, 1901; Fitzgerald, 1969; Imber, 1985).

The intestinal mucosa contains villi and crypts of Lieberkühn (Fitzgerald, 1969; Hill, 1971; Ziswiler, 1985; Bezuidenhout and Van Aswegen, 1990). Epithelial cells of the villi have about 10^5 microvilli per square millimeter on their apical surface, increasing the absorbing surface area 15-fold. The villi contain a rich capillary bed, which picks up the absorbed nutrients and transfers them to the portal blood vessels, which go to the liver. In some species (e.g., quail), but not all (e.g., chicken, Ostrich), the villi have a central lacteal, similar to that in mammals, which facilitates the collection of interstitial fluids. The thickness of the intestinal mucosa decreases gradually along the length of the intestine, as the villi become shorter and the crypts decrease in depth. The avian mucosal epithelium does not contain an equivalent of mammalian Brünner's glands, but numerous goblet cells secrete copious mucus, which protects the epithelium from digestive enzymes and abrasion by the digesta. The mucus is particularly thick along the anterior duodenum, where it protects the villi from the excessive acidity of the digesta leaving the gizzard. Two muscle layers, the inner circular and outer longitudinal, surround the intestine and are responsible for the mixing and movement of the digesta.

Intestinal epithelial cells are formed in the base of the crypts and migrate up the villi while attaining digestive and absorptive capabilities (Imondi and Bird, 1966; Klob and Starck, 1993; Starck, 1997). After reaching the apex of the villi, the cells are shed into the lumen of the intestine resulting in an appreciable amount of endogenous nutrients entering the small intestine. The turnover time of epithelial cells is in the range of 2–4 days in growing chickens, Common Starling, and Japanese Quail, but the rate of division slows by a factor of two or more in adults. The length of the villi is adaptable, being determined by the rates of cell division within the crypts versus the rate of loss of cells at the apex.

Across species, the most variable aspects of the small intestine are its overall length, the size and shape of the villi, and the degree to which microorganisms populate the ileum. All three of these factors are subject to some variation within a single bird, depending upon the composition of its diet (Chapter 4). Relative to body size, the small intestine tends to be long in herbivores and granivores and relatively short in carnivores, nectarivores, and frugivores (Herpol and van Grembergen, 1967; Ziswiler and Farner, 1972). This is because the enzymatic digestion of meat or fruit is proficient and rapid compared with the digestion of seeds or vegetation, which are rich in cell walls. The relationship between diet and intestine length is somewhat confounded by taxonomic position and holds better within orders than between orders. The surface area of the intestine increases at 0.72 power of body mass, which keeps it proportional to the bird's metabolic rate (see Fig. 2.3). The villi of carnivorous birds are more well developed and finger-like than the flatter villi of herbivores, partially compensating for the decreased length of their small intestine.

In some herbivorous species, the posterior region of the small intestine is richly populated with microorganisms and may be involved in the fermentation of nutrients that have withstood digestion in the more anterior intestine. Although the ileum is rarely the major site for microbial fermentation, it may function as a transition between the digestive functions that use enzymes produced by the bird and the microbial-based digestion that occurs in the more posterior tract.

The ceca originate from the rectum at a location immediately posterior to the junction of the small intestine and the rectum (see Fig. 2.4). There is a very large degree of variation in the size of ceca among species, ranging from voluminous paired ceca to a single cecum to complete absence. For those species that have them, the most common arrangement is two ceca of approximately equal length with separate openings into the rectum. The ceca are highly developed in herbivores and omnivores, where they serve as a site for microbial fermentation of complex carbohydrates that resist digestion in the small intestine. In some species they may serve water- and nitrogen-absorption or immunosurveillance functions. Within closely related species, the size of the ceca often correlates with the diet type. For example, within North American Anatidae, the carnivorous duck species have much smaller ceca than omnivorous species, and the ceca of herbivores are largest of all (Kehoe and Ankney, 1985).

Various classification schemes have been used for the avian ceca. Table 2.1 shows a classification based strictly on the size of the ceca. When a combination of size and histological criteria is used, ceca can be divided into four general types: intestinal; glandular; lymphoepithelial; and vestigial (Mitchell, 1901). Intestinal ceca are well developed and histologically similar to the intestines (e.g., some species of ducks, geese, grebes, and ratites). In some species they are relatively large, and rich in lymphoid tissue (e.g., many Galliformes, such as chickens, turkeys, and quail). Herbivorous birds, such as ratites, screamers, tinamou, and many species of grouse, have large ceca that are highly sacculated or have spiral out-pouches, which provide an area for microbial fermentation (Bezuidenhout, 1993). Glandular ceca are usually long and have conspicuous goblet cells and secretory glands in the epithelium (e.g., owls). The function of this type of ceca appears to be related to water absorption and nitrogen excretion, but is not well characterized. Lymphoepithelial ceca are found in Passeriformes and some species of doves and pigeons. These ceca are usually small, have negligible fermentation function, and serve as a secondary lymphoid tissue involved in immunosurveillance of the ileum and rectum. Vestigial (or absent) ceca are found in penguins, petrels, hawks, parrots, lorikeets, woodpeckers, swifts, humming-birds, and many doves and pigeons. Upon histological examination, a nodule of lymphatic tissue is commonly seen within the wall of the anterior rectum, but no distinct ceca are visible macroscopically. More than two-thirds of all species of birds either lack ceca completely or their ceca are very small and lympho-epithelial.

The chicken has three distinct regions in its ceca: proximal, middle, and distal (Strong *et al.*, 1989, 1990). The proximal region has well-developed villi with large numbers of microvilli, lymphoid cells in the lamina propria, and many goblet cells in the epithelium. In the proximal region are large aggregated lymphoid nodules, the cecal tonsils, which are visible macroscopically. Each tonsil extends over much of the circumference of the cecum and protrudes into the lumen. The tonsils have longer villi than the adjacent tissue and they actively sample the cecal contents, permitting immunosurveillance of the lower gastrointestinal tract. The cecum walls are thin in the middle region and have well-developed

Table 2.1. Cecal characteristics generalized by families as summarized by Clench and Mathias (1995).

Family	Examples	Size			
		Large	Moderate	Vestigial	Absent
Struthioniformes	Ostrich	X			
Rheiformes	Rheas	X			
Casuariiformes	Cassowaries		X		
	Emus				
Apterygiformes	Kiwis	X			
Tinamiformes	Tinamous	X			
Sphenisciformes	Penguins			X	
Gaviiformes	Loons		X		
Podicipediformes	Grebes		X	X	
Procellariiformes	Albatrosses			X	
	Shearwaters				
	Petrels				
Pelecaniformes	Cormorants			X	
	Boobies				
	Pelicans				
Ciconiformes	Herons			X	
	Storks				
	Ibises				
Anseriformes	Geese	X	X		
	Ducks				
	Swans				
Falconiformes	Vultures			X	
	Eagles				
	Hawks				
Galliformes	Grouse, quail	X			
	Turkeys				
	Chicken				
Gruiformes	Cranes	X			
	Rails, coots				
	Bustards				
Charadriiformes	Plovers	X	X	X	
	Snipes				
	Stilts, auks				
Columbiformes	Pigeons			X	X
	Doves				
Psittaciformes	Lories				X
	Parrots				
	Macaws				
Cuculiformes	Turacos	X	X		X
	Cuckoos				
	Roadrunners				
Strigiformes	Owls	X			

Table 2.1. Continued.

Family	Examples	Large	Moderate	Vestigial	Absent
Caprimulgiformes	Potoos Frogmouths Goatsuckers	X			
Apodiformes	Swifts Hummingbirds				X
Coliiformes	Mousebirds				X
Trogoniformes	Trogons Quetzals		X		
Coraciiformes	Kingfishers Bee-eaters Hornbills		X	X	X
Piciformes	Barbets Toucans Woodpeckers				X
Passeriformes	Larks, finches Thrushes Jays, wrens			X	

The header over Large/Moderate/Vestigial/Absent is "Size".

longitudinal folds in the mucous membrane. In the distal region of the cecum, the folds are less developed and the villi are short and blunt, with few goblet cells. There is considerable variation in the morphology of the epithelium in the middle and distal regions across other species of Galliformes.

The junction of the ileum, paired ceca, and the rectum is surrounded by a continuous layer of muscle, which forms three sphincters in the duck and probably most birds that have large ceca (Mahdi and McLelland, 1988). These sphincters regulate the rate and direction of flow of digesta leaving the ileum. Contraction of the ileal sphincter, which surrounds the ileum at its junction with the rectum, is important in regulating the rate at which digesta leaves the ileum and in preventing the reflux of rectal contents into the ileum. The cecal sphincters determine the aperture of the entrance into the ceca. The luminal epithelium directly under the cecal sphincters has very long villi. Partial contractions of the sphincters create a narrow orifice, which causes the villi to interdigitate and act like a filter. This arrangement permits the entrance of the nutrient-rich liquid fraction of the intestinal contents into the ceca while keeping out the relatively indigestible solid fibrous fraction (Fig. 2.9). This meshwork of villi at the ileocecal junction also exists in species with very small ceca, such as the House Sparrow, but its function is not known (Klem *et al.*, 1983).

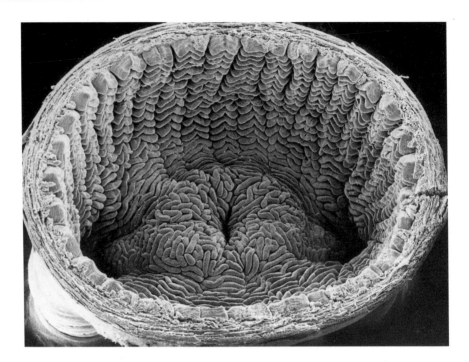

Fig. 2.9. The ileal–cecal–rectal junction of the domestic duck viewed from the rectum towards the ileum (center) and the two cloaca openings. The long ileal and cecal villi (papillae) protrude into the lumen of the rectum. Constriction of the cecal sphincters narrows the openings and permits the long villi to filter out particulate matter, while admitting the liquid fractions. (Used with permission from Mahdi and McLelland, 1988.)

Rectum

The length of intestine between the ileocecal junction and the cloaca is called the rectum. A frequently used synonym for the avian rectum is 'colon.' Typically it is very short and small in diameter compared with the large intestine of mammals and thus is not appropriately referred to as a 'large intestine.' In fact, the rectum of birds is usually smaller in diameter than the duodenum and in the Mistle Thrush, pigeon, chickens, and Japanese Quail the rectum is only 1, 3, 4, and 5%, respectively, of the total intestinal length. The rectum of herbivorous birds does not have the sacculation necessary for significant microbial fermentation that it has in many mammals. An exception is the Ostrich, in which the rectum is more than 50% of the total intestinal length and is sacculated. Histologically, the rectum is similar to the small intestine, except that the villi are shorter and the lamina propria is richer in lymph follicles in most species (Fitzgerald, 1969; Griminger, 1983; King and McLelland, 1984).

Cloaca

The rectum empties into the cloaca, which has a much larger diameter than the rectum (Fig. 2.10.). The cloaca serves as a storage area for urine and feces and it

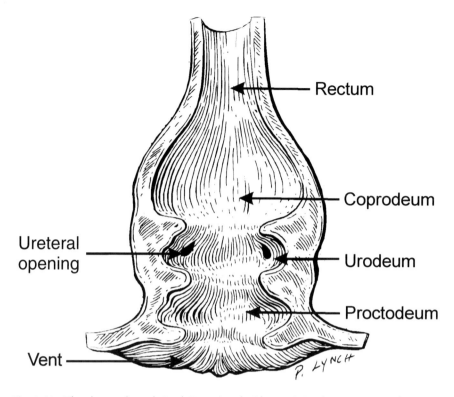

Fig. 2.10. The cloaca of a male Rock Dove (used with permission from Proctor and Lynch, 1993).

receives the ureters and the exit ducts of the reproductive system (King and McLelland, 1984; Duke *et al.*, 1995). Two folds of the mucosa divide the cloaca into three compartments: coprodeum, proctodeum, and urodeum. The anterior region of the cloaca (coprodeum) receives the rectum, and the midregion (urodeum) receives the ureters and reproductive system. The most posterior region (proctodeum) opens externally through the vent. The mucosal fold that separates the coprodeum from the urodeum can act as a diaphragm to prevent the movement of feces into the urodeum. Upon defecation, this diaphragm can be everted through the vent, permitting the expulsion of feces without passing through and contaminating the urodeum and proctodeum. In the chicken, this fold closes during egg laying by the female and ejaculation by the male to prevent fecal contamination of the egg or semen, respectively. The cloaca may also act as a bladder in a few species, e.g., some sea birds. The expandable bladder-like proctodeum of the Ostrich permits urination separate from defecation. The bursa of Fabricius is a prominent diverticulum of the dorsal cloaca. It serves as a location for the differentiation of B lymphocytes in the immature chick and becomes a secondary lymphoid organ involved in immunosurveillance of the lower gastrointestinal tract later in life.

The vent is usually a transverse slit, which includes lips on both the dorsal and ventral sides, which project inward into the proctodeum. The proctodeum and the vent are surrounded by voluntary muscles, which form a sphincter. The voluntary nature of these muscles gives the bird some control over the timing of its defecation. During defecation, the dorsal and ventral lips are partly everted, forming a circular orifice for passage of the feces and urine.

Blood drainage

In chickens, blood from venules draining the duodenum and parts of the ileum and ceca joins the gastropancreaticoduodenal vein. Blood from venules in the jejunum and the rest of the ileum and ceca drains into the cranial mesenteric vein. Blood is drained from the rectum by the caudal mesenteric vein. These three veins form the right hepatic portal vein. The left portal vein receives blood from parts of the proventriculus and gizzard. The two portal veins transport nutrients from the gastrointestinal tract to the liver (McLelland, 1975).

Accessory organs

The liver, gallbladder, and pancreas are important accessory organs of the digestive system. The liver has two primary lobes and is usually larger in piscivores and insectivores than in florivores. At hatching, the liver has a yellow color, caused by the uptake of large amounts of yolk lipids and pigments. The characteristic dark red color is obtained gradually over the next weeks but becomes yellowish again in laying hens. The primary nutritional role of the liver is metabolism of absorbed nutrients (Chapters 6–10) and the production of bile acids and bile salts. Among a limited sample of wild birds, chenodeoxycholic acid is the predominant bile acid, except among faunivores, where cholic and allocholic acids are more common. β-Phocacholic acid is a major component in the bile of ducks, geese, and flamingos. Bile acids are usually conjugated to taurine, but glycine conjugation occurs in some species. Bile salts, along with cholesterol and phospholipids, are secreted into the bile canaliculi and collected by the bile ducts. Depending on the species, the duct from the right side of the liver may enlarge into or branch into a gallbladder. In some species, a gallbladder is absent (e.g., Ostrich, hummingbirds, and many species of passerines, doves, pigeons, and parrots). In ducks, bile from the left side of the liver can reach the gallbladder through a common sinus. In the chicken, bile from the left duct drains directly into the duodenum without being stored in the gallbladder. The location of the entrance for the bile ducts into the duodenum varies among species. In pigeons, one bile duct enters the anterior duodenal loop and the second enters in the posterior duodenum (Crompton and Nesheim, 1972; Sturkie, 1976; Hagey *et al.*, 1990, 1994).

The pancreas lies within the duodenal loop. The digestive enzymes produced in the tubuloacinar glands are collected into one, two, or three ducts that enter the duodenum, usually near the entrance of the bile duct. Avian pancreatic juice contains enzymes similar to those of mammals, including amylase, lipases, trypsin, chymotrypsin, carboxypeptidases A, B, and C, deoxyribonucleases, ribonucleases, and elastases. The pancreas also produces bicarbonate (HCO$_3$), which buffers the intestinal pH (Krogdahl and Sell, 1989; Pubols, 1991).

References

Akester, A.R. (1986) Structure of the glandular layer and koilin membrane in the gizzard of the adult domestic fowl (*Gallus gallus domesticus*). *Journal of Anatomy* 147, 1–25.

Belman, A.I. and Kare, M.R. (1961) Character of salivary flow in the chicken. *Poultry Science* 40, 1377–1388.

Berkhoudt, H. (1985) Special sense organs: structure and function of the avian taste receptors. In: King, A.S. and McLelland, J. (eds) *Form and Function in Birds*, Vol. 3. Academic Press, New York, pp. 462–496.

Bezuidenhout, A.J. (1993) The spiral fold of the caecum in the ostrich (*Struthio camelus*). *Journal of Anatomy* 183, 587–592.

Bezuidenhout, A.J. and Van Aswegen, G. (1990) A light microscopic and immunocytochemical study of the gastrointestinal tract of the Ostrich (*Struthio camelus*). *Onderstepoort Journal of Veterinary Research* 57, 37–48.

Bock, W.J. (1964) Kinetics of the avian skull. *Journal of Morphology* 114, 1–42.

Bock, W.J. (1966) An approach to the functional analysis of bill shape. *Auk* 83, 10–51.

Bonser, R.H.C. and Witter, M.S. (1993) Indentation hardness of the bill keratin of the European starling. *Condor* 95, 736–738.

Burton, P.J.K. (1974) *Feeding and the Feeding Apparatus in Waders.* British Museum, London.

Buyse, J., Adelsohn, D.S., Decuypere, E. and Scanes, C.G. (1993) Diurnal nocturnal changes in food intake, gut storage of ingesta, food transit time and metabolism in growing broiler chickens – a model for temporal control of energy balance. *British Poultry Science* 34, 699–709.

Clench, M.H. and Mathias, J.R. (1995) The avian cecum – a review. *Wilson Bulletin* 107, 93–121.

Crompton, D.W.T. and Nesheim, C. (1972) A note on the biliary system of the domestic duck and a method for collecting bile. *Journal of Experimental Biology* 56, 545–550.

Degen, A.A., Duke, G.E. and Reynhout, J.K. (1994) Gastroduodenal motility and glandular stomach function in young ostriches. *Auk* 111, 750–755.

Dorst, J. (1974) *The Life of Birds*, 1st edn. Columbia University Press, New York.

Duke, G.E., Place, A.R. and Jones, B. (1989) Gastric emptying and gastrointestinal motility in Leach's storm petrel chicks. *Auk* 106, 80–85.

Duke, G.E., Degen, A.A. and Reynhout, J.K. (1995) Movement of urine in the lower colon and cloaca of ostriches. *Condor* 97, 165–173.

Dumont, J.N. (1965) Prolactin-induced cytologic changes in the mucosa of the pigeon crop during crop 'milk' formation. *Zeitschrift fuer Zellforschung* 68, 755–782.

Fisher, H.I. and Dater, E.E. (1961) Esophageal diverticula in the redpoll, *Acanthis flammea*. *Auk* 78, 528–531.

Fitzgerald, T. (1969) *The Coturnix Quail: Anatomy and Histology*, 1st edn. Iowa State University Press, Ames, 306 pp.

Fowler, M.E. (1991) Comparative clinical anatomy of ratites. *Journal of Zoo and Wildlife Medicine* 22, 204–227.

Gionfriddo, J.P. and Best, L.B. (1996) Grit-use patterns in North American birds: the influence of diet, body size, and gender. *Wilson Bulletin* 108, 685–696.

Goodman, D.C. and Fisher, H.I. (1962) *Functional Anatomy of the Feeding Apparatus in Waterfowl.* Southern Illinois University Press, Carbondale.

Grajal, A., Strahl, S.D., Parra, R., Dominguez, M.G. and Neher, A. (1989) Foregut fermentation in the hoatzin, a neotropical leaf-eating bird. *Science* 245, 1236–1238.

Griminger, P. (1983) Digestive system and nutrition. In: Abs, M. (ed.) *Physiology and Behaviour of the Pigeon.* Academic Press, London.

Hagey, L.R., Schteingart, C.D., Ton-Nu, H.T., Rossi, S.S., Odell, D. and Hofmann, A.F. (1990) β-Phocacholic acid in bile: biochemical evidence that the flamingo is related to an ancient goose. *Condor* 92, 593–597.

Hagey, L.R., Schteingart, C.D., Ton-Nu, H.T. and Hofmann, A.F. (1994) Biliary bile acids of fruit pigeons and doves (Columbiformes): presence of 1-beta-hydroxychenode-oxycholic acid and conjugation with glycine as well as taurine. *Journal of Lipid Research* 35, 2041–2048.

Hainsworth, F.R. and Wolf, L.L. (1972) Crop volume, nectar concentration and hummingbird energetics. *Comparative Biochemistry and Physiology* 42A, 359–366.

Healy, S. and Guilford, T. (1990) Olfactory-bulb size and nocturnality in birds. *Evolution* 44, 339–346.

Herpol, C. and van Grembergen, G. (1967) L'activité protéolytique du système digestif de *Gallus domesticus. Zeitschrift fuer Vergleichende Physiologie* 57, 1–6.

Hill, K.J. (1971) The physiology of digestion. In: Bell, D.J. and Freeman, B.M. (eds) *Physiology and Biochemistry of the Domestic Fowl,* Vol. 1. Academic Press, London, pp. 1–49.

Horseman, N.D. and Buntin, J.D. (1995) Regulation of pigeon cropmilk secretion and parental behaviors by prolactin. *Annual Review of Nutrition* 15, 213–238.

Imber, M.J. (1985) Origins, phylogeny and taxonomy of the gadfly petrels *Pterodroma* spp. *Ibis* 127, 197–229.

Imondi, A.R. and Bird, F.H. (1966) The turnover of intestinal epithelium in the chick. *Poultry Science* 45, 142–147.

Kehoe, F.P. and Ankney, C.D. (1985) Variation in digesta organ size among five species of diving ducks (*Aythya* spp.). *Canadian Journal of Zoology* 63, 2339–2342.

King, A.S. and McLelland, J. (1984) *Birds: Their Structure and Function,* 2nd edn. Baillière Tindall, London..

Klem, D.J., Finn, S.A. and Nave, J.H. (1983) Gross morphology and general histology of the ventriculus, intestinum, caeca and cloaca of the house sparrow (*Passer domesticus*). *Proceedings of the Pennsylvania Academy of Science* 58, 151–158.

Klob, E. and Starck, J.M. (1993) Immunohistochemischer Nachweis der Zellproliferation und Zellwanderung im Dermepithel des Zebrafinken (*Taeniopygia guttata*). *Verhandlungen der Deutscher Zoologischen Gesellschaft* 86, 165–172.

Krogdahl, A. and Sell, J.L. (1989) Influence of age on lipase, amylase and protease activities in pancreatic tissue and intestinal contents of young turkeys. *Poultry Science* 68, 1561–1568.

McLelland, J. (1975) Aves: digestive system. In: Getty, R. (ed.) *The Anatomy of the Domestic Animals,* Vol. 2. Saunders, Philadelphia, pp. 1857–1882.

Mahdi, A.H. and McLelland, J. (1988) The arrangement of the muscle at the ileo-caeco-rectal junction of the domestic duck (*Anas platyrhynchos*) and the presence of anatomical sphincters. *Journal of Anatomy* 161, 133–142.

Mitchell, P.C. (1901) On the intestinal tract of birds, with remarks on the valuation and nomenclature of zoological characters. *Transactions of the Linnean Society of London, Zoology* 8, 173–275.

Navarro, J.L. and Bucher, E.H. (1992) Capsicin effects on consumption of food by cedar waxwings and house finches. *Wilson Bulletin* 104, 549–551.

Obst, B.S. and Nagy, K.A. (1993) Stomach oil and the energy budget of Wilson's storm-petrel nestlings. *Condor* 95, 792–805.

Pernkopf, E. and Lehner, J. (1937) *Vorderdarm. Vergleichende Beschreibung des Vorderdarm bei den einzelnen Klassen der Kranoiten.* Urban und Schwarzer, Berlin, pp. 349–376.

Piersma, T., Koolhaas, A. and Dekinga, A. (1993) Interactions between stomach structure and diet choice in shorebirds. *Auk* 110, 552–564.

Place, A.R., Stoyan, N.C., Ricklefs, R.E. and Butler, R.G. (1989) Physiological basis of stomach oil formation in Leach's storm petrel, *Oceanodroma leucorhoa. Auk* 106, 687–699.

Poulin, B. and Lefebvre , G. (1995) Additional information on the use of tartar emetic in determining the diet of tropical birds. *Condor* 97, 897–902.

Prevost, J. and Vilter, V. (1963) Histologie de la sécrétion œsophagienne du manchot empereur. *Proceedings International Ornithological Congress* 2, 1085–1094.

Pritchard, P.J. (1972) Digestion of sugars in the crop. *Comparative Biochemistry and Physiology* 43A, 195–205.

Proctor, N.S. and Lynch, P.J. (1993) *Manual of Ornithology.* Yale University Press, New Haven.

Pubols, M.H. (1991) Ratio of digestive enzymes in the chick pancreas. *Poultry Science* 70, 337–342.

Ricklefs, R.E. (1996) Morphometry of the digestive tracts of some passerine birds. *Condor* 98, 279–292.

Soler, J.J., Soler, M. and Martinez, J.G. (1993) Grit ingestion and cereal consumption in five corvid species. *Ardea* 81, 143–149.

Spitzer, G. (1972) Jahreszeitliche Aspekte der Biologie der Bartmeise (*Panurus biarmicus*). *Journal für Ornithologie* 113, 241–275.

Starck, J.M. (1997) Intestinal growth in altricial European starling (*Sturnus bulgaris*) and precocial Japanese quail (*Coturnix coturnix japonica*). *Acta Anatomica* 156, 289–306.

Stevens, C.E. and Hume, I.D. (1995) *Comparative Physiology of the Vertebrate Digestive System.* Cambridge University Press, Cambridge.

Stresemann, E. (1927–1934) *Handbuch der Zoologie. Sauropsida: Aves.* W. de Gruyter, Berlin.

Strong, T.R., Reimer, P.R. and Braun, E.J. (1989) Avian cecal microanatomy – a morphometric comparison of 2 species. *Journal of Experimental Zoology,* Suppl. 3, 10–20.

Strong, T.R., Reimer, P.R. and Braun, E.J. (1990) Morphometry of the galliform cecum – a comparison between Gambel quail and the domestic fowl. *Cell and Tissue Research* 259, 511–518.

Sturkie, P.D. (1976) *Avian Physiology,* 3rd edn. Springer-Verlag, New York.

Turk, D.E. (1982) The anatomy of the avian digestive tract as related to feed utilization. *Poultry Science* 61, 1225–1244.

Wackernagel, H. (1964) Was futtern die Flamingos ihren Jungen? *Internationale Zeitschrift fuer Vitaminforschung* 34, 141–143.

Ziswiler, V. (1985) Function and structure of the alimentary tract as an indicator of evolutionary trends. *Fortschritte der Zoologie.* 130, 295–303.

Ziswiler, V. (1990) Specialisation in extremely unbalanced food: possibilities and limits of its investigation exclusively by functional morphology. *Netherlands Journal of Zoology* 40, 299–311.

Ziswiler, V. and Farner, D.S. (1972) Digestion and the digestive system. In: Farner, D.S., King, J.R. and Parkes, K.C. (eds) *Avian Biology,* Vol. 2. Academic Press, New York, pp. 343–430.

CHAPTER 3
Digestion of Food

The uptake of nutrients is dependent on their prehension, digestion, and absorption from the gastrointestinal tract and requires the coordinated effort of all of the digestive organs described in the previous chapter. As in mammals, digesta move in the direction of mouth to vent and the food is denatured and hydrolyzed along this path. However, in birds this posterior flow of digesta is often interrupted by refluxes in the opposite direction, known as retrograde flow. Retrograde movement of digesta occurs between: (i) the proventriculus and the gizzard; (ii) the small intestine and the gizzard; (iii) the rectum and the ceca; and (iv) the cloaca and the rectum. As food passes through the gastrointestinal tract, two types of digestion may occur, autoenzymatic and alloenzymatic. Autoenzymatic digestion is due to the action of enzymes of bird origin, which are produced in the proventriculus, small intestine, pancreas, and possibly other organs of the tract (Table 3.1). Alloenzymatic digestion is due to enzymes of microbial origin, usually in the lumen of the ceca or rectum. Because of the low oxygen tension in these areas, fermentation is the primary microbial activity, and fermentation and alloenzymatic digestion are often used synonymously. The physiological and biochemical aspects of digestion follow similar patterns among different avian species and between birds and mammals. The following description details the processes as they occur in a typical omnivorous bird and also gives general adaptations found in a wide variety of birds. Differences due to adaptation to more specialized diets (e.g., carnivory, herbivory, frugivory) are detailed in Chapter 4.

The complete digestion and assimilation of complex foods requires substantial time, and there is considerable debate about the trade-offs between maximizing the efficiency of digestion and maximizing the rate of digestion. These questions exist at both the physiological level and the ecological level. At the physiological level, complete digestion and absorption of nutrients follows the principle of diminishing returns. Even the simplest food nutrients, such as monosaccharides and free amino acids, require an investment in intestinal transporters and a large intestinal surface area of very high structural integrity. Complex nutrients, such as starch, protein, and fat, require considerably more investment, including mechanical grinding (muscle contraction), hydrolytic enzymes (protein synthesis), and other digestive factors, such as bile acids and hydrochloric acid (HCl). Digestion of plant structural components requires additional mechanical grinding and substantial investments for microbial assistance, including their housing (ceca), husbandry (warmth, moisture, nitrogen), and immunosurveillance. Quantitatively, these costs may not always be worth the returns and maximum digestive efficiency may not always be appropriate. In birds, the cost–benefit ratio of high digestive efficiency is skewed by the demands of flight, because large digestive organs and the digesta contained in these organs compromise flight. This trade-off is important at the ecological level, because acrobatic flight is often necessary to obtain food and avoid predators. Conversely, obtaining more food to compensate for nonoptimal digestion is associated with costs, including increased foraging time and energy expenditure; more competition with others; increased exposure to predators; higher consumption of toxins in foods; and the threat of exhausting food supplies. Chapter 4 provides a framework for the diverse patterns of digestive strategies that reflect these compromises.

Movement and Autoenzymatic Digestion of Food

Prehension and swallowing

A wide variety of food-prehension techniques are used by birds, including pecking, gleaning, scooping, spearing, grasping, and filtering. The beak is well adapted for the prehension of preferred dietary items and often functions together with the tongue in processing food by sorting, hulling, tearing, or crushing.

The anticipation of eating or the presence of food in the mouth stimulates the secretion of saliva. Saliva may contain low amounts of amylase in some granivorous birds, such as the House Sparrow, but not in others, such as the chicken. Due to the low level of enzymes, minimal mechanical action, and short time prior to swallowing, little breakdown of food occurs in the mouth (Jerret and Goodge, 1973).

Swallowing usually occurs quickly after procurement of food. The tactile stimulation of the tongue by food causes a series of rapid posterior tongue movements, which push the food to the pharynx. The choanal opening closes to prevent food from entering the nasal cavity and the tongue occludes the glottis to keep food from entering the larynx. The posterior movement of food is guided by the posteriorly directed papillae on the tongue and the roof of the mouth, in combination with the posterior thrusting of the laryngeal mound. In many species, a quick forward thrust of the head aids in the propulsion of the food posteriorly. The esophagus moves forward to receive the food and propels it downward by a constriction of muscles behind the bolus of food and dilatation of muscles ahead. This peristaltic action delivers the food bolus to the proventriculus or crop. The movement of food is lubricated by saliva and mucus secreted within the esophagus (Suzuki and Nomura, 1975).

Drinking water requires similar coordination of the tongue, laryngeal mound, and esophagus as swallowing food. To initiate the process, water is trapped in the oral cavity and, in most species, elevation of the head results in the downward flow into the esophagus. Backward movement of water may also be accomplished by lapping, as in parrots.

Table 3.1. Autoenzymatic digestion in birds (modified from Stevens and Hume, 1995).

Substrate	Extracellular enzymes and intermediates	Intestinal mucosal enzymes	End products
Carbohydrates			
Amylose Amylopectin Glycogen	α–Amylase → Maltose, Isomaltose, α-1, 4-Oligosaccharides	Maltase Isomaltase	Glucose
Chitin	Chitobiose	Chitobiase	Glucosamine
Sucrose		Sucrase	Glucose Fructose
Lactose	⊘	Lactase	Glucose Galactose
Trehalose	?	Trehalase	Glucose
Lipids			
Triglycerides	Lipase Colipase		Monoglyceride 2 Fatty acids
Phospholipids	Phospholipase	Phosphatase	Alcohol Fatty acids Phosphate
Cholesterol esters	Cholesterol esterase		Cholesterol Fatty acid
Waxes	Lipase Esterases		Monohydric alcohol Fatty acid

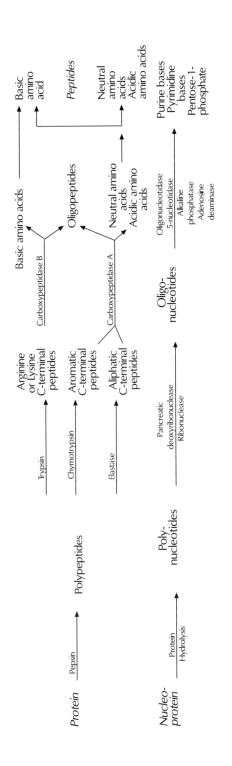

⊘ not found in Aves.

Crop and gizzard

When the gizzard is empty, food enters it immediately (Levey and Duke, 1992). When the gizzard is full, food is stored in the esophagus or crop. For example, Cedar Waxwings can consume sufficient fruit in one foraging trip for two fillings of the gizzard. The fruits consumed first move quickly into the gizzard and the second 'meal' of fruits remains in the esophagus until the gizzard finishes processing the first. In birds with a crop, the entrance of the food bolus is controlled by the distention of the gizzard. When the gizzard is filled with digesta, the relaxing of the esophagus permits the food bolus to be deflected into the crop. The addition of saliva to food during swallowing, mucus secreted into the crop, and water consumed following a meal help to soften the food during its residence in the crop. It has been suggested that some species may begin to digest starches in the crop, due to physiologically relevant secretions of amylase into the crop or due to salivary amylase. Glucose present in the food or released by amylase can be absorbed from the crop. More significant digestion may occur due to the presence of enzymes in the food itself or from microbial action in the warm moist environment of the crop (Soedarmo *et al.*, 1961; Ziswiler and Farner, 1972).

Emptying of the crop or esophagus plays an important role in regulating the rate of passage of digesta through the entire tract. The rate of emptying depends on a number of factors, including the capacity of the crop and the fullness of the gizzard. From the crop or esophagus, the food bolus moves quickly through the proventriculus (except in birds such as penguins and petrels that store food in their proventriculus). Passage through the proventriculus coats food particles with acid and pepsin. The secretions of the proventriculus have a pH of about 2, but the digesta usually buffers some of the acid, resulting in a slightly higher pH within the proventriculus and gizzard (Fig. 3.1). The act of feeding initiates the release of proventricular secretions, with high-protein foods inducing the highest quantity of secretion. Control of gastric secretions and emptying is coordinated by vagal input and the action of the hormones gastrin, pancreatic polypeptide, cholecystokinin, and secretin. Studies in a variety of species have shown a complex cycle of proventricular, gizzard, and duodenal contractions that propels the food in alternate directions between these three organs. This refluxing of digesta permits cycles of grinding, followed by fresh pepsin and HCl addition, which promote protein digestion, initiate the breakup of large lipid globules, and homogenize the food. This process also serves to retain lipid components in the anterior region of the tract, so that they can be digested more slowly than the protein and carbohydrate components of the meal (Duke, 1986).

The gizzard contracts rhythmically at a rate of about three contractions per minute in the chicken and turkey (Hill, 1971; Duke, 1986). The frequency and the amplitude of gizzard contractions increase with the coarseness of the food. A contraction sequence in chickens can be divided into three distinct phases. Phase 1 begins with the gizzard full of digesta. During this phase, maximal contraction of the pair of thin muscles of the gizzard forces some of the more liquid digesta into the duodenum. There is no sphincter at the junction between the gizzard and the duodenum, but large pieces of digesta are retained in the gizzard due to the filtering action of the gizzard's mucosal ridges at the entrance to the duodenum.

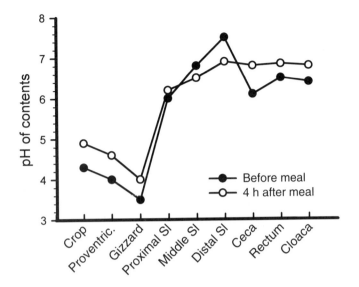

Fig. 3.1. The pH of gastrointestinal contents of a domestic goose. Samples were taken from the indicated areas either prior to a meal or 4 h after consumption of a meal. SI = small intestine; proventric. = proventriculus. (From Clemens *et al.*, 1975.)

In the second phase, a wave of contraction starts in the small intestine and moves forward through to the proventriculus. When this wave passes through the gizzard, the asymmetric contraction of the thick pair of muscles grinds the cuticle surfaces together. Food is ground into smaller particles during this period and some is also pushed back into the proventriculus. The anterior direction of this contraction wave prevents food from being forcefully expelled into the duodenum. In the third phase, a final wave of contraction begins in the proventriculus and moves digesta back into the gizzard. Repetition of this cycle serves to retain coarse material for repeated grinding while expressing fluid material into the small intestine for subsequent digestive actions. The cycle also continuously mixes the digesta with the gastric fluid. Protein denaturation by acid and partial hydrolysis by pepsin constitute the primary digestive activity in the gizzard.

Phase 2 contractions can be amazingly strong and, in conjunction with grit, serve to crush the food and grind the edges of the resulting particles together (Kato, 1914; Stresemann, 1927–34). Intraluminal pressures have been reported to be in the following ranges: Common Buzzard, 20 mmHg; duck, 178 mmHg; chicken, 100-200 mmHg; and goose, 257 mmHg. In the turkey, the pressure is sufficient to completely crush 24 walnuts (in the shell) in less than 4 hours. The length of time that food material spends in the gizzard depends largely on its size. Small particles and liquid components pass through in minutes, whereas hard grains may remain in the gizzard for several hours. Especially strong reverse peristalsis of digesta from the duodenum back into the gizzard occurs about every

15–20 min in turkeys and almost continuously in chickens. These contractions permit the remixing of duodenal contents with gastric secretions to maximize protein digestion. Experiments in geese have demonstrated preferential reflux of large particles (> 5 mm) over smaller particles and fluid digesta. In Leach's Storm-Petrel the frequent duodenal reflux may be helpful in digestion of a high-fat diet (Dziuk and Duke, 1972; Clemens *et al.*, 1975; Sklan *et al.*, 1978).

Small intestine

The vast majority of enzymatic digestion and absorption of nutrients occurs in the small intestine. Enzymes are secreted by the intestine or arrive from the pancreas and proventriculus (Table 3.1). Release of pancreatic and intestinal secretions is stimulated by duodenal distention, HCl, vagal stimulation, secretin, vasoactive intestinal peptide, and cholecystokinin (Hill, 1971; Satoh *et al.*, 1995). Cholecystokinin release is induced by peptides, amino acids, and fat. The interplay of these regulatory factors ensures that the amount of individual digestive enzymes released approximates the type and amount of substrate in the digesta. For example, high-carbohydrate, low-fat meals result in greater amounts of amylase and lower levels of lipase secretion (Hulan and Bird, 1972).

Many pancreatic enzymes are secreted as inactive zymogens, which must be activated in the lumen of the small intestine by enterokinase (Vonk and Western, 1984). Enterokinase is initially bound to intestinal epithelial cells and is released into the lumen of the intestine by the action of bile salts and pancreatic secretions. It then hydrolyzes a specific peptide bond in trypsinogen, giving active trypsin. Trypsin in turn catalyzes hydrolysis of a variety of other pancreatic enzymes, including chymotrypsinogen, procarboxypeptidase, proelastase, and prophospholipase. Some enzymes, such as amylase, deoxyribonuclease, ribonuclease, elastase and the lipases, do not require proteolytic activation. In general, hydrolysis of food macromolecules in the lumen of the small intestine by pancreatic enzymes results in smaller oligomers. These are further hydrolyzed at the enterocyte brush border to constitutive molecules, such as sugars, amino acids, and nucleotides, which are then absorbed. The spectrum of enzyme activities in the brush border of the small intestine of birds is similar to that in mammals and includes phosphatases, disaccharidases, and peptidases. A notable difference is that lactase has yet to be found in the intestine of those birds examined. Trehalase activity has been reported in turkeys, but its presence in other species remains to be confirmed (Zoppi and Shmerling, 1969; Sell *et al.*, 1991).

Intestinal pH ranges from about 5.6 to 7.2 and increases posteriorly, due to pancreatic secretions and buffers secreted by the intestinal epithelium (Herpol and van Grembergen, 1967; Hurwitz and Bar, 1968). Pancreatic and intestinal digestive enzymes usually have pH optima in this range and are fully active. The movement of digesta through the intestine of birds is due to peristaltic and segmenting contractions. In the ileum of the turkey, contractions occur at a mean frequency of four per minute, punctuated by occasions of more frequent and intense contractions. In addition to moving the digesta along, intestinal contractions probably serve to mix the digestive fluids and digesta. Viscosity of the digesta increases during passage from the duodenum to the terminal ileum, due to the loss of water and soluble carbohydrates and the relative concentrating of cellulose, pectins, and other fibers. From studies in Galliformes and ducks, it appears that there is a massive secretion and reabsorption of water through the intestines, which may amount to two to three times the volume of water consumed each day. Daily secretion and absorption of sodium (Na) approaches the bird's total pool size (Duke *et al.*, 1975; Thomas and Skadhauge, 1989a,b).

All of the digestive enzymes, mucosal secretions, and exfoliated cells that enter the lumen of the digestive tract are subject to digestion along with dietary nutrients. These endogenous sources of nutrients are surprisingly high in quantity but are recovered with little loss. In fasting chickens and Ring-neck Pheasants, strong retrograde contractions move endogenous secretions from the posterior ileum anteriorly, apparently to digest and recycle these nutrients (Clench and Mathias, 1995).

Ceca and rectum

The fate of digesta leaving the small intestine is variable, depending on the species, age, diet, and state of hydration of the bird. In birds with functional ceca, the digesta moves through the ileocecal junction and water, electrolytes, small particles, and the water-miscible components are directed into the ceca, leaving the large particles to continue through the rectum (Fig. 3.2). The capacity of the ceca to absorb some nutrients may approach or even surpass that of the intestine in some species (Obst and Diamond, 1989). In birds with lymphoepithelial or vestigial ceca, all of the digesta enters the rectum, which absorbs much of the remaining electrolytes and water and some nutrients.

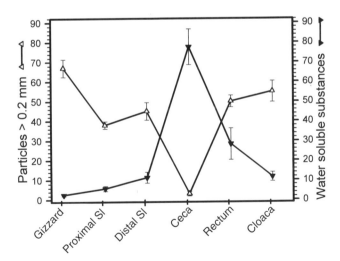

Fig. 3.2. Water-soluble substances in the digesta are concentrated in the ceca of turkeys, but large particles are excluded from this organ (Björnhag and Sperber, 1977). SI, small intestine.

Motility within the ceca results from peristaltic (filling) and reverse peristaltic (emptying) contractions, which occur approximately once every 50 s in the turkey (Duke *et al.*, 1980). More frequent contractions of lower amplitude serve to mix the cecal contents. A series of major reverse peristaltic contractions occasionally evacuate the ceca, resulting in the voiding of cecal feces, called cecotropes. In many species, removal of the ceca is easily compensated for by other physiological systems, illustrating redundancy and reserve capacity of the digestive and excretory systems. In an omnivore, such as Gambels' Quail, the ceca perform subtle water absorption, fiber digestion, and nitrogen recycling functions but the relative contribution to any single function is not dominant (Anderson and Braun, 1984). Table 3.2 lists the primary functions of the ceca in other birds.

Components of digesta that bypass the ceca move through the rectum and are either defecated or moved back to the ceca (Duke, 1986; Clench and Mathias, 1995). Peristaltic waves of long duration move the digesta posteriorly through the ceca towards the cloaca. In many species, small waves of short duration occur almost continuously and are reverse peristaltic, moving urine from the cloacal urodeum into the rectum and the ceca. This antiperistalsis has been observed in Emus, herons, ducks, geese, hawks, many Galliformes, gulls, roadrunners, owls and crows. It probably occurs in a variety of other species that have not yet been examined; however, it does not occur in the Ostrich and some parrots. The rectum and ceca are sites for water and electrolyte absorption from digesta and from the urine in those species with antiperistalsis. Nutrients, such as volatile

Table 3.2. Functions of the avian ceca.

Cecal function	Examples	Rationale
Water resorption primarily	Owl	Carnivore that forms dilute urine
Fiber digestion primarily	Ostrich	Herbivore with well-developed renal urine concentration
	Many ducks and geese	Herbivores with salt glands
Both functions	Chicken, turkey, quail, guinea fowl, ptarmigan, Emu, ducks and geese	Omnivores and florivores that must be able to adapt to many foods and environments
Nitrogen homeostasis	All species with ceca	Conservation of dietary and endogenous nitrogen
Immunosurveillance of posterior regions of tract	All species with ceca	Site of lymphatic tissue positioned to monitor and control gut microflora
None, ceca are absent or residual	Hawks, parrots, woodpeckers, swifts, hummingbirds, passerines	Cecal functions are carried out by other organs (kidney, rectum)

fatty acids and glucose, may also be absorbed in the rectum.

During defecation, there is an intense peristaltic contraction, beginning in the anterior rectum, which propels the fecal material through the entire rectum and cloaca in less than 4 s. The contents of the rectum are frequently evacuated, whereas the cecal contents are less frequently defecated. In gallinaceous birds, one or two cecotropes are eliminated daily, compared with 15 or more rectal droppings per day. In geese the number of rectal droppings can exceed more than 15 h^{-1}. In most species, cecal droppings have a light brown color and homogeneous texture, which distinguishes them from rectal droppings. For nutritional study, the components of feces can be divided into three categories: dietary; endogenous; and microbial. Dietary components include undigested macromolecules, especially fiber, and unabsorbed minerals. The endogenous components include epithelial cells, mucus, residual digestive enzymes, bile acids, and excreted metabolites, such as uric acid. The uric acid can be observed as white crystals, usually on the surface of the feces. The microbial component is especially high in cecotropes. The cecotropes are sometimes consumed by birds, a practice known as cecotrophy. The consumption of either rectal or cecal feces is known as coprophagy.

Alloenzymatic Digestion

The extent to which avian species utilize microbial assistance to digest their food (alloenzymatic digestion) is extremely variable. Alloenzymatic digestion is essentially absent in most faunivores, nectarivores, and frugivores, but species that consume large amounts of plant fiber may extensively utilize alloenzymatic digestion.

The capacity to ferment complex macromolecules, such as cellulose, pectin, and hemicellulose, markedly contributes to the energy nutrition of herbivores and, to a lesser extent, graminivores and omnivores (dietary fiber is explained in more detail in Chapter 8). Further, the location of a fermentation area has profound nutritional implications (Fig. 3.3). In general, the location of the fermentation area in the posterior part of the digestive tract is beneficial when very high-quality vegetation is consumed, and an anterior location is beneficial when poor-quality vegetation is consumed. With only a few exceptions, the fermentation area in florivorous birds is always located posterior to the gastric stomach, typically in the ceca, but occasionally in the rectum or ileum. This postgastric location permits a bird to first autoenzymatically digest food and obtain essential amino acids, fatty acids, and vitamins, prior to fermentation of the remaining fiber (strategy 2, Fig. 3.3). This strategy optimizes the nutritional value obtained from highly digestible foods (e.g., animal matter, seeds) and from vegetative matter that has a low content of cell walls, which is usually the more tender plant components, such as very young leaves, buds, and grasses. However, this strategy does not permit efficient use of high-fiber foods because the abundant cell walls are impervious to alloenzymatic digestion and they physically inhibit autoenzymatic digestion of the nutritious cell contents. In some

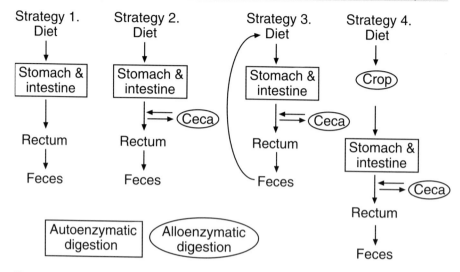

Fig. 3.3. Four strategies for handling dietary fiber in birds. Strategy 1: little or no alloenzymatic digestion and therefore little fiber utilization. Strategy 2: autoenzymatic digestion followed by selective alloenzymatic digestion of the most fermentable components of the remainder. Strategy 3: same as strategy 2, but cecotropes are reingested so that the nutrients produced by cecal microflora can be autoenzymatically digested. Strategy 4: alloenzymatic digestion followed by autoenzymatic digestion of the microbes produced during fermentation — additional alloenzymatic digestion of remaining fiber may occur in the ceca following autoenzymatic digestion (e.g., Hoatzins).

birds, the location of a fermentation area prior to the true stomach, or pregastric, permits the efficient utilization of plant materials that are high in cell walls by utilizing microbial enzymes to hydrolyze the cellulose and lignin (strategy 4, Fig. 3.3). This strategy is especially effective for obtaining energy from high-roughage plant materials, such as mature leaves, stems, branches, etc., which make up the bulk of the biomass on land. However, few birds have adopted a pregastric strategy because of the incompatibility between the weight required for these digestive structures and the demands of flight (see Chapter 4). A pregastric alloenzymatic digestive strategy also results in considerably lower nutritional returns for protein, starch, and lipids found in highly nutritious foods compared with autoenzymatic digestion. This is because fermentation of essential amino acids diminishes their nutritional value and fermentation of highly digestible starch and lipids is energetically inefficient. Thus, most birds use their capacity of flight to harvest foods that are easiest to digest with simple digestive systems, even though these foods are considerably less abundant and sometimes patchily distributed.

Pregastric fermentation

Microbial fermentation may occur in the crop of some species. This is taken to the furthest extreme in the Hoatzin, which has a very large crop, containing a rich microbial flora that make a substantial contribution to the digestion of foliage (strategy 4, Fig. 3.3). A limited amount of pregastric fermentation may also occur in less specialized species. For example, lactobacilli extensively colonize the epithelium of the crop in pigeons, quail, and chickens. When these birds are fed large meals, food remains in the crop for several hours and microbial digestion results in high levels of lactic acid production. The contribution of this microbial action to digestion of the food is minor and the major nutrients consumed by microbes are the readily available carbohydrates and amino acids, which are digested easily by the bird without microbial help. However, microbial synthesis of vitamins, which can then be digested and absorbed in the small intestine, may make a small but important contribution in deficiency situations.

Postgastric digestion

In most species, the major contribution of microbial action to the digestive process occurs in the ceca, although some fermentation may also occur in the posterior ileum (Barnes, 1979; Mead, 1989). The ceca provide a relatively stable environment for microorganisms and contain a large and complex biota. The predominant organisms present are bacteria, especially obligate anaerobes, which occur in the lumen of the chicken ceca at 10^{11} g^{-1} wet weight. At least 38 different types of anaerobic Gram-negative and Gram-positive nonsporing rods and cocci have been isolated from the chicken cecum. These bacteria are often attached to the intestinal epithelium where they exclude transient populations. The cecal mucosa of the chicken has a layer of Gram-negative bacteria approximately 200 cells deep. These bacteria are attached to the mucus secreted by the epithelium; the type of food consumed alters the type and amount of mucus secreted and thus the types of bacteria present.

The large intestine of the Ostrich has a high concentration of facultative anaerobes, which resemble the flora of mammalian ruminants in microbial ecology. Some of the bacteria population inhabiting the ceca, rectum, and ileum produces the enzyme, β-(1–4)-glycosidase, which is necessary for the hydrolysis of the 1–4 glucose bonds of cellulose. The bacteria also produce enzymes capable of at least partially digesting hemicellulose, pectin, lignin, gums, mucilages, and other complex molecules, if given sufficient time. The microbes cannot oxidize the resulting sugars, due to a lack of oxygen in the lumen of the ceca and rectum. Instead, sugars are fermented to the volatile fatty acids acetate, propionate, and butyrate. Of these, acetate production predominates, but all three can be readily absorbed from ceca and rectum (Mackie, 1987; Savory and Knox, 1991; Swart *et al.*, 1993a, b).

The contribution of microbial fermentation to the nutrition of a bird is a function of the volume of digesta present in the fermentation area and the length of time it spends there. High volumes and slow rates of passage favor microbial fermentation. In most birds, the ceca are the primary area in the digestive tract with this combination. Because a cecum is blind-ended, its contents can be retained for longer periods of time than is possible in the small intestine or rectum, where digesta moves through relatively rapidly in most birds. Figure 3.4 shows the quantity of volatile fatty acids along the gastrointestinal tract of a goose. In Emu and possibly geese, the posterior small intestine may retain digesta

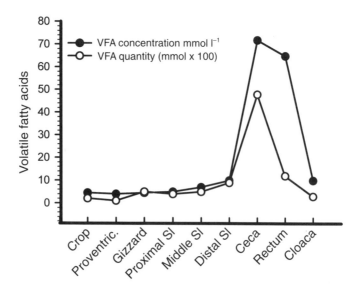

Fig. 3.4. Volatile fatty acid (VFA) levels along the gastrointestinal tract of a goose expressed both as the concentration (mmol l^{-1}) and the quantity, which equals the concentration times the volume of contents. SI, small intestine; proventric, proventriculus. (From Clemens *et al.* 1975.)

long enough to make a significant contribution to total fermentation. In Ostriches, the rectum is a major site for fermentation. The contribution of microbial fermentation to the energy requirements of some herbivores (e.g., Ostriches) may approach 50%. In the omnivorous chicken, cecal fermentation only contributes about 3–4% of the energy needs and this amount does not change with the level of dietary fiber. Apparently, the cecal capacity is met at fairly low dietary fiber levels and increments above this threshold are not diverted to the ceca but are excreted (Herd and Dawson, 1984; Buchsbaum *et al.*, 1986; Jorgensen *et al.*, 1996).

Uric acid arriving from retrograde flow of urine can be utilized by cecal anaerobes and stimulates their growth, presumably by providing a nitrogen and energy supply (Mackie, 1987). Consequently, the cecal flora provides energy-rich fermentation end products (volatile fatty acids) to the host bird, but they require nitrogen, which is supplied by the bird. This symbiotic relationship likely developed as a mutual detoxification system. The nutrients within the cellular structure of bacteria are not directly available to the bird and most are eventually excreted. Some birds consume part of their feces, especially the cecotropes, permitting the digestion of the bacteria and the absorption of the nutrients by the upper gastrointestinal tract (see strategy 3, Fig. 3.3). The extent to which this phenomenon occurs in birds and the nutritional implications are not well characterized.

The nutritional impact of bacteria in the digestive tract is normally positive, but in some situations bacteria may compete with the host for nutrients, especially vitamins, or may produce metabolites that are harmful. For example, the decarboxylation of lysine to cadaverine and histidine to histamine can induce pathology. Some of the microbial organisms present in the digestive tract have the capacity to enter the epithelium and cause deleterious infections. Birds that have large areas dedicated to fermentation must also invest considerable effort in immunosurveillance.

Digestion of Specific Nutrients

The digestion of nutrients by the gastrointestinal tract of avian omnivores is generally very similar to that in their mammalian counterparts. The complement of enzymes secreted by the gastrointestinal tract and the pancreas is also similar. As in mammals, carbohydrate and protein digestion begins in the lumen of the tract due largely to the action of pancreatic enzymes and is finished by enzymes attached to the intestinal villi. The final hydrolysis of disaccharides and peptides by enzymes attached to the villi focuses the resulting monosaccharides, dipeptides and amino acids at the absorptive surface for efficient transport. Lipid digestion is largely a luminal event and requires enzymes from the pancreas and bile acids from the liver. The digestion and absorption of protein, lipids, and carbohydrates is very efficient, usually above 85% for low-fiber diets. In most species, this efficiency is not quantitatively affected by the nutritional state of the bird but remains maximal even when supplies and stores are high, such as with obesity. This is not the case with water, electrolytes, and many vitamins and minerals. With these nutrients, the efficiency of absorption changes with metabolic need, being high when body levels are low, but decreasing when body levels are high.

Carbohydrates

In omnivorous and florivorous birds, carbohydrates are the predominant form of dietary energy intake. Glucose, disaccharides, such as sucrose, and polysaccharides, such as amylose and amylopectin, are the most common forms of dietary carbohydrates of plant origin that are digested autoenzymatically. Amylose is a glucose polymer with α-1,4 linkages, and amylopectin consists of α-1,6-linked amylose helices. Carbohydrates associated with the cell walls of plants (fiber) typically do not have α-linked glucose and require the presence of bacteria for digestion. Foods of animal origin typically contain nutritionally minor amounts of glucose, mostly as glycogen.

MONOSACCHARIDES

Glucose can be transported throughout the small intestine, the proximal ceca, and the rectum of most species (Amat *et al.*, 1996). It is transported into the enterocyte by active carrier-mediated absorption, probably by the type II transporter isoform. In some species, glucose may also be transported between the junctions of enterocytes by diffusion and solvent drag. About 80% of glucose absorption in the nectarivorous Rainbow Lorikeet and omnivorous Northern Bobwhite Quail is thought to be by passive absorption. In the chicken, the sum of passive and active transport is greatest in the duodenum and jejunum (Karasov and Cork, 1994; Levey and Cipollini, 1996).

The capacity for glucose absorption generally matches the glucose content of the natural diet across avian species. For example, glucose uptake in the nectarivorous Rufous Hummingbird is 15 times higher than in the omnivorous chicken and 50 times higher than in the carnivorous Loggerhead Shrike (Diamond and Karasov, 1987; Karasov and Levy, 1990).

In addition to glucose, galactose is a common dietary hexose. Pentoses (arabinose and xylose) and uronic acids (glucuronic and galacturonic) are also found in avian diets, especially those high in fiber. In the chicken, absorption rates of hexoses are twice as great as those of pentoses in the small intestine, but these two classes of monosaccharides are absorbed at similar rates in the ceca (Savory and Mitchell, 1991).

DISACCHARIDES

Disaccharidases present on the enterocyte membrane hydrolyze dietary di-saccharides to their constituent monosaccharides. For example, dietary sucrose is hydrolyzed at the enterocyte membrane by sucrase, and the resulting glucose and fructose are absorbed. However, several species of birds in the Sturnidae (starlings), Muscicapidae (thrushes) and Mimidae (catbirds, mockingbirds, and thrashers) families lack sucrase. When fed sucrose, they display sucrose intoler-ance, including diarrhea and conditioned aversion (Martinez del Rio, 1990; Malcarney *et al.*, 1994). Unlike mammals, the avian enterocytes do not have lactase activity. In most birds, this results in lactose intolerance when fed large amounts of milk or other lactose-containing foods.

STARCH

Starch is considerably more difficult to digest than disaccharides and requires the sequential participation of a variety of enzymes (Moran, 1985; Biviano *et al.*, 1993). Hydration of food during storage in the esophagus or crop facilitates subsequent hydrolysis by pancreatic α-amylase in the duodenum. This enzyme attaches to the amylose helix at aqueous surfaces and then sequentially hydrolyzes interior maltose units until the nonreducing end is reached, leaving a maltotriose. Production of dextrins occurs at the α-1,6 linkages of amylopectin. Maltose, isomaltose, and dextrins diffuse through the unstirred layer of the glycocalyx and are hydrolyzed by the multifunctional maltases-isomaltases attached to the membranes of enterocytes in the villi. Maltase and sucrase activities are greatest in the proximal to mid jejunum and the glucose that results from their action is absorbed in these intestinal segments (Fig. 3.5).

Within a species, the capacity to digest and absorb carbohydrates such as starch is not fixed but is highly adaptable. In chickens, the amount of pancreatic amylase secreted varies directly with the amount of starch consumed and adaptations can be seen within a few days. Maltase and sucrase also vary directly with the amount of substrate arising from starch or sucrose consumption.

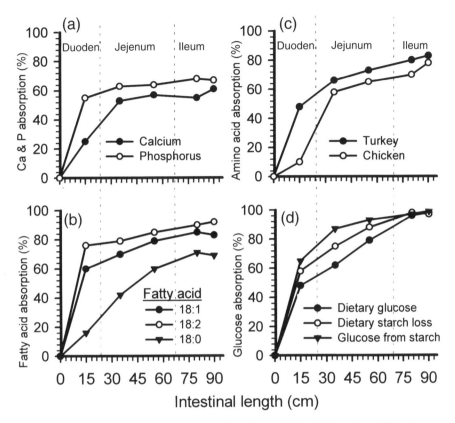

Fig. 3.5. Absorption of amino acids, fatty acids, calcium (Ca), and phosphorus (P) by the small intestine of the turkey; duoden, duodenum (Hurwitz *et al.*, 1979; Riesenfeld *et al.*, 1982). (a) Disappearance of dietary Ca and P along the turkey intestine. The primary dietary sources of Ca and P were calcium carbonate, calcium monophosphate, sodium monophosphate, and potassium monophosphate. (b) Disappearance of dietary fatty acids along the turkey intestine. The primary dietary source of these fatty acids was soybean oil. (c) Disappearance of dietary protein from the intestinal tracts of turkeys and chickens. The primary dietary protein sources were soybean meal and sorghum. (d) Disappearance of dietary glucose monohydrate, dietary starch, and the glucose resulting from the hydrolysis of dietary starch.

Selection for high growth rates and high efficiency of feed utilization in domestic poultry may have resulted in a greater reserve in absorptive capacity compared with wild birds. Presumably there is high selective pressure against excess absorptive capacity in wild birds, due to the cost of maintaining this capacity compared with the size of the occasional returns (Hulan and Bird, 1972; Blum *et al.*, 1979; Biviano *et al.*, 1993; Jackson and Diamond, 1995).

Digestion of grain starches occurs in the upper jejunum, but tuber and legume starches are usually more resistant to hydrolysis, and digestion occurs over the entire length of the small intestine. In the plant, starch can occur in small granules that have a core of starch surrounded by a network of protein. Amylopectin forms a very orderly crystalline structure within the granule and is resistant to digestion. The presence of starch granules in seeds, tubers, or fruits, such as bananas, may prevent the digestion of up to 50% of the starch when consumed in the raw state. Cooking or pelleting results in hydration of the starch molecules, causing swelling and gelatinization of the starch granules. This increases the susceptibility of the starch to enzymatic hydrolysis, with a consequent increase in digestibility.

FIBER

Plant fiber contains variable amounts of hexoses (glucose and galactose), pentoses (arabinose and xylose), and uronic acids (glucuronic and galacturonic). These sugars are normally components of the polymers (cellulose, hemicellulose, pectins, gums, and lignins) found in the cell walls. Birds are incapable of producing the hydrolytic enzymes that digest dietary fiber, but microbial action in the ceca or enzymes in the food can result in the release of monosaccharides from the complex fiber polymers (alloenzymatic digestion). Those monosaccharides that are not directly fermented by microbes can be absorbed throughout the intestines and in the ceca.

CHITIN

Chitin is a mucopolysaccharide that serves a structural role in many invertebrates. For example, it is the primary constituent of the exoskeleton of arthropods. Chitin is the predominant carbohydrate in the diets of insectivores and other microfaunivores, and proventricular enzymes with chitinase activity have been identified in Common Starling, raptors, and a variety of sea birds. Chitinase hydrolyzes chitin to chitobiose, the β-1,4-linked dimer of N-acetyl-D-glucosamine. Chitobiose is hydrolyzed by chitobiase to N-acetyl-D-glucosamine. The digestibility of chitin is low compared with that of starches but still presents a useful energy source. In Red-billed Leiothrix, King Penguins, and Leach's Storm-Petrels, the apparent digestibility of chitin is 57, 85, and 35% respectively. Although the N-acetyl-D-glucosamine and glucosamine produced upon the hydrolysis of chitin are not always efficiently absorbed, an important effect of chitinolysis may be the mechanical breakdown of the exoskeleton, permitting particle-size reduction, access of digestive enzymes, and a faster rate of passage. Chitinase is low in chickens and absent in African Grey Parrots and pigeons (Jeuniaux and Cornelius, 1978; Leprince *et al.*, 1979; Jackson *et al.*, 1992).

Proteins

Dietary proteins are very diverse in their amino acid sequences and tertiary structures. In general, they may be classified based on their solubility properties in various solutions. Classes include globular proteins, such as albumins, globulins, gliadins, and glutelins; fibrous proteins, such as collagens, elastins, and keratins; and conjugated proteins, such as mucoproteins, glycoproteins, lipoproteins, and nucleoproteins. Both the tertiary structure and the amino acid sequence of a protein influence its digestibility. Protein hydrolysis results in 20 different L-amino acids. Usually very few free amino acids are found in foodstuffs but they may be added to prepared diets.

Protein digestion in birds is very similar to that in nonruminant mammals (Esumi *et al.*, 1980; Austic, 1985; Tarvid, 1995). The hydrolysis of protein to its constituent amino acids requires the sequential action of a large number of enzymes. Protein digestion begins in the proventriculus and gizzard, where the acid environment denatures the three-dimensional structure to a linearized form, with individual peptide bonds exposed. Cooking also denatures protein and facilitates enzymatic digestion. Pepsin hydrolyzes protein at a variety of peptide bonds, resulting in polypeptides. The wide specificity of pepsin ensures that at least some of the exposed bonds in any protein will be hydrolyzed, resulting in greater denaturation and solubility. Pepsin is secreted as an inactive precursor, pepsinogen, which is hydrolyzed by HCl or previously activated pepsin, to become a catalytically active endopeptidase with a pH optimum of 3. Pepsin from quail, chicken, and ducks shows much greater activity and stability at higher pH than mammalian pepsin.

Polypeptides arriving in the duodenum stimulate the secretion of pancreatic enzymes. Pancreatic trypsin hydrolyzes at peptide bonds with basic amino acids (arginine and lysine), chymotrypsin hydrolyzes at bonds with aromatic amino acids (phenylalanine and tyrosine), and elastase hydrolyzes between a variety of amino acids with small uncharged R groups (e.g., glycine, alanine, serine). Pancreatic carboxypeptidase removes terminal carboxyl amino acids. The combined action of proteolytic enzymes results in oligopeptides of about two to six amino acids and free amino acids. Oligopeptides are further hydrolyzed by aminopeptidase, an exopeptidase, and various dipeptidases associated with the brush border of the villi. A large portion of di- and tripeptides are transported intact into the enterocyte and hydrolyzed in the cytoplasm to free amino acids; however, little is known about oligopeptide transport systems in Aves. Less than half of the absorbed amino acids are transported as free amino acids. Free amino acids traverse the enterocyte membrane by diffusion when at high concentrations, and at lower concentrations specific active transport systems, which recognize the charge and the size of the amino acids, are important. At least four transport systems are involved in amino acid transport in the small intestine: leucine and other neutral amino acids; proline, β-alanine, and related imino and amino acids; basic amino acids; and acidic amino acids. Amino acids are also absorbed in the rectum and ceca, but at low rates relative to the small intestine. The rate of amino acid absorption is extremely variable across species, with the chicken having a greater rate than turkey, Japanese Quail, Chuckar Partridge, Ring-necked Pheasant, or Elegant Crested Tinamou. Presumably, selection for high feed efficiency has resulted in a very high capacity for amino acid absorption in chickens. The transporter systems usually show more affinity for L-amino acids, resulting in slower absorption of the D isomers. In chickens, competition between amino acids sharing a transport system can occur. Some amino acids are absorbed more rapidly as peptides than as free amino acids. Most amino acids and peptides are absorbed in the duodenum and upper jejunum of the chicken

(Fig. 3.5). Once in the enterocyte, most amino acids are transported into the blood as free amino acids, although some may be metabolized in the enterocyte (Lerner and Kratzer, 1976; Riley *et al.*, 1989; Tarvid, 1995).

Lipids

Lipids are a heterogeneous group of compounds that share the common property of relative insolubility in water but solubility in organic solvents, such as ether and chloroform. Lipids in the diet can be divided into two groups: saponifiable and nonsaponifiable. Those lipids, such as triglycerides, phospho-lipids, cholesterol esters, and waxes, that contain fatty acids are known as saponifiable lipids, since they form soaps (salts of fatty acids) upon alkaline hydrolysis. The nonsaponifiable fatty acids include terpenes, carotenoids, and steroids, such as cholesterol. Triglycerides are used for energy storage in both plants and animals and are the most common form of dietary lipid. A triglyceride that is solid at room temperature is called a fat; one that is liquid at room temperature is called an oil. The melting temperature depends on the length and saturation of the component fatty acids. The longer the chain length and the more highly saturated the fatty acids, the higher the melting temperature. Triglycerides that have melting temperatures well above a bird's body tem-perature are often more difficult to digest than those with lower melting temperatures. Free fatty acids are not usually found in high concentrations in most food items but may build up upon storage or processing.

Lipid digestion begins with emulsification by the mechanical action of the gizzard and is facilitated by the presence of peptic digests of dietary proteins. In some species that eat high-fat diets, emulsification in the gizzard may be greatly facilitated by bile acids, cholesterol, and monoglycerides from digesta refluxed from the duodenum (Place, 1992a,b). The conjugation of bile salts to taurine in birds permits their solubility in the acid environment of the gizzard. Emulsifica-tion is accelerated in the small intestine by the action of bile acids, pancreatic lipase, phospholipase, and esterase. Lipases act at the lipid–water interface and their attachment to the bile-covered lipid droplets requires the previous attach-ment of colipase. Pancreatic lipase specifically releases the fatty acids esterified to glycerol in the 1 and 3 positions. Following hydrolysis, free fatty acids, 2-monoglycerides, and lysophospholipids enter into mixed micelles formed spontaneously with bile salts. Phospholipids from the diet or from bile also enter into the micelles. The hydrophobic cores of the micelles attract diglycerides, long-chain saturated fatty acids, cholesteryl esters, carotenoids, and fat-soluble vitamins (A, D, E, and K).

Micelles facilitate fat absorption by providing a high concentration of lipids in the unstirred water layer adjacent to the epithelial cells of the villi. Passage of lipids through the enterocyte membrane is passive and the free fatty acids are bound to a fatty acid-binding protein in the cytoplasm. In chickens, the concen-tration of fatty acid-binding proteins is highest in the duodenum and decreases along the length of the intestine (Katongole and March, 1979; Sklan *et al.*, 1984). About 30% of the free fatty acids are absorbed into the bloodstream without further metabolism. The rest are resynthesized into triglycerides in the endoplasmic reticulum, utilizing both the monoglyceride and the glycerol-3-phosphate pathways. The monoglyceride pathway involves the esterification of fatty acids to preexisting monoglycerides, whereas the glycerol-3-phosphate pathway begins with free glycerol. The newly synthesized triglycerides and phospholipids coalesce with apoproteins to form lipoproteins, which are incorporated into secretory vesicles. The secretory vesicles fuse with the basal cell membrane and release lipoproteins into the interstitial fluid. The lipoproteins diffuse through spaces between cells of the lamina propria and enter the capillaries. Thus, the lipoproteins absorbed in the chicken's small intestine enter the portal blood and not the lymphatics as in mammals. For this reason, the lipoproteins are called portomicrons and not chylomicrons. In turkeys fed a cereal-based diet, most of the fat absorption occurs in the duodenum (Fig. 3.5). The presence of lipids in the ileum decreases the rate of lipid passage out of the gizzard and increases the refluxing of digesta from the duodenum back to the gizzard. This ileal brake ensures matching of the rate of passage of lipid to the completeness of its digestion (Bensadoun and Rothfeld, 1972; Martinez *et al.*, 1995).

The capacity of various lipids to form micelles influences the efficiency of their digestion (Hurwitz *et al.*, 1979; Leeson and Atteh, 1995). Saturated fatty acids (e.g., stearic, palmitic, and myristic acids), either as free fatty acids or in triglycerides (at the 1 or 3 position), do not form micelles well unless triglycerides containing unsaturated fatty acids (e.g., linoleic, linolenic acids) are also present in the meal. Unabsorbed saturated fatty acids can form soaps with calcium or magnesium and are fermented in the ceca and rectum or are excreted.

When lipids are absorbed at the microvilli, bile salts reenter the lumen of the intestine and can form new micelles or they can be efficiently absorbed (Sklan, 1980; Place, 1992a,b). In chickens and turkeys most bile salts are absorbed in the duodenum and jejunum, but in Leach's Storm-Petrel the gizzard is also a primary site, presumably following reflux from the duodenum. The release and consequent absorption of bile salts represent an enterohepatic circulation that permits efficient reutilization.

Waxes found in invertebrates, plankton, and fruits can be important energy sources for those birds that can digest them (Diamond and Place, 1988; Jackson *et al.*, 1992; Place, 1992a,b; Place and Stiles, 1992). Pelagic sea birds, such as Procellariiformes (albatrosses, shearwaters, petrels, and storm petrels), consume krill and other zooplankton, which have much of their energy content as wax esters. These wax esters are assimilated with an efficiency of greater than 90% in Leach's Storm-Petrel. High concentrations of triglycerides in bile, high frequencies of duodenal reflux of bile and chyme back to the proventriculus, and the presence of a pancreatic carboxyl ester lipase facilitate the emulsification, digestion, and absorption of these waxes. Some passerines can also utilize wax esters with high efficiency. Wax esters containing saturated long-chain fatty acids coat some types of fruit (e.g., bayberry, mistletoe, and poison ivy) and can be a major component of the diet of seasonal frugivores, such as Yellow-rumped Warblers, Tree Swallows, and many woodpeckers. The capacity to digest wax is also found in honeyguides, which digest beeswax. However, many avian species, including the chicken, cannot utilize waxes effectively.

Vitamins and minerals

The digestion of vitamins and minerals is usually a process of freeing them from other food components and solubilizing them so that they may interact with specific receptors or transporters on the enterocytes of the intestinal tract. The denaturing and hydrolysis of proteins, carbohydrates, and nucleic acids releases associated vitamin and mineral cofactors. The acid environment of the proventriculus and gizzard solubilizes many mineral salts. Dietary minerals and most other micronutrients are absorbed in the duodenum and upper jejunum (see Fig. 3.4). Discussion of the absorption of individual vitamins and minerals may be found in Chapters 10 and 11.

Microflora in the gastrointestinal tract have the capacity to synthesize vitamins from carbohydrates and nitrogen. The B vitamins, including thiamin, niacin, riboflavin, pyridoxine, pantothenic acid, biotin, folic acid, and vitamin B_{12}, are synthesized in the ceca of many birds (Coates *et al.*, 1968). These vitamins are bound as complexes within the bacterial cells and are essentially unavailable to chickens, with the possible exception of folic acid. A significant contribution of microbial vitamin synthesis to nutritional requirements in birds that practice coprophagy is theoretically possible but has not been quantified. Also, the contribution of microbial vitamin synthesis in the crop of species that store food for extended periods of time could be nutritionally significant. Vitamins synthesized in the crop are digestible and absorbable in the small intestine.

Water and electrolytes

Water and electrolytes may come from the diet or from drinking. Further, the digestive process results in the secretion of large amounts of water and electrolytes, particularly in the anterior half of the gastrointestinal tract. Regardless of source, net absorption of water takes place in the ileum and is proportional to sodium chloride (NaCl) absorption. Water diffuses across the enterocyte membrane to maintain osmolarity between the intestinal lumen and the epithelium, and this process may be facilitated by the maintenance of local hypertonicity within the epithelium. The absorption of NaCl, and consequently water, is stimulated by the absorption of monosaccharides and amino acids. The active transport of these metabolites requires the cotransport of sodium across a lumen-to-enterocyte concentration gradient. In the rectum and cloaca, water absorption is not greatly influenced by glucose or amino acid absorption, but is regulated by the Na balance of the bird through aldosterone. Injections of aldosterone augment NaCl and water absorption, while NaCl loading decreases aldosterone and impairs NaCl absorption in birds that do not have salt glands. In diets with low levels of Na, absorption of potassium and ammonia provide the osmotic force for water absorption in the rectum and coprodeum (Skadhauge, 1989; Thomas and Skadhauge, 1989a,b; Wilson, 1989).

Urine excreted by the kidneys also contributes to water in the posterior intestinal tract (Gasaway *et al.*, 1975; Bjornhag and Sperber, 1977; Anderson and Braun, 1984; Thomas and Skadhauge, 1988). For example, 64% of the urine undergoes retrograde flow and enters the ceca in the hydrated turkey. One advantage of mixing together urine with digesta arriving from the small intestine is that the rectum and ceca become an effective integrator of renal and intestinal functions for purposes of osmotic balance. In many species, little net absorption of water and electrolytes (except potassium) occurs in the small intestine and the kidney secretes a dilute urine, so the rectum and ceca (if present) are responsible for regulating water balance. For example, around 80% of small-intestinal water and Na enter the ceca of Japanese Quail and Rock Ptarmigans and the ceca plays a primary role in regulating absorption of water, electrolytes, and ammonia for homeostasis. The quantitative importance of this organ can be seen in the Rock Ptarmigan, where 98% of the water reabsorption that occurs in the posterior intestines takes place in the ceca.

The cecal absorption of water is Na-dependent, so birds lacking salt glands that become dehydrated due to a high Na intake greatly diminish their reflux of urine into the rectum and increase renal concentration (Thomas *et al.*, 1984). This change in excretion strategy avoids the counterproductive action of Na-dependent water absorption. Birds with functional salt glands, such as the domestic duck or Glaucous-winged Gulls, do not have strong water absorption capacity in the rectum and ceca. In these birds, the small intestine appears to have this responsibility and the Na that is cotransported can be excreted by the salt gland. Some species, such as the Ostrich, do not have retrograde flow of urine and have well-developed kidneys that are solely responsible for water balance (Goldstein, 1989; Duke *et al.*, 1995).

Rate of Passage

The rate of passage is a measure of how long portions of digesta spend in the gastrointestinal tract being subjected to the processes of mechanical mixing, digestion, microbial fermentation, absorption, etc. Dietary-dependent changes in the rate of passage can occur rapidly and differences can be observed between meals of different composition. Even different components of the same meal may move at greatly different rates through the digestive tract (e.g., the retention of soluble fibers in the ceca or grit in the gizzard). Slower rates of passage improve nutrient acquisition by increasing the time that digesta contacts enzymes and absorptive cells. On the other hand, slower rates of passage may limit the quantity of food that can be consumed (Herd and Dawson, 1984; Washburn, 1991; Afik and Karasov, 1995).

Dietary fiber and fat require relatively long times for digestion and can have marked impacts on the rate of passage in some species. Different species have adopted varying strategies to deal with these dietary components in order to maximize net nutrient gain (see Chapter 4). For example, many herbivorous species, such as geese, have adopted a digestive strategy that does not maximize digestive efficiency, but instead maximizes food intake and results in relatively high rates of passage. Others, such as Ostriches, have adopted a strategy of more complete alloenzymatic digestion of fiber, resulting in slow rates of passage. Actual interspecies comparisons of passage rates for specific components of foods reveal the radically different approaches to digestion. For example, fruit

Table 3.3. Terminology used to describe the movement of food through the gastrointestinal (GI) tract. Details on measurement of these rates can be found in Warner (1981).

Term	Definition	Units of measure
Velocity	Rate of movement of digesta through the GI tract	$mm\ s^{-1}$
Flow rate	Amount of material that moves through the GI tract over time	$kg\ digesta\ h^{-1}$
Rate of passage	General term describing the time required for digesta to move through the GI tract	h^{-1}
Retention time	General term describing the length of time digesta spends in the GI tract	h
Mean retention time	The average time required for digesta to move through the GI tract	h

seeds pass through the digestive tract of small frugivores in less than 10 min but may be retained by ratites for several weeks. Pectin, hemicellulose, and other plant cell-wall components pass through the digestive tract of a frugivore in less than half an hour, but are retained and fermented in the ceca for 12 h or more in many omnivores and herbivores.

A variety of terms (Table 3.3) can be used to describe the movement of food through the gastrointestinal tract (Warner, 1981; Herd and Dawson, 1984; Place *et al.*, 1989; Karasov and Levey, 1990). The time that food is stored in the crop or esophagus prior to movement to the proventriculus is extremely variable and is usually not included in estimates of passage rates. A period of fasting prior to measurement is commonly used experimentally to eliminate errors due to variable residence times in the crop or esophagus. In petrels, chickens, turkeys, and geese, about 50% of the mean residence time is spent in the proventriculus and gizzard. For those food components that move into the ceca, the residence time in the digestive tract is at least three times as long as that of components that are directly excreted. For this reason, it is important to know what components of digesta are represented in a given measurement of retention time. Food tends to clear the digestive tract in an exponential fashion, so the time required to clear 98% of a food is equal to about four times the time required to clear 50% (mean retention time). For comparative purposes, measurements that give the rate of passage of the fastest or slowest food components are not as useful as mean retention times. Table 3.4 shows mean retention times for various consumption categories using markers of different sizes and solubilities. Small frugivores and nectarivores have the shortest mean retention times of about 45 min. An Ostrich has a mean retention time of 24 h when fed a high-fiber diet. Although food characteristics and digestive anatomy are primary determinants of retention time, larger birds have longer gastrointestinal tracts and consequently greater retention times. When examined over a variety of species, the relationship between retention time and body weight scales to the 0.21 power (calculated from data of Karasov, 1990).

Table 3.4. Typical mean retention times in birds.

Consumption category	Retention time (min)
Carnivore	360–600
Piscivore	360–780
Insectivore	30–90
Nectarivore	30–50
Granivore	40–100
Frugivore	15–60
Herbivore	
Flying	50–300
Nonflying	300–1440

In general, gastrointestinal motility is under the control of the nervous system, with nerve fibers emanating from several ganglia along the tract. Stimulation is under parasympathetic control. Several hormones (e.g., gastrin, cholecystokinin, and secretin) also regulate motility, particularly by affecting transit between organs, including: crop to esophagus; esophagus to proventriculus; proventriculus to gizzard; gizzard to duodenum; and ileum to rectum to ceca. Finally, the frequency and amplitude of contractions is dependent upon the composition of the digesta. Low pH, large particle sizes, high osmotic pressure, or significant amounts of lipid decrease motility.

Digestibility

Measurements of digestibility are essential in order to define the efficiency of utilization of nutrients within foods, to classify the nutritional quality of food items, and to formulate diets for captive birds. Indexes of digestibility measure the sum of activities within the gastrointestinal tract, including size reduction, autoenzymatic hydrolysis, alloenzymatic hydrolysis, microbial synthesis, enterocyte transport capacity, rate of passage, and, in some instances, endogenous losses. For a given food item, the digestibility is determined by factors inherent in its chemical and structural makeup and by the digestive physiology of the bird consuming it. Birds often do not maximize the digestibility of the foods they consume but choose a strategy that compromises efficiency for low digesta volumes and high total rates of nutrient extraction.

Digestibility may be expressed in terms of apparent or true. Apparent digestibility is the relationship between the amount of nutrients consumed in the diet and the amount that disappears from the gastrointestinal tract: (nutrient intake − nutrient in feces)/nutrient intake. 'Apparent' indicates that the measurement is biased by the amount of nutrient that was absorbed but then excreted back into the digestive tract, as well as by endogenous nutrient losses, such as those from the shedding of the intestinal epithelia and mucous secretions.

True digestibility corrects for those components of the excreted nutrients that were not originally in the food. In other words, it corrects for the portion of nutrient in the feces that is of endogenous origins: (nutrient intake − nutrient in feces + endogenous fecal losses)/nutrient intake. Thus, the values obtained for true digestibility are always greater than those for apparent digestibility. The separation of endogenous losses arising from the digestive tract from the metabolic losses excreted in urine is difficult in birds, due to the simultaneous voiding of feces and urine. This results in estimates of digestibility that are often confounded by metabolic losses contributed by the urine (uric acid, electrolytes, etc.). Consequently, nutrient digestibility values are not commonly used in avian nutrition. Instead, nutrient metabolizable values are used to evaluate foods: apparent metabolizability = (nutrient intake − nutrient in feces and urine)/ nutrient intake). In practical avian nutrition, this approach is acceptable when the metabolic losses are not an artifact of the measuring procedure. This assumption does not hold for foods that are low in energy or protein, have poor amino acid balance, or have toxins. Measurements of amino acid and vitamin digestibility are further compromised by the microflora in the posterior tract, which may synthesize some of the amino acids and vitamins found in the excreta. Alternatively, they may metabolize unabsorbed amino acids and vitamins, so that these unabsorbed nutrients are not observed in the excreta. In both cases, an estimate of metabolic and microfloral contributions must be made to give accurate metabolizability values.

The most common metabolizability measurement made on foods fed to birds is that for energy. The apparent metabolizable energy value of an avian food is often expressed as a fraction of the gross energy of the food (AME/GE) and referred to as the apparent metabolizable energy coefficient (*MEC). The *MEC has been determined for a very wide variety of foods in a vast number of avian species. Although generalizations across species are inappropriate for most nutritional uses, average *MEC values illustrate the important differences in the value of general classes of food items (Fig. 3.6). Digestibility or metabolizability measurements for amino acids, vitamins, and minerals are technically much more difficult to make than those for energy and are not generally known except for poultry. In poultry, the digestibility of amino acids, vitamins, and minerals of nonprocessed feed ingredients is roughly correlated with that of energy.

When comparing birds that naturally eat similar foods, the composition of the food has a greater impact on *MEC than the species of bird consuming it (Castro *et al.*, 1989; Karasov, 1990; van Tets and Sanson, 1996). Among foods of plant origin, the primary factor affecting the *MEC is the amount of fiber associated with cell-wall structures. Among foods of animal origin, the primary factor is the amount of chitin, bones, scales, fur, or feathers. Foods high in fiber have considerably greater digestibilities in animals that have large fermentation areas for alloenzymatic digestion. For example, the digestibility of neutral detergent fiber fed to an adult Ostrich is tenfold greater than its digestibility in an Ostrich chick or in adult chickens, which have relatively small fermentation areas. This difference is reflected in a 40% greater *MEC of an alfalfa based-diet for adult Ostriches compared with chicks (Angel, 1993).

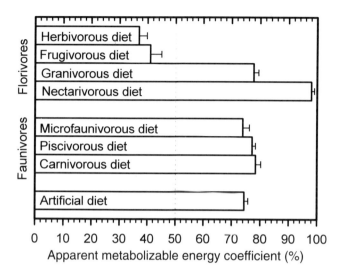

Fig. 3.6. Apparent metabolizable energy coefficients (*MEC) for common food categories. Values represent averages for a variety of species from several families that typically consume each food category. Error bars indicate the standard error of means. Data from Castro *et al.* (1989), except nectarivore value, which is from Karasov (1990).

For many nutrients, such as amino acids, vitamins, and minerals, the specific chemical form impacts the degree to which it can be utilized to support important metabolic processes. For example, some stereoisomers of amino acids (e.g., D-amino acids) are efficiently digested and absorbed, but are not efficiently used for protein synthesis and are degraded. Thus, their nutritional value as amino acids is low, even though their digestibility and energy metabolizability may be very high. Similarly, some forms of vitamins (e.g., vitamin D_2) are readily absorbed but are poorly utilized and are degraded. The term bioavailability is often used in nutrition to describe the degree to which a nutrient in a particular source can be absorbed and utilized for its required purpose by an animal. Bioavailability is usually expressed as a percentage of an established standard. Frequently, the standard reference is the same chemical form as that used in experiments to determine a bird's requirement (Ammerman *et al.*, 1995).

Fate of Absorbed Nutrients

Monosaccharides, amino acids, lipids, volatile fatty acids, and most other nutrients leave the intestinal enterocytes through the basal–lateral membrane and enter the rich capillary bed of the villi. Nutrients and blood are transported to the liver by the left and right portal veins. These portal veins branch repeatedly within the liver, ultimately forming a three-dimensional network of fenestrated sinusoids between sheets of hepatic cells, where nutrients can be selectively taken up. Fine rami of the hepatic arteries join the sinusoids and form into branches of the hepatic veins, which eventually drain into the caudal vena cava. Dietary nutrients may be taken up by the liver and metabolized or they may exit the hepatic circulation through the vena cava and enter the general circulation. The tissue-specific metabolism of individual nutrients will be considered in later chapters.

Ontogeny of Digestive Capacity and Strategy

Many avian species undergo dramatic changes in their diet as they develop (Downhower, 1976; Sedinger and Raveling, 1984). As embryos, all birds begin life on a diet of lipid and protein (yolk and albumen), but essentially no carbohydrate. Faunivores maintain this intake pattern posthatch and throughout life. Nestlings of florivorous or omnivorous adults often consume a faunivorous diet also. This is particularly true of altricial birds, which grow very quickly immediately posthatching and have the luxury of being fed by parents capable of capturing arthropods and other fauna. In many cases, the adults continue florivory while feeding their offspring a more nutritious faunivorous diet. As the chicks grow and mature, they are gradually switched to a diet more similar to that of their parents. Most precocial chicks are also fed a highly faunivorous diet after hatching, but some (e.g., geese) switch to a high-carbohydrate, florivorous diet immediately following hatching. Domestic ducks, quail, turkeys, and chickens are fed a cereal-based diet that is low in fat and high in carbohydrate immediately following hatching. They must quickly make the appropriate digestive and enzymatic adaptations to facilitate this abrupt switch. The dramatic homeostatic changes in hormone levels and the rapid induction of enzymes needed for carbohydrate catabolism and lipid synthesis have been described in remarkable detail, including molecular characterization of the promoters and modifiers of the responding genes (Hillgartner et al., 1995).

Altricial versus precocial development

There are important temporal differences between precocial and altricial birds in the development and maturation of their digestive systems into a completely functional organ system (Konarzewski et al., 1989; 1990; Starck, 1992). Altricial hatchlings are incapable of life without considerable parental assistance, because of closed eyes, poor muscle development, and limited down. However, the allometric growth rate of the embryonic digestive tract is faster than that of the rest of the body, so that at hatching it is relatively large and functionally developed. This gives the young altricial chick the capacity to receive and efficiently digest the large quantities of feed supplied by its parents and permits very rapid postnatal growth. For example, in the Budgerigar the absorptive area of the intestines relative to body weight is maximum at hatching and decreases later as growth slows. Also, the parents of altricial chicks may contribute to the digestion of food by softening it with saliva and sometimes by enzymatic action prior to regurgitation. Conversely, precocial chicks have well-developed eyes, musculature, and coordination and plentiful down but the digestive tract is relatively immature at hatching and develops digestive capacity over a period of weeks. This may correlate with food intake, which typically is low immediately posthatching due to inexperience in finding food and minimal parental assistance. Following hatching, the intake of food stimulates rapid growth in intestinal size and markedly increases the size and surface area of villi. In the young chicken, turkey, and quail, the gastrointestinal tract grows at a much greater rate than other organs, especially during the first 7–10 days (Baranylova and Holman, 1976; Lilja *et al.*, 1985).

Prenatal development

The embryo receives the vast majority of its nutrients via the yolk-sac membrane and the developing digestive system does not play an important part in nourishing the embryo. However, the digestive tract is anatomically complete prior to the end of embryonic development. For example, villi in the small intestine have rudimentary microvilli by 16 days of incubation in chickens. During the last third of incubation, the embryo begins to ingest amniotic fluid. This first 'meal' stimulates the growth of intestinal villi and microvilli. Chicken embryos can begin to digest carbohydrates at day 18 of incubation, when pancreatic α-amylase appears and the brush-border disaccharidases begin to increase in activity. The quantity of pancreatic enzymes is high immediately prior to hatching, demonstrating some reserve for use after hatching (Dautlick and Strittmatter, 1970; Chambers and Grey, 1979).

An important component of development is the uptake of two forms of immunoglobulins that play an important role in the selective colonization of the gastrointestinal tract following hatching. Immunoglobulin Y (IgY) is the avian homologue of IgG and is passed from the hen into the yolk during follicular development. It is taken up by the embryo during yolk-sac resorption and is the primary immunoglobulin that circulates in the blood. Immunoglobulin A is passed from the hen to the albumen in the shell gland. In the chicken, IgA is taken up beginning on day 14 of incubation, concurrently with the consumption of amniotic fluid, and it is particularly important in determining the microflora that will be permitted to colonize the gastrointestinal tract. The differential absorption of albumen and yolk immunoglobulins by birds may be analogous to the uptake of immunoglobulins through the placenta (IgG) and via colostrum (IgA) in mammals (Rose *et al.*, 1974; Loeken and Roth, 1983).

Postnatal development

The rate of passage of digesta increases with age posthatch, with the most dramatic increase occurring during the first week in chickens and turkeys (Krogdahl and Sell, 1989; Sell *et al.*, 1991; Pinchasov, 1995; Noy and Sclan, 1995). This is accompanied by a dramatic increase in secretion of pancreatic enzymes and activity of brush-border enzymes (Fig. 3.7). The net result of these changes is the matching of the capacity to digest carbohydrate and protein to their intake. During the first week posthatch, the digestibility of dietary lipid is low, due to limiting rates of bile-salt production and a lag in the developmental increase in

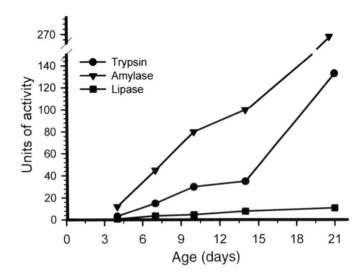

Fig. 3.7. Net daily secretion of amylase, trypsin, and lipase into the duodenum of chickens at different ages. One unit of enzyme activity equals an amount of enzyme that hydrolyzes 1 μmol of substrate min^{-1} (Noy and Sklan, 1995).

pancreatic lipase secretion. This developmental limitation of fat absorption is particularly pronounced for saturated fats. In the Ostrich, fat digestibility increases from 44 to 86% between 3 and 10 weeks of age (Angel, 1993).

The gizzard is relatively small and weak following hatching and increases in strength and resistance to abrasion as the chick begins to consume food. This developmental pattern constrains the size and hardness of foods that many young chicks can consume. In particular, hard seeds are poorly ground by the gizzard. For example, almost half of the grass seeds consumed by young Grey Partridge chicks are excreted with intact seed coats, but this proportion quickly decreases in the second week of life (Green *et al.*, 1987).

Birds hatch from the egg with a completely sterile digestive tract. Microbes are quickly picked up within the nest or incubator. The source of this microflora depends on the hygienic condition of the environment and the presence or absence of the parents. The process of microbial seeding occurs both orally and through the vent. Precocial chicks begin pecking and consuming nest materials or shell debris soon after hatching. Food and water also contribute to their flora. Altricial chicks are fed by their parents and the food material is contaminated with a robust population of microflora from the parents' anterior digestive tract. Spontaneous sucking movements of the vent (cloacal drinking) facilitate the uptake of microflora from the environment for colonization of the posterior digestive tract. Additionally, young chicks of many species have been observed eating the feces of their parents. Coprophagy may help seed the digestive tract with beneficial microflora from the established flora of the parents' posterior digestive tract. Some of the microflora ingested by the chick will not find the conditions suitable and are either killed by digestive secretions or eliminated by the immune system (e.g., IgA), or they are unable to attach to the epithelium and are defecated. The remaining microflora proliferate and compete for space and nutritional resources. This process continues until the 'normal' flora develops. The small intestine of the chicken develops a stable bacterial population, similar to that of adult birds, within the first 2 weeks, but the ceca acquire a succession of predominant organisms over a period of 6 weeks (Mead, 1989). The composition of the normal flora is dependent on the species, age, and diet. Most young florivorous birds are relatively inefficient at fermenting fiber and typically consume low amounts. In Ostrich chicks, neutral detergent fiber digestibility increases from 6.5% at 3 weeks of age to 51% at 10 weeks of age. Poor fiber digestibility is reflected in cecum size, which is small relative to body weight but grows four times faster than the rest of the body during the first few weeks in Red Jungle Fowl chicks (Angel, 1993; Jackson and Diamond, 1995).

References

Afik, D. and Karasov, W.H. (1995). The trade-offs between digestion rate and efficiency in warblers and their ecological implications. *Ecology* 76, 2247–2257.

Amat, C., Planas, J.M., and Moreto, M. (1996) Kinetics of hexose uptake by the small and large intestine of the chicken. *American Journal of Physiology – Regulatory Integrative and Comparative Physiology* 40, R1085–R1089.

Ammerman, C.B., Baker, D.H., and Lewis, A.J. (1995) *Bioavailability of Nutrients for Animals.* Academic Press, San Diego.

Anderson, G.L. and Braun, E.J. (1984) Cecae of desert quail: importance in modifying the urine. *Comparative Biochemistry and Physiology* 78A, 91–94.

Angel, R.C. (1993) Research update: age related changes in digestibility of nutrients in ostriches and nutrient profiles of eggs of the hen and chick. *Proceedings of the Association of Avian Veterinarians* 3, 275–281.

Austic, R.E. (1985) Development and adaptation of protein digestion. *Journal of Nutrition* 115, 686–697.

Baranylova, E. and Holman, J. (1976) Morphological changes in the intestinal wall in fed and fasted chickens in the first week after hatching. *Acta Veterinaria Branoi* 45, 151–158.

Barnes, E.M. (1979) The intestinal microflora of poultry and game birds during life and after storage. *Journal of Applied Bacteriology* 46, 407–419.

Bensadoun, A. and Rothfeld, A. (1972) The form of absorption of lipids in the chicken, *Gallus domesticus. Proceedings of the Society of Experimental Biology and Medicine* 141, 814–817.

Biviano, A.B., Delrio, C.M. and Phillips, D.L. (1993) Ontogenesis of intestine morphology and intestinal disaccharidases in chickens (*Gallus gallus*) fed contrasting purified diets. *Journal of Comparative Physiology B – Biochemical Systemic and Environmental Physiology* 163, 508–518.

Björnhag, G. and Sperber, I. (1977) Transport of various food components through the digestive tract of turkeys, geese and guinea fowl. *Swedish Journal of Agriculture Research* 7, 57–66.

Blum, J.C., Gauthier, A. and Guillaumin, S. (1979) Variations of intestinal maltase and sucrase activities in chicks according to age and diet. *Annales de Biologie Animale Biochemie, Biophysique* 19, 807–812.

Buchsbaum, R., Wilson, R. and Valiela, I. (1986) Digestibility of plant constituents by Canada geese and Atlantic brant. *Ecology* 67, 386–393.

Castro, G., Stoyan, N. and Nyers, J.P. (1989) Assimilation efficiency in birds: a function of taxon or food type? *Comparative Biochemistry and Physiology* 92A, 271–278.

Chambers, C. and Grey, R.D. (1979) Development of the structural components of the brush border in absorptive cells of the chick intestine. *Cell Tissue Research* 204, 387–405.

Clemens, E.T., Stevens, C.E. and Southworth, M. (1975) Sites of organic acid production and pattern of digesta movement in the gastrointestinal tract of geese. *Journal of Nutrition* 105, 1341–1350.

Clench, M.H. and Mathias, J.R. (1995) Motility responses to fasting in the gastrointestinal tract of three avian species. *Condor* 97, 1041–1047.

Coates, M.E., Ford, J.E. and Harrison, G.F. (1968) Intestinal synthesis of vitamins of the B complex in chicks. *British Journal of Nutrition* 22, 493–500.

Dautlick, J. and Strittmatter, C.F. (1970) Developmental and hormone-induced changes in chicken intestinal disaccharidases. *Biochimica et Biophysica Acta* 222, 444–454.

Diamond, A.W. and Place, A.R. (1988) Wax digestion by black-throated honey-guides, *Indicator indicator. Ibis* 130, 558–561.

Diamond, J.M. and Karasov, W.H. (1987) Adaptive regulation of intestinal nutrient transporters. *Proceedings of the National Academy of Sciences of the USA* 84, 2242–2245.

Downhower, J.F. (1976) Darwin's finches and the evolution of sexual dimorphism in body size. *Nature* 263, 558–563.

Duke, G.E. (1986) Alimentary canal: anatomy, regulation of feeding, and motility. In: Sturkie, P.D. (ed.) *Avian Physiology*. Springer-Verlag, New York.

Duke, G.E., Kostuch, T.E. and Evanson, O.A. (1975) Electrical activity and intraluminal pressure in the lower small intestine of turkeys. *American Journal of Digestive Diseases* 10, 1040–1046.

Duke, G.E., Evanson, O.A. and Huberry, B.J. (1980) Electrical potential changes and contractile activity of the distal cecum of turkeys. *Poultry Science* 59, 1925–1934.

Duke, G.E., Degen, A.A. and Reynhout, J.K. (1995) Movement of urine in the lower colon and cloaca of ostriches. *Condor* 97, 165–173.

Dziuk, K.E. and Duke, G.E. (1972) Cineradiographic studies of gastric motility in turkeys. *American Journal of Physiology* 222, 159–166.

Esumi, H., Yasugi, S., Mizuno, T. and Fujiki, H. (1980) Purification and characterization of a pepsinogen and its pepsin from proventriculus of the Japanese quail. *Biochimica et Biophysica Acta* 611, 363–370.

Gasaway, W.C., Holleman, D.F. and White, R.G. (1975) Flow of digesta in the intestine and cecum of the rock ptarmigan. *Condor* 77, 467–474.

Goldstein, D.L. (1989) Absorption by the cecum of wild birds – is there interspecific variation? *Journal of Experimental Zoology* Suppl. 3, 103–110.

Green, R.E., Rands, M.R.W. and Moreby, S.J. (1987) Species differences in the diet and the development of seed digestion in partridge chicks, *Perdix perdix* and *Alectoris rufa. Ibis* 129, 511–514.

Herd, R.M. and Dawson, T.J. (1984) Fiber digestion in the emu, *Dromaius novaebollaniae*, a large bird with a simple gut and high rates of passage. *Physiological Zoology* 57, 70–84.

Herpol, C. and van Grembergen, G. (1967) L'activité protéolytique du système digestif de *Gallus domesticus. Zeitschrift fuer Vergleichende Physiologie* 57, 1–6.

Hill, K.J. (1971) The physiology of digestion. In: Bell, D.J. and Freeman, B.M. (eds) *Physiology and Biochemistry of the Domestic Fowl*, Vol. 1. Academic Press, London, pp. 1–49.

Hillgartner, F., Salati, L.M. and Goodridge, A.G. (1995) Physiological and molecular mechanisms involved in nutritional regulation of fatty acid synthesis. *Physiological Reviews* 75, 47–76.

Hulan, H.W. and Bird, F.H. (1972) Effect of fat level in isonitrogenous diets on composition of avian pancreatic juice. *Journal of Nutrition* 102, 459–466.

Hurwitz, S. and Bar, A. (1968) Regulation of pH in the intestine of the laying fowl. *Poultry Science* 47, 1029–1035.

Hurwitz, S., Eisner, U., Dubrov, D., Sklan, D., Risenfeld, G. and Bar, A. (1979) Protein, fatty acids, calcium and phosphate absorption along the gastrointestinal tract of the young turkey. *Comparative Biochemistry and Physiology* 62A, 847–850.

Jackson, S. and Diamond, J. (1995) Ontogenetic development of gut function, growth, and metabolism in a wild bird, the red jungle fowl. *American Journal of Physiology – Regulatory Integrative and Comparative Physiology* 38, R1163–R1173.

Jackson, S., Place, A.R. and Seiderer, L.J. (1992) Chitin digestion and assimilation by seabirds. *Auk* 109, 758–770.

Jerret, S.A. and Goodge, W.R. (1973) Evidence for amylase in avian salivary glands. *Journal of Morphology* 139, 27–33.

Jeuniaux, C. and Cornelius, C. (1978) Distribution and activity of chitinolytic enzymes in the digestive tract of birds and mammals. In: Muzzarelli, R.A.A. and Pariser, E.R. (eds) *Proceedings of the First International Conference on Chitin/Chitosan*. MIT Press, Cambridge, Massachusetts, pp. 542–549.

Jorgensen, H., Zhao, X., Knudsen, K.E. and Eggum, B.O. (1996) The influence of dietary fibre source and level on the development of the gastrointestinal tract, digestibility and energy metabolism in broiler chickens. *British Journal of Nutrition* 75, 379–395.

Karasov, W.H. (1990) Digestion in birds: chemical and physiological determinants and ecological implications. *Studies in Avian Biology* 13, 391–415.

Karasov, W.H. and Cork, S.J. (1994) Glucose absorption by a nectarivorous bird – the passive pathway is paramount. *American Journal of Physiology* 267, G18–G26.

Karasov, W.H. and Levey, D.J. (1990) Digestive system trade-offs and adaptations of frugivorous passerine birds. *Physiological Zoology* 63, 1248–1270.

Kato, T. (1914) Druckmessungen im Muskelmagen der Vogel. *Pfluegers Archiv Gisamte Physiologie des Menschen und der Tiere* 159, 6–26.

Katongole, J.B. and March, B.E. (1979) Fatty acid binding protein in the intestine of the chicken. *Poultry Science* 58, 372–375.

Konarzewski, M., Kozlowski, J. and Ziolko, M. (1989) Optimal allocation of energy to growth of the alimentary tract in birds. *Functional Ecology* 3, 589–596.

Konarzewski, M., Lilja, C., Kozlowski, J. and Lewonczuk, B. (1990) On the optimal growth of the alimentary tract in avian postembryonic development. *Journal of Zoology* 222, 89–101.

Krogdahl, A. and Sell, J.L. (1989) Influence of age on lipase, amylase and protease activities in pancreatic tissue and intestinal contents of young turkeys. *Poultry Science* 68, 1561–1568.

Leeson, S. and Atteh, J.O. (1995) Utilization of fats and fatty acids by turkey poults. *Poultry Science* 74, 2003–2010.

Leprince, P., Dandrifosse, G. and Schoffeniels, E. (1979) The digestive enzymes and acidity of the pellets regurgitated by raptors. *Biochemical Systematics and Ecology* 7, 223–227.

Lerner, J. and Kratzer, F.H. (1976) A comparison of intestinal amino acid absorption in various avian and mammalian species. *Comparative Biochemistry and Physiology* 53A, 123–127.

Levey, D.J. and Cipollini, M.L. (1996) Is most glucose absorbed passively in northern bobwhite? *Comparative Biochemistry and Physiology A – Physiology* 113, 225–231.

Levey, D.J. and Duke, G.E. (1992) How do frugivores process fruit? Gastrointestinal transit and glucose absorption in cedar waxwings (*Bombycilla cedrorum*). *Auk* 109, 722–730.

Lilja, C., Sperber, I. and Marks, H.L. (1985) Postnatal growth and organ development in Japanese quail selected for high growth rate. *Growth* 49, 51–62.

Loeken, M.R. and Roth, T.F. (1983) Analysis of maternal IgG subpopulations which are transported into the chicken oocyte. *Immunology* 49, 21–28.

Mackie, R.I. (1987) Microbial digestion of forages in herbivores. In: Hacher, J.B. and Tenouth, J.H. (eds) *The Nutrition of Herbivores*. Academic Press Australia, Sydney, pp. 233–265.

Malcarney, H.L., Delrio, C.M. and Apanius, V. (1994) Sucrose intolerance in birds – simple nonlethal diagnostic methods and consequences for assimilation of complex carbohydrates. *Auk* 111, 170–177.

Martinez, V., Jimenez, M., Gonalons, E. and Vergara, P. (1995) Intraluminal lipids modulate avian gastrointestinal motility. *American Journal of Physiology – Regulatory Integrative and Comparative Physiology* 38, R445–R452.

Martinez del Rio, C.M. (1990) Dietary, phylogenetic, and ecological correlates of intestinal sucrase and maltase activity in birds. *Physiological Zoology* 63, 987–1011.

Mead, G.C. (1989) Microbes of the avian cecum – types present and substrates utilized. *Journal of Experimental Zoology*, Suppl. 3, 48–54.

Moran, E.T. (1985) Digestion and absorption of carbohydrates in fowl and events through prenatal development. *Journal of Nutrition* 115, 665–674.

Noy, Y. and Sklan, D. (1995) Digestion and absorption in the young chick. *Poultry Science* 74, 366–373.

Obst, B.S. and Diamond, J.M. (1989) Interspecific variation in sugar and amino acid transport by the avian cecum. *Journal of Experimental Zoology*, Suppl. 3, 117–126.

Pinchasov, Y. (1995) Early transition of the digestive system to exogenous nutrition in domestic post-hatch birds. *British Journal of Nutrition* 73, 471–478.

Place, A.R. (1992a) Bile is essential for lipid assimilation in Leach's storm petrel, *Oceanodroma leucorhoa*. *American Journal of Physiology* 263, R389–R399.

Place, A.R. (1992b) Comparative aspects of lipid digestion and absorption – physiological correlates of wax ester digestion. *American Journal of Physiology* 263, R464–R471.

Place, A.R. and Stiles, E.W. (1992) Living off the wax of the land – bayberries and yellow-rumped warblers. *Auk* 109, 334–345.

Place, A.R., Stoyan, N.C., Ricklefs, R.E. and Butler, R.G. (1989) Physiological basis of stomach oil formation in Leach's storm petrel, *Oceanodroma leucorhoa*. *Auk* 106, 687–699.

Riesenfeld, G., Geva, A. and Hurwitz, S. (1982) Glucose homeostasis in the chicken. *Journal of Nutrition* 112, 2261–2266.

Riley, W.W., Welch, C.C., Nield, E.T. and Austic, R.E. (1989) Competitive interactions between the basic amino acids in chicken intestine *in situ*. *Nutrition Reports International* 40, 383–393.

Rose, M.E., Orlans, E. and Buttress, N. (1974) Immunoglobulin classes in the hen's egg: their segregation in yolk and white. *European Journal of Immunology* 4, 521–523.

Satoh, S., Furuse, M. and Okumura, J. (1995) Factors influencing the intestinal phase of pancreatic exocrine secretion I: the turkey. *Experientia* 51, 249–251.

Savory, C.J. and Knox, A.I. (1991) Chemical composition of caecal contents in the fowl in relation to dietary fibre level and time of day. *Comparative Biochemistry and Physiology A – Comparative Physiology* 100, 739–743.

Savory, C.J. and Mitchell, M.A. (1991) Absorption of hexose and pentose sugars *in vivo* in perfused intestinal segments in the fowl. *Comparative Biochemistry and Physiology A – Comparative Physiology* 100, 969–974.

Sedinger, J.S. and Raveling, D.G. (1984) Dietary selectivity in relation to availability and quality of food for goslings of cackling geese. *Auk* 101, 295–306.

Sell, J.L., Angel, C.R., Piquer, F.J., Mallarino, E.G. and Albatshan, H.A. (1991) Developmental patterns of selected characteristics of the gastrointestinal tract of young turkeys. *Poultry Science* 70, 1200–1205.

Skadhauge, E. (1989) An overview of the interaction of kidney, cloaca, lower intestine, and salt gland in avian osmoregulation. In: Hughes, M.R. and Chadwick, A. (eds) *Progress in Avian Osmoregulation.* Leeds Philosophical and Literary Society, Leeds, pp. 333–346.

Sklan, D. (1980) Site of digestion and absorption of lipids and bile acids in the rat and turkey. *Comparative Biochemistry and Physiology* 65A, 91–95.

Sklan, D., Shachaf, B., Baron, J. and Hurwitz, S. (1978) Retrograde movement of digesta in the duodenum of the chick: extent, frequency and nutritional implications. *Journal of Nutrition* 108, 1485–1490.

Sklan, D., Geva, A., Budowski, P. and Hurwitz, S. (1984) Intestinal absorption and plasma transport of lipids in chicks and rats. *Comparative Biochemistry and Physiology* 78A, 507–510.

Soedarmo, D., Kare, M.R. and Wasserman, R.H. (1961) Observations on the removal of sugar from the mouth and crop of the chicken. *Poultry Science* 40, 123–141.

Starck, J.M. (1992) Evolution of avian ontogenies. In: Power, D.M. (ed.) *Current Ornithology*, Vol. 10. Plenum Press, New York, pp. 275–367.

Stevens, C.E. and Hume, I.D. (1995) *Comparative Physiology of the Vertebrate Digestive System.* Cambridge University Press, Cambridge.

Stresemann, E. (1927–1934) *Handbuch der Zoologie. Sauropsida: Aves.* W. de Gruyter, Berlin.

Suzuki, M. and Nomura, S. (1975) Electromyographic studies on the deglutition movement in the fowl. *Japanese Journal of Veterinary Science* 37, 289–295.

Swart, D., Mackie, R.I. and Hayes, J.P. (1993a) Influence of live mass, rate of passage and site of digestion on energy metabolism and fibre digestion in the ostrich (*Struthio camelus* var. *domesticus*). *South African Journal of Animal Science* 23, 119–126.

Swart, D., Siebrits, F.K. and Hayes, J.P. (1993b) Utilization of metabolizable energy by ostrich (*Struthio camelus*) chicks at 2 different concentrations of dietary energy and crude fibre originating from lucerne. *South African Journal of Animal Science* 23, 136–141.

Tarvid, I. (1995) The development of protein digestion in poultry. *Poultry and Avian Biology Reviews* 6, 35–54.

Thomas, D.H. and Skadhauge, E. (1988) Transport function and control in bird caeca. *Comparative Biochemistry and Physiology* 90A, 591–596.

Thomas, D.H. and Skadhauge, E. (1989a) Functions of the flow of urine and digesta in the avian lower intestine. *Acta Veterinaria Scandinavica*, Suppl. 86, 212–218.

Thomas, D.H. and Skadhauge, E. (1989b) Water and electrolyte transport by the avian ceca. *Journal of Experimental Zoology*, Suppl. 3, 95–102.

Thomas, D.H., Pinshow, B. and Degen, A.A. (1984) Renal and intestinal contributions to the water economy of desert-dwelling phasianid birds: comparison of wild and captive chukars and sand partridges. *Physiological Zoology* 57, 128–136.

van Tets, L. and Sanson, G. (1996) Use of the metabolisable energy coefficient in bird studies

– statistical power in taxa and food comparisons. *Australian Journal of Zoology* 44, 1–7.

Vonk, H.J. and Western, J.R.H. (1984) *Comparative Biochemistry and Physiology of Enzymatic Digestion.* Academic Press, San Diego.

Warner, A.C.I. (1981) Rate of passage of digesta through the gut of mammals and birds. *Nutrition Abstracts and Reviews B* 51, 789–820.

Washburn, K.W. (1991) Efficiency of feed utilization and rate of feed passage through the digestive system. *Poultry Science* 70, 447–452.

Wilson, J.X. (1989) The renin–angiotensin system in birds. In: Hughes M.R. and Chadwick, A. (eds) *Progress in Avian Osmoregulation.* Leeds Philosophical and Literary Society, Leeds, pp. 143–162.

Ziswiler, V. and Farner, D.S. (1972) Digestion and the digestive system. In: Farner, D.S., King, J.R., and Parkes, K.C. (eds) *Avian Biology,* Vol. II. Academic Press, New York, pp. 343–430.

Zoppi, G. and Shmerling, D.H. (1969) Intestinal disaccharidase activities in some birds, reptiles and mammals. *Comparative Biochemistry and Physiology* 29, 289–294.

CHAPTER 4
Nutritional Strategies and Adaptations

The behavioral, morphological, functional, and biochemical adaptations that give a bird the capacity to consume, digest, and metabolize the nutrients inherent in its foods are referred to as the nutritional strategy. Radically diverse nutritional strategies have evolved over time, due to multiple ecological and physiological factors, including: (i) the potential food sources available; (ii) the compatibility between the digestive structures needed to utilize potential food sources and the demands of flight; (iii) morphological and physiological limitations of digestive-tract design that preclude efficient utilization of foods of markedly different compositions; (iv) critical events, such as 'bottlenecks' in food availability or periodic surges in nutritional demands; (v) competitors and predators associated with each nutritional niche; and (vi) coevolution of foods, including chemical defenses, physical defenses, and rewards for seed dispersal or pollination. Ancestral birds were predominantly faunivorous, but during the last 50 million years the radiation of flowering plants has driven a rapid adaptive radiation in avian nutritional strategies. During this time period, the appearance of seeds and fruits facilitated the adoption of diverse nutritional strategies for the digestion and metabolism of plant nutrients (Wing and Tiffney, 1987; Chivers and Langer, 1994).

The majority of avian species use their capacity for flight to procure the most digestible and nutritious food items. This strategy sometimes requires migration over large distances to ensure a steady supply of appropriate foods. A smaller number of species have adopted a more sedentary strategy, often requiring seasonal adaptation to a food supply that is highly variable in quantity and quality. For these birds, the available food may be of low nutritional quality and require a considerable investment in digestive-tract size and complexity to be utilized. Through convergent evolution, distantly related birds eating similar diets have very similar nutritional strategies and can extract similar nutritional value from their diet. At the morphological level, all birds have their digestive tracts built out of the same units in the same order, but with wide variations in design depending upon the type of diet typically consumed. In general, birds consuming diets that are digestible autoenzymatically (faunivores, granivores, nectarivores) possess tracts with relatively small ceca and rectums but large proventriculi and small intestines. Conversely, birds that consume foods that require alloenzymatic digestion using symbiotic microflora have tracts with relatively large ceca and rectums relative to their proventriculi and small intestines. More subtle functional

71

adaptations are required to accommodate the physical and nutritional character-istics of the food. Adaptations found in various birds include: the capacity to egest the exoskeleton of arthropods, the bones of vertebrate prey, or seeds of fruits; the ability to concentrate dietary lipids in the proventriculus; the capacity to sort fermentable from refractory cell-wall components; and the ability to modulate the rate of passage of digesta to match the type of feedstuff consumed. Nutritional adaptations at the biochemical level include a wide variety of catabolic, anabolic, and detoxifying pathways and will be discussed in later chapters.

The nutritional strategy determines the type of foods that may be consumed without digestive or metabolic complications. Without these adaptations, many foods are unattainable or indigestible or cannot be metabolized appropriately. For example, an owl lacks the beak and tongue morphology necessary to obtain nectar. Bypassing these morphological inadequacies by intubating nectar causes severe osmotic diarrhea due to glucose malabsorption and would fail to meet an owl's amino acid requirements. Similarly, a lorikeet, which normally consumes nectar, cannot catch or swallow mice. If it were forced to consume a mouse, blockage of the upper gastrointestinal tract by bones and fur and hyper-ammonemia due to excessive protein would occur. Although these examples border on the ridiculous, they illustrate the importance of nutritional adapta-tions.

As discussed in Chapter 1, most birds do not fit neatly into a single food-consumption category and classification often requires major oversimplifications and generalization. During a bird's lifetime, it is rare that a food source is sufficiently steady in supply and invariant in content that adjustments in the nutritional strategy are not required. Populations within a species often have subtle differences in food selection due to variability in morphology and experience. The following nutritional strategies adopted by each food-consumption category illustrate convergent morphological and physiological adaptations. Some species are extremely specialized for a specific food category (obligate consumers) and the generalizations present in the following sections do not do justice to their unique and exquisite adaptations. For many other species, the primary consumption category changes seasonally (facultative consumers) or a variety of foods are consumed continually (omnivory). For these birds, compromises in digestive morphology are often apparent. Adaptations that permit switching between food categories are discussed in the final section of this chapter.

Faunivores

Faunivores possess distinct morphological traits designed for pursuing and catching prey, including flight, beak, and talon modifications. The general similarity in the chemical composition of foods of animal origin permits a relatively similar digestive strategy across the various subcategories (e.g., piscivores and insectivores). All faunivores are reliant on very competent autoenzymatic digestive capacity. Alloenzymatic digestion is not advantageous and large areas for fermentation of food components are absent. Perhaps a defining adaptation of most faunivores is the capacity to separate the highly digestible soft tissues from relatively indigestible components, such as exoskeletons, bones, fur, feathers, fins, scales, and shells. This separation is sometimes accomplished with the beak prior to ingestion, but it also occurs in the gizzard, followed by egestion of the indigestible components back out of the mouth. When these components are not separated, digestibility is markedly impaired. Often the probability of blockage of the gastrointestinal tract by indigestible food components precludes the consumption of animal prey items by many nonfaunivores.

Classification of faunivores into more specialized consumption categories is somewhat arbitrary and ambiguous. Piscivores primarily eat fish but may also consume some amphibians, small mammals, and birds. Likewise, carnivores occasionally eat fish, and members of both groups may eat large arthropods when the opportunity presents. In general, insectivores consume mostly insects, but they usually also consume other terrestrial invertebrates (worms, spiders, and crustacea) and small vertebrates (lizards, amphibians). Consumers of aquatic invertebrates usually eat fewer insects and larger amounts of crustacea and mollusks. Thus, classifications are imprecise, but they are useful for making some generalizations.

Animal tissues are readily digestible and have a very reliable level and balance of nutrients. In particular, animal foods are high in protein and have a balance of essential amino acids that is similar to the bird's requirement. The nutritional completeness of animal prey is exemplified by the absence of species in which parents feed their nestlings a more florivorous diet. If dietary accommodations are made, nestlings are provided different types or sizes of animal prey, or the prey is partially predigested prior to feeding.

Insectivores

Insectivory is the most common pattern of food consumption in birds. Many species of birds are primarily insectivorous and eat little plant material (see Chapter 1, Table 1.2). Other birds are highly insectivorous during part of the year, usually spring and summer, and switch to seeds or fruits during the remainder of the year. Some birds specialize in the consumption of a few types of insects, whereas others opportunistically prey on a large variety. Of the passerines, 46% of the families are primarily insectivorous, as are 51% of the families of small terrestrial nonpasserines. Almost 80% of bird families include some insects in their diet (Morse, 1975).

The value of insects as a complete food is exemplified by the Golden-crowned Kinglet (Heinrich and Bell, 1995). These birds are among the smallest of birds (5 g, which is smaller than many species of hummingbirds), and yet they winter in an environment that routinely reaches −30°C by feeding almost exclusively on hibernating insect larvae. More prevalent but less dramatic is the fact that most granivorous, frugivorous, nectarivorous, and herbivorous birds feed their fast-growing young a diet of insects, spiders, and other invertebrates.

Despite being the most common dietary strategy, our understanding of the nutrition of insectivores lags behind that of less populous granivores, herbivores, and frugivores.

INSECTS AS FOOD

The composition of insects is variable, depending on the species and the stage of the life cycle (Redford and Dorea, 1984; Landry *et al.*, 1986; Robel *et al.*, 1995). Adult insects are high in protein (50–75%, by Kjeldahl's method) and lipid (5–35%), with low levels of nonchitin carbohydrate. Larvae, such as those from Lepidoptera, are also high in protein (40–70%, by Kjeldahl's method) and lipid (10–40%) and have variable levels of nonchitin carbohydrate (3–30%). In larvae, much of the nonchitin carbohydrate is glycogen. The amino acid balance of insects relative to the requirement of birds is almost as good as vertebrate prey and much better than plant proteins. Insects are also a good source of phosphorus, most of the trace minerals, and vitamins, but are low in calcium. The dominant aspect of insects that negatively affects their digestibility is a chitinous exoskeleton. Other invertebrates consumed by insectivores, such as spiders, mollusks, and earthworms, are roughly similar to insects in nutritional properties (see next section).

PHYSICAL ADAPTATIONS

The general body morphology of insectivores usually reflects their foraging method. Insects may be obtained during highly maneuverable flight (e.g., swifts and swallows), by sallying out from exposed perches (hawking or flycatching), by picking insects from the surface of leaves or limbs (gleaning), by climbing vertically on trees and excavating prey from the bark or wood, and by foraging on the ground. Insectivores tend to be smaller than their more omnivorous or frugivorous relatives. The beak of many insectivorous birds is longer and narrower than that of granivorous relatives. Those that obtain insects by probing and drilling tend to have longer beaks than birds that catch flying insects or obtain them from the surface of plants or the ground. Tropical insectivores tend to have longer beaks than their temperate counterparts, probably due to the larger size of insects. Insectivorous ducks (e.g., Ruddy Ducks), have numerous and closely spaced lamellae in their beak for filtering small insects from the water.

Compared with granivores or frugivores, insectivorous birds usually have a less expansive esophagus or crop, presumably because the rate of capture of insects is relatively slow (Wooller *et al.*, 1990). The proventriculus of insectivores is large compared with that of closely related granivorous or herbivorous species. A large proventriculus is an adaptation necessary to provide copious amounts of pepsin and hydrochloric acid (HCl) for protein digestion. Species that consume soft-bodied insects have particularly small gizzards, but those that consume hard-bodied insects have gizzards that are larger and stronger in order to break open the exoskeleton. In birds that switch seasonally between seeds and insects, the size and musculature of the gizzard declines when insects are consumed. Insectivores typically have ceca that are small and lymphoepithelial or vestigial.

NUTRITIONAL STRATEGIES

Insectivores autoenzymatically digest the nonchitin components of their food during its passage through their relatively simple gastrointestinal tracts. The moderate rate of passage of insects is comparable to that of other nutrient-dense foods, such as seeds, and much slower than nutrient-dilute foods, such as fruits. The efficiency of digestion of the nonchitin components of insects is very high and probably approaches 100% in some situations. For example, blowfly larvae, excluding their exoskeleton, have an apparent metabolizable energy coefficient (*MEC) of 90% in Redshanks, and correction for excreted uric acid gives an apparent digestibility close to 100%. The chitin content of insects is the major characteristic that affects their digestibility. This is because chitin is relatively indigestible and physically blocks the access of digestive enzymes to lipid and protein. The *MEC of intact insects, including their exoskeletons, is usually in the range of 50–80% and averages about 75% (Speakman, 1987; Bryant and Bryant, 1988; Castro et al., 1989; Karasov, 1990).

The proportion of chitin in arthropods is extremely variable, ranging from 18 to 60%, and is dealt with in a variety of ways (Kaspari, 1991; Kaspari and Joern, 1993):

1. Birds often select insects that contain low amounts of chitin.
2. Many birds prepare insects by removing the parts highest in chitin before swallowing or prior to feeding their young. For example, Grasshopper Sparrows usually remove the wings and tibia from grasshoppers prior to consumption. These parts contain more than 50% chitin, compared with less than 10% in the abdomen, which is always consumed.
3. Birds may mandibulate (crush in their beak and oral cavity) insect parts to extract the muscle tissue and then discard the exoskeleton.
4. Some birds eat insects whole but have the capability to egest the undigested exoskeletons (Kestrels, flycatchers, swallows). For example, European Bee-eaters regurgitate in their daily pellets 24% of the weight of the honeybees or dragonflies that they consume (Krebs and Avery, 1984).
5. Some species produce a chitinase in the proventriculus (see Chapter 3). Even in species that produce useful levels of chitinase, the energy value of chitin is very low, due to poor absorption. However, chitinase is useful because it reduces chitin's physical hindrance of the digestion of other nutrients.

Insectivorous adults often feed their chicks insects that are larger than they select for themselves. This may be due to the decline in chitin content with increasing insect size, or because fewer trips back to the nest are required when larger insects are transported. The high concentration and digestibility of essential amino acids, vitamins, and most minerals found in insects are exemplified by the observation that 40% of the insect diet of House Martin chicks can be replaced with pure fat without affecting weight at fledging (Johnston, 1993a,b). Malnutrition results from such a swap in the diet of frugivores or granivores, resulting in slow growth and mortality. Many insects have low concentrations of calcium or a low ratio of calcium to phosphorus relative to the requirement of growing chicks and egg-laying females. This deficiency is often corrected by consumption of snail shells, bone fragments, or egg shells.

OTHER CONSIDERATIONS

In the tropical environment, insects are usually available through the seasons, but their density is often variable or patchy in distribution. In temperate climates, insect populations are often restricted in seasonal availability. In many habitats, a flush of insect abundance during the growing season (usually spring) is followed by a dearth of active insects in the nongrowing season. Many insectivorous birds in temperate zones solve this problem by seasonal migrations. Others are facultatively insectivorous and switch to other foods, such as fruit or grain, when insect numbers decline. A few species at high northern and southern latitudes do not migrate and maintain insectivory by finding insects hidden in wood (woodpeckers) or in bark crevices (tits, creepers, and nuthatches).

Whether or not birds have a controlling influence on the numbers of insects in natural or agricultural ecosystems has been a longstanding issue. Several major scientific efforts have been launched to evaluate this possibility, such as the US Biological Survey in the early 1900s. In general, results indicate that the very large numbers of insects consumed by birds rarely exert any primary controlling effect on population outbreaks of economically important agricultural pests. This may be because birds typically prey on nonpest species, which are endemic and have more stable population numbers. Insectivorous birds have a major impact on the population dynamics of their preferred prey items, which can be observed in the evolutionary adaptations of these insects, including size, shape, colors, toxins, and behavioral characteristics (Holmes, 1990).

Other microfaunivores

A variety of freshwater and saltwater birds feed primarily on invertebrates taken from the water (sea birds, flamingos, and some ducks) or from the bottom (e.g., plovers, sandpipers, stilts, Limpkins and some rails, ducks, and storks). The invertebrates that are consumed by birds include worms, echinoderms (e.g., starfish and sea urchins), mollusks (e.g., squid, octopus, snails, and bivalves), and arthropods, such as crustacea and arachnids. Some species of sea birds feed primarily on a single type of invertebrate. More commonly, birds eat a very wide variety of organisms from these diverse phyla, specializing according to the size range and location of invertebrates that can be acquired with their foraging strategy and beak structure. Zooplankton, the suspended invertebrates of salt or fresh water, are targeted by some birds (e.g., flamingos, some ducks, penguins, petrels, and small alcids). Other birds target benthic invertebrates, which are those found on the bottom of aquatic systems. Sea birds (e.g., some large penguins, petrels, alcids) that consume diets high in the larger invertebrates, such as krill or squid, may also consume small fish when available. Similarly, birds of the shore and land that primarily eat noninsect invertebrates often eat small vertebrates (amphibians, rodents, fish) and insects when available.

INVERTEBRATES AS FOOD

Crustacea, such as shrimp, crayfish, krill, crabs, water fleas, copepods, and barnacle larvae, dominate the plankton and the benthos of marine and freshwater environments. Their sizes vary from microscopic, with more than 10,000 l^{-1} of water, to some which are too large for birds to consider as prey. Like insects, crustacea have a chitinous exoskeleton, but it may also be partly calcified and therefore higher in ash content. Krill, the generic name for a variety of euphausiids, are typical of the crustacea consumed by birds. On a dry-matter basis, krill are about 10% chitin, 45% protein, 30% lipid, 15% ash, and less than 3% carbohydrate (Clark, 1980; Roby, 1991). Lipid is the most variable component, both in amount and composition. Wax esters are the main type of lipid in some species of crustacea, while triglycerides predominate in others. In many ways, crustacea are similar to insects in nutritional value, but they often have a lower chitin content. They are also reasonably similar to fish in protein and energy content. Squid are one of the most common mollusks consumed by planktonivores. Squid contain less fat than most krill and small fish, giving them a lower energy content. They also have a third of the calcium content of fish or krill. Bivalves and snails are commonly consumed by shore birds. These mollusks are low in chitin, but the calcium carbonate shell makes up a very large percent of their weight. For example, blue mussels are about 63% shell and are often consumed whole by ducks and wading birds. The shells must be crushed by a very strong gizzard and contribute to a high digesta weight relative to body weight, which impacts the flying capacity of consumers, such as Common Eiders (Guillemette, 1994).

PHYSICAL ADAPTATIONS

Wading shore birds (Charadriiformes), such as plovers, sandpipers, avocets, oystercatchers, and curlews, have beaks of a variety of lengths and strengths for selecting and manipulating invertebrates (Burton, 1974). The bill may be long and thin for probing deep into the wet substrate, short and thin for picking up surface prey, or long and stout for finding and opening bivalves. For example, the strong beak of oystercatchers is used to open large bivalves, such as mussels and cockles, so that the meat can be consumed without the shell. This is done by stabbing the beak into a partially open bivalve and forcing it between the shells. The adductors are severed and the shell is pried open by widely opening the beak. For closed bivalves, a hole is made in one of the shells by a series of hammering blows with the bill tip. The beak is inserted into the hole, the adductors are cut, and the meat is extracted. Shore birds that have thin beaks typically must select bivalves and snails sufficiently small to consume in the shell.

Planktonivores must mechanically separate their prey from the water before swallowing it (Zweers *et al.*, 1977; Crome, 1985; Rubega and Obst, 1993). Many sea birds (e.g., penguins, albatrosses, petrels) individually select large items, such as krill of 1–5 cm in length. Phalaropes seize individual small prey items (e.g., *Daphnia*) with the forceps-like tip of their thin beak and transport them to their mouth using the capillary action resulting from the surface tension of water surrounding the prey. Waterfowl have a highly specialized beak for filtering and sorting potential food items. Sorting food from nonfood items is done by utilizing touch and taste receptors in coordination with the papillae of the tongue, which interdigitate with the lamellae of the beak. The tongue acts as a piston to create suction, which draws water inward through the tip of the beak. Water is forced posteriorly and laterally through the mouth and particles are strained out by the lamellae as the water exits. Filtered invertebrates are brushed into a food-transport groove by the action of papillae on the tongue, forming a bolus of food, which is swallowed. A more extensive array of lamellae in the beak of flamingos provides for efficient filtration of plankton (Fig. 4.1). Several species of prions also have well-developed lamellae in their beak, which permit these pelagic birds to selectively consume small crustacea. Sympatric species of planktonivorous ducks, prions, and flamingos often have different spacing of their lamellae, which minimizes interspecific competition for food by allowing different size ranges of invertebrates to be sieved (Jenkin, 1957; Prince, 1980).

The gastrointestinal tract, especially the small intestine and ceca (lympho-epithelial or vestigial), is typically similar to that of insectivores in length and morphology. Petrels, albatrosses, and penguins have a very large and distensible proventriculus and a comparatively small gizzard. Their proventriculus is used for storing prey accumulated during long foraging trips. The proventriculus and gizzard of many pelagic sea birds is also adapted for accumulating dietary lipids for provisioning of their young. Those molluscivores that swallow snails or bivalves whole have large well-muscled gizzards for crushing the strong shells. The gizzard of Red-necked Stints and Red Knots is more than ten times larger than the proventriculus. The size of the gizzard is adaptable in these shore birds, becoming atrophied when soft food items, such as worms, are consumed and increasing in size and muscularity following prolonged consumption of snails, cockles, or mussels (Piersma *et al.*, 1993).

NUTRITIONAL STRATEGIES

The foraging strategies and beak adaptations for procuring prey are well studied but, with the exception of digestive efficiency, other aspects of the nutritional strategy of microfaunivores have not been examined in detail (Adams, 1984; Heath and Randall, 1985; Jackson, 1986; Jackson and Place, 1990). The digestibility of krill and squid is typically high, suggesting that the protein and fat are highly digestible or that a substantial portion of the chitin is assimilated. For example, krill have an *MEC of about 75% in White-chinned Petrels and in a variety of penguins, and squid has an *MEC of 81% in King Penguins. However, squid is not efficiently digested by some sea birds, such as Jackass Penguins. Petrels and albatrosses digest the waxes found in krill and other crustacea with an efficiency of greater than 85% (see Chapter 3). Production of a chitinase for the hydrolysis of chitin is important for birds that eat crustacea and some mollusks, and has been demonstrated in a variety of petrels, penguins, and an albatross. The highly chitinous beak of squid may be regurgitated by some birds (e.g., albatrosses, cormorants).

When oystercatchers remove mussels from their shells, they obtain an *MEC of 82% from the flesh. However, when oystercatchers consume snails without

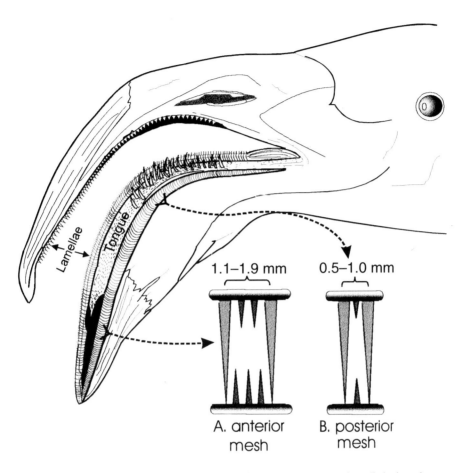

1.1–1.9 mm   0.5–1.0 mm

A. anterior mesh   B. posterior mesh

Fig. 4.1. The beak of a flamingo is adapted to gather large quantities of small plankton by filtration. During feeding, the beak is closed and a mesh is formed by the juxtaposition of lamellae on the margins of the upper and lower beak. The size of the mesh is larger anteriorly (A) and smaller posteriorly (B). Water is moved in and out of the beak by the pumping action of the tongue. When water is brought into the beak, the gape is widened so that food material can freely flow in. When water is forced out, the firmness with which the gape is closed and the area along the beak where the water is directed determines the size of food organisms that are selectively retained (0.5–4 mm). The filtered food organisms are swept posteriorly for swallowing by the large papillae of the tongue. (Redrawn with permission from Zweers *et al.*, 1995.)

removing the shell, the *MEC of the flesh is only 64%. Apparently the high ash content (59%) physically limits access of digestive enzymes and interferes with digestion. Thus, it is often nutritionally advantageous for a bird to expend the energy to remove the shell and to wash the contents to remove sand and other nondigestible material prior to consumption (Speakman, 1987).

OTHER CONSIDERATIONS

Most noninsect invertebrates live in water, especially the oceans. Although oceans cover 60% of the earth's surface, less than 3% of the world's birds exploit these foods (Croxall, 1987). Nevertheless, sea birds are major consumers of zooplankton in the oceans at northern latitudes. The availability of prey, especially in the winter, is a limiting factor on the size of sea-bird populations. Competition for food probably exists between birds and other major zoo-plankton consumers, such as humans, whales, and pinnipeds, although the extent of this competition is not well characterized.

Zooplankton is distributed very unevenly, often in patches more than 100 km apart, requiring extensive traveling in order to locate and utilize this food. Further complicating the location of food, much of the zooplankton follows a diurnal rhythm, rising at night towards the surface and moving deeper during the day. Many birds that catch zooplankton are strong divers (e.g., penguins, alcids) and can feed efficiently during the day. Often, the morphological adaptations required for diving make flight either difficult or impossible and mobility is reduced. Highly mobile sea birds (e.g., albatrosses, gulls, petrels, and prions) are better suited to locating patches of prey but are generally limited to the top few meters of the ocean. These birds often feed at night, when the zooplankton rises, or opportunistically during the day on dead or injured prey. The patchy distribution of zooplankton requires long foraging trips, resulting in very infrequent feeding of young chicks. Thus chicks may be subjected to a feast–famine consumption regimen.

Carnivores and piscivores

Carnivores include most falcons, hawks, kites, eagles, vultures, owls, and shrikes. Piscivorous species include a variety of falcons, hawks, eagles, owls, kingfishers, grebes, loons, wading birds (e.g., herons, egrets), sea birds (e.g., terns, pelicans, some penguins, gulls, auks), and mergansers among the ducks. Many of the small carnivores also consume larger insects. Many piscivores also consume mollusks and larger crustacea. In general, raptors are more intensely studied than the rest, due largely to their prevalence in captivity.

VERTEBRATES AS FOOD

The nutrient composition of vertebrates is relatively constant, with the main differences being the amounts of fat and thus the caloric density. Mammals tend to have more fat than birds or lizards, but in all vertebrates the amount is highly variable, depending on the age, species, and individual. For example, fish can be subdivided by their fat content. Oily fish, which store fat in their muscles, can contain more than 20% fat prior to spawning and less than 1% fat afterward. Nonoily fish store their fat in the liver and normally contain less than 2%. Cold-water species have higher fat contents than warm-water species. The protein content of vertebrate foods varies inversely with the fat content, but the protein quality is uniformly excellent. The balance of essential amino acids closely resembles the requirement of birds. In general, vertebrate foods are also very high in readily digestible vitamins and minerals. Hair, feathers, skin, scales, and bone are refractory to autoenzymatic digestion and are the primary limitation to efficient utilization. The lighter, hollow bones of birds make them somewhat more digestible than those of mammalian prey. The water content of vertebrate foods ranges between 50 and 80%, which is sufficient to fulfill the water needs of birds in many environments and a drinking-water source is not absolutely required.

PHYSICAL ADAPTATIONS

While vertebrate foods are nutritionally complete, considerable physical adaptations are needed for catching and digesting them. The beak of carnivores and piscivores is designed for procuring and sometimes tearing apart prey. The beak tip is usually pointed and the edges are sharp for tearing flesh. In raptors, the hooked tip aids in holding and ripping prey. Piscivorous birds may have long thin beaks for spearing their prey (e.g., anhingas) or long broad beaks for scooping and holding prey (e.g., puffins, pelicans). Some piscivorous birds have dentate patterns on the edges of their beak (e.g., mergansers) or rough tongues to keep slippery fish from escaping their grasp. Vultures routinely scavenge large carcasses and have minimal feathering on their head. This adaptation apparently minimizes contamination of the head with tissue debris during consumption of entrails.

Carnivores and piscivores have a very expandable esophagus, which is enlarged at the junction with the proventriculus (Herpol, 1964; Rhoades and Duke, 1975; Duke et al., 1989). In some species, a distinct crop is evident (e.g., Falconiformes). The proventriculus and gizzard of carnivores are large in size to accommodate large meal sizes with high protein content. There is some evidence for the production of pepsin in the gizzard of the Kestrel, Common Buzzard, and Little Owl. The gizzard of many carnivores and piscivores lacks distinct pairs of thin and thick muscle and functions to massage and mix the contents rather than grind them. The pancreas is small relative to that of granivores and herbivores, presumably due to the high digestibility of flesh and the lack of need for amylase. In many species, the pancreas occupies only half or less of the duodenal loop. The small intestine is usually short relative to most other birds, except frugivores, but piscivores usually have longer small intestines than carnivores. Among Falconiformes, scavengers, such as vultures, have longer intestine length relative to body size compared with species that capture live prey. Those Falconiformes that attack prey, such as small birds, by aerial pursuit have intestines that are 20–30% shorter than those that hunt with soaring flight or by pouncing from a perch. Short intestine length may lighten the body to facilitate rapid acceleration and maneuverability for the capture of flying prey. The shorter intestinal tract decreases the efficiency of digestion, but this loss may be compensated by a much larger increase in successful hunting due to increased agility (Barton and Houston, 1993a,b, 1994).

Many carnivores and piscivores have vestigial ceca, but large ceca are found in a few species (Duke et al., 1981, 1997; Chaplin, 1989). The large ceca of the Great Horned Owl have mostly water-absorbing functions, with little effect on food digestibility. Loss of cecal function by surgical removal can be compensated for by increased water intake at thermoneutral temperatures. At hot environmental temperatures, high rates of water loss due to evaporative cooling make the ceca essential for preventing dehydration. American Kestrels have a uniquely long rectum, which may aid in water resorption. The presence of extensive lymphatic cells and nodes in the epithelium of the ceca and rectum of carnivorous birds suggests vigilant immunosurveillance. Infected birds, rodents, and other prey are more likely to be consumed by predators, and the large numbers of lymphatic follicles along the gastrointestinal tract aid in the control of these potential pathogens.

NUTRITIONAL STRATEGIES

When consumed whole, vertebrate prey represent a complete nutritional package that is not deficient in any nutrient, including water. Many carnivores that consume large prey often consume one meal or less each day. In contrast, many piscivores that eat small prey have a more continuous consumption pattern. Small prey items are often consumed headfirst with no preparation. Sometimes, the prey must be crushed to break large bones, such as the pelvis and pelvic girdle, in order to be swallowed (e.g., by roadrunners and kingfishers). In some cases the prey is larger than the bird's gape and must be torn into smaller pieces to be consumed. Poorly digestible components, such as the head or skin, may not be consumed. The prey's gastrointestinal tract may also be discarded when adequate food is available. Of the consumed components, bones, claws, hair, beaks, feathers, and scales are not very digestible. These materials are massaged into a pellet in the gizzard, and mucus is used to make them adhere together. Regurgitation and egestion of the pellet has been well studied in Great Horned Owls and is distinct from vomiting or ruminating in mammals. Egestion occurs after gastric digestion is complete and is stimulated by the presence of bulk in the gizzard and the absence of amino acids and other nutrients in the lumen of the gastrointestinal tract. The anterior propulsion of the pellet is accomplished by a series of unique contractions of the gizzard and abdominal wall, and the process requires about 4 min. When fed one meal a day, there is usually one pellet egested per day in Falconiformes, Strigiformes, and American Bitterns. In some species, pellets may be egested less frequently (e.g., Bald Eagle) or more frequently (Long-eared Owl). If a second meal becomes available prior to egesting a pellet, an owl will consume it and egest only one pellet after the second meal is digested. Such skipping of an egestion does not decrease digestive efficiency. Piscivorous species that consume many small fish egest pellets composed of the indigestible components from many meals. Circadian cues, such as sunrise, are a contributory stimulus of an egestion reflex in raptors. The pellets that accumulate below a roost give an accurate dietary history of a bird (Duke *et al.*, 1975, 1976, 1993; Duke and Rhoades, 1977).

Carnivores and piscivores rely solely on autoenzymatic digestion and a slow rate of passage to efficiently digest their food (Laugksch and Duffy, 1986; Barton and Houston, 1993a,b). Cape Gannets and Jackass Penguins require 12.8 and 10.3 h, respectively, to excrete 50% of a meal of fish. The mean retention time of a meal of chicks in a variety of raptors is about 8 h, with the time being positively correlated with the length of the small intestine. The slow rate of passage of digesta results in efficient and complete digestion of the flesh and organs of the prey. Averaged across a variety of species, the *MEC of vertebrate foods, including bones, feathers, scales, fur, teeth, etc., is about 75%. Relatively small variations in digestibility of vertebrate foods are due to a bird's species, age, metabolic demand, and meal size and the type of prey. Whole fish are usually more digestible than whole birds or rodents, because they contain less indigestible bone. However, when regurgitated components are excluded, the utilization of fish, birds, and mammals is similar. For example, the *MEC of catfish and shad is 89% in Double-crested Cormorants when regurgitated bones are excluded. Great Horned Owls digest the nonegested components of chicks, turkey poults, hamsters, and white mice with an *MEC of about 90%. About 20% of the energy of dietary protein is lost in excreted uric acid, so an *MEC of 90% represents an apparent digestibility of almost 100%. The complete digestion of most protein and fat has been confirmed by feeding Common Buzzards a meal of rabbits or pigeons that have had bones, skin, and fur or feathers removed (Fig. 4.2). The remaining components have an apparent digestibility of 96% (Kirkwood, 1979; Wijnandts, 1984; Campbell and Koplin, 1986; Jackson and Place, 1990; Brugger, 1993; Tabaka et al., 1996).

Carnivorous and piscivorous birds secrete large quantities of HCl and pepsin from their proventriculus (Herpol, 1964; Duke et al., 1975). The pH of gastric secretions of hawks and owls is 1.7 and 2.4, respectively. Presumably, this difference in acidity accounts for the difference in digestive efficiency of the

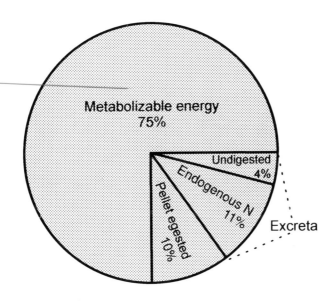

Fig. 4.2. Partitioning of energy contained in the vertebrate prey of carnivores and piscivores. Typical values were selected from the literature. The proportion of energy lost in the pellet is highly variable, depending upon the specific avian consumer and the prey item. See text for references. N, nitrogen.

bones. When fed mice, bones make up 6.5% of the pellet of Falconiformes and 45.8% of the pellets of Strigiformes (owls). The difference in bone digestibility translates into a pellet mass of 5.2% of the food consumed in Falconiformes, versus 12.5% in Strigiformes. It is important that bones are consumed and partially digested, because the ratio of calcium to phosphorus in soft tissue is very poor (1:44 for beef liver and 1:15 for horse meat). Without the partial digestion of bones, a calcium deficiency is probable (Chapter 10).

The diet of carnivores and piscivores has the lowest carbohydrate content of any consumption category (< 2%). This anomaly impels a variety of metabolic adaptations. The paucity of dietary carbohydrate and excess of protein necessitate high rates of amino acid catabolism and use of the carbon skeleton for gluconeogenesis. The disaccharidase activity and glucose-transport capacity of the intestine is low and, when a high-glucose meal is fed, the glucose that is absorbed is cleared slowly from the circulation (Diamond and Karasov, 1987; Meyers, 1995).

OTHER CONSIDERATIONS

Most carnivorous and piscivorous birds raise altricial young. Presumably, precocial development patterns are limited by the physical prowess and skill required to locate and kill vertebrate prey. Parents often forage very widely to find food for themselves and their chicks. Sea birds, in particular, travel great distances to exploit prey that is patchily distributed. Sea birds have evolved various systems for transporting fish to their young, including storage in a gular pouch (e.g., some auklets), storage in the proventriculus (e.g., gulls, cormorants, boobies, gannets, petrels, and penguins), or carrying in the beak (e.g., terns and some auks). The proventricular transport results in partial digestion of the prey items. In some species, this partial digestion leaves behind a lipid-enriched meal for the chick.

Carnivores can have a significant impact on populations of rodents and are highly appreciated in agricultural regions. Though given little credit, vultures provide an important service by efficiently locating and consuming dead animals, thereby preventing the stench and potential diseases that would otherwise occur. Large carnivores are disparaged for predation on agricultural animals, but their economic impact is minimal. Though faunivores have very little impact on modern production of poultry and livestock, piscivores, such as cormorants, egrets, and herons, can impact fish populations and their presence is often discouraged by aquaculturists and fishermen.

Florivores

Foods of plant origin are considerably more diverse in their chemical and nutritional composition than foods of animal origin. Consequently, there is extreme variability in the digestive strategies across the florivorous consumption categories. A comparison of nectarivores and herbivores illustrates this diversity. Nectarivorous birds have very simple, small digestive tracts, which digest nectar quickly and efficiently, whereas herbivores typically possess large, complex digestive tracts, which digest vegetation slowly and inefficiently. Similarly, body morphology is diverse, with the smallest and most acrobatic of birds (e.g., hummingbirds) consuming easily digested but relatively rare foods, such as nectar, and the largest of birds (e.g., Ostriches) forsaking the capacity of flight to accommodate a digestive system capable of utilizing ubiquitous leaves and grasses. Anthropomorphically, the food of nectarivores and frugivores 'desires' to be consumed and advertises this fact with conspicuous packages of digestible nutrients. Nectar is sporadically distributed in space and time and nectarivores, because of their great mobility, are well adapted to exploit these nectar packages. Conversely, the food of herbivores may be protected by structural components and toxins to thwart consumption and digestion, but is typically available throughout the year in all but the coldest climates.

The names given to several florivore consumption categories are well worn but are not botanically correct. For example, fruit consumption of frugivores is confined to succulent, fleshy fruits and does not include hard, dry fruits, such as grains, nuts, and legumes. Granivory is not limited to the consumption of grains but also includes other hard, dry fruits, such as nuts and beans. Likewise, herbivory includes the consumption of herbaceous plants as well as grasses and other vegetation. Thus, consumption categories are nutritionally relevant operational categories with relatively loose botanical boundaries.

Nectarivores

A variety of birds have become reliant on nectar from flowers as a major component of their diet. The most prominent of these are hummingbirds (Trochilidae of northern and neotropical America), sunbirds (Nectariniidae of Africa), sugarbirds (Promeropidae of Africa), honeyeaters (Meliphagidae of Australia), lorikeet parrots (Loriinae of Australia), honeycreepers and flowerpeckers (neotropical Emberizidae). No bird is completely nectarivorous, as nectar alone cannot meet the amino acid requirements of any bird. Other foods, usually insects, are supplemented to provide essential amino acids. Representative species from about 5% of the avian families include nectar as an important part of their diet. Nectarivorous species are found predominantly in the tropics. Flowers are seasonal in the temperate zone, so it is not surprising that fewer species are found there and these are highly migratory (Morse, 1975).

NECTAR AS FOOD

Nectar is a very dilute solution of various sugars. The nectar of some flowers also contains low concentrations of polysaccharides, amino acids, and lipids; however, nectar is probably the most nutrient-dilute food consumed by birds. Flowers that are pollinated by hummingbirds produce nectar that contains mostly sucrose, whereas plants pollinated by most other birds produce nectars that are enriched in glucose and fructose. Thus, the distribution of sugar composition of flower nectars appears to be the result of the evolutionary response of plants to the preferences of their particular avian pollinators (Hainsworth and Wolf, 1976; Baker and Baker, 1990).

Several foods are nutritionally similar to nectar, in that they have high concentrations of sugars and low levels of amino acids. These include: honey from beehives; honeydew, composed of the secretions of nymphal stages of aphids and other sap-sucking insects; and manna, the fluid that exudes from damaged plants and later crystallizes. These foods require a similar digestive strategy as nectar and are commonly consumed by birds classified as nectarivores, especially lorikeets and honeyeaters. During times of low nectar availability, these foods can substitute as the primary dietary energy source (Cannon, 1979; Paton, 1980).

PHYSICAL ADAPTATIONS

Most nectarivores have relatively long narrow beaks and tongues, which are adapted for retrieving nectar from flowers. The length and shape vary across the taxa and according to the types of flowers visited. Beak length and curvature largely determine the range of floral lengths and shapes that can be probed. Some species (e.g., flowerpiercers, flowerpeckers) that are primarily or partially nectarivorous have shorter, sharper beaks, which are used to rip or pierce flowers to obtain nectar. The tongues of nectarivores contain grooves, bristles, and papillae, which increase the surface area of the tongue and facilitate the collection of nectar by capillary attraction. The tongue may be extended beyond the tip of the beak to permit licking of nectar from the base of the flower. Licking rates for hummingbirds, sunbirds, and honeyeaters range from 6 to 17 s^{-1}; the speed changes with sugar concentration of the nectar and the corolla length. Sucking does not occur because fast licking speeds and the morphology of the tongue make it impossible (Snow, 1981; Paton and Collins, 1989; Richardson and Wooller, 1990).

Nectar is often abundant in flowers, and birds can consume it quickly. In the case of hummingbirds, a large crop permits the consumption of high quantities of nectar during a foraging bout. Then the bird can rest while emptying the crop and digesting its contents. Compared with related granivorous or insectivorous psittacines and passerines, nectarivorous species have smaller, less muscular gizzards and shorter intestines. The proventricular and pyloric openings of most nectarivores lie in the same plane, so that nectar can bypass the lumen of the gizzard. Ceca are vestigial and the rectum is short (Richardson and Wooller, 1986, 1990).

NUTRITIONAL STRATEGY

The digestive strategy of nectarivores is simple relative to other florivore categories. The composition of nectar allows almost complete digestion, with a fast rate of passage. In hummingbirds, the rate of intestinal hydrolysis of sucrose is extremely high and the capacity for carrier-mediated glucose transport is the highest reported among vertebrates. Intestinal glucose transporters are adapted to a high glucose concentration in the lumen by exhibiting an exceptionally high V_{max}. This results in an *MEC of glucose, fructose, and sucrose of > 95%. The Rainbow Lorikeet does not have remarkably high carrier-mediated glucose transport but apparently is capable of efficient absorption by the passive pathway (Diamond *et al.*, 1986; Martinez del Rio, 1990a,b; Karasov and Cork, 1994).

Digestion of nectar does not require participation of the gizzard or extensive enzyme action, permitting fast transit through the gastrointestinal tract relative to that in birds consuming other highly digestible foods, such as seeds or insects, but not compared with frugivores. Following a meal of nectar, hummingbirds begin excreting the first components after 15 min and have a mean retention time of 49 min. Rainbow Lorikeets and New Holland Honeyeaters have similarly high rates of passage. The rate of passage of insects consumed by nectarivores is much slower, due to the need for considerably greater proventricular and gizzard action for digestion of protein and lipid (Paton, 1982; Diamond *et al.*, 1986; Karasov and Cork, 1996).

Hummingbirds and honeyeaters have very low protein requirements when expressed as percent of the diet (Paton, 1981, 1982; Brice and Grau, 1989, 1991). The low requirement on a percentage of the dry-matter basis (< 3%) is due to the very high rates of food intake driven by the exceptional energy demands of these small active birds. The protein requirement for maintenance on a mg kg^{-1} body weight day^{-1} basis is not remarkably low relative to other florivores (see Chapter 6). Nectar has such a low level and poor balance of amino acids that its sole consumption meets less than 15% of the needs for essential amino acids, especially methionine. Soft-bodied arthropods, such as small flies, wasps, mosquitoes, and spiders, are frequently consumed throughout the day to meet amino acid needs. Adult hummingbirds at maintenance often spend more than 25% of their time foraging for arthropods and increase this effort when breeding and feeding their young or when nectar is scarce. The reduced size and musculature of the gizzard of most nectarivores make seeds and hard-bodied arthropods an unlikely supplement, unless these items are first fragmented by stabbing them with the beak prior to consumption (Guentert, 1981; Stiles, 1995).

Pollen may be consumed in the process of obtaining nectar and is high in protein (7–40% dry weight), minerals, and vitamins (Brice *et al.*, 1989). The rate of passage through the digestive tract is the limiting factor for pollen digestibility and is dependent on the bird's age and species. Slow passage rates give longer exposure to digestive enzymes in the proventriculus, gizzard, and small intestine and improve pollen digestibility. The digestibility of eucalyptus pollen in adult hummingbirds and adult Rainbow Lorikeets is only 7%. Young lorikeets consuming insects as a predominant component of their diet digest pollen more efficiently (24%) because of slower rates of passage. Pollen probably contributes little to the amino acid needs of many nectarivores, given the relatively small quantity consumed and its poor digestibility.

The consumption of nectar results in considerable water intake. This is particularly true for small nectarivores, such as hummingbirds, which have the highest metabolic rates of any animal and commensurate rates of nectar intake. When consuming maximum amounts of nectar, hummingbirds must excrete a daily volume of urine equivalent to about three times their body mass. Even though this urine is very dilute (< 100 mosmol), the daily excretion of electrolytes is nutritionally relevant. At this excretion rate, about 15% of the bird's total body electrolytes must be replaced daily. Dietary electrolytes may come from insect consumption, but hummingbirds have been observed drinking sea water, eating sand, or licking ashes, presumably to obtain electrolytes (Calder and Hiebert, 1983; Beuchat *et al.*, 1990; des Lauriers, 1994).

The altricial young of nectarivores have high protein requirements to support fast growth and low energy requirements relative to adults, due to parental brooding and low activity. Consequently, nectar does not meet the nutrient needs of chicks, and adult nectarivores must feed their young a predominantly insect-based diet. As the chick matures, nectar increases in dietary importance.

OTHER CONSIDERATIONS

Flowers produce nectar to attract a variety of different pollinators. Flowers that attract hummingbirds by nectar production are typically scentless and often red in color or have adjacent bracts or leaves that are red. The nectar from these flowers typically has much lower concentrations of sugars (20–25 g 100 ml^{-1}) and amino acids than those pollinated by insects, apparently to reduce the viscosity and permit fast acquisition through capillary action of the tongue.

Nectar is offered by plants as a nutritional reward for pollination, so its nutritional deficiencies are somewhat of an ecological enigma (Brice, 1992). A comparison of the amino acid levels in nectars commonly consumed by hummingbirds with the estimated dietary requirements of these birds indicates that essential amino acids are particularly low. It may be that the amino acid deficiency inherent in nectar compels birds to forage more widely for insects, increasing the radius of pollen dispersal in a manner similar to that suggested for frugivores (see below).

The supplementation of wild hummingbirds with sugar-water feeders has been questioned from a nutritional viewpoint by concerned ornithologists. This composition of sugar-water is not very different from nectar. Therefore, birds feeding on either sugar-water feeders or flowers still must find supplemental protein sources. The main difference between natural nectar and artificial preparations is the quantity and timing of their availability. In this regard, sugar–water feeders may support early migration and breeding of hummingbirds on the west coast of North America and allow their winter residence at subzero temperatures (Pimm *et al.*, 1985; Taylor and Kamp, 1985).

Frugivores

Birds that eat fleshy fruits as their primary food are known as frugivores. Fruits contain nutrients in the pulp to entice consumption by birds, so that their indigestible seeds might be widely dispersed. Fruits are especially common in the tropics, where there are numerous species of birds adapted to the consumption of fruit as their primary or sole food (primary or obligatory frugivory). Fruit availability in temperate climates is highly seasonal and very few birds demonstrate obligatory frugivory. Nevertheless, many species of temperate birds consume fruit as the primary component of their diet when they are seasonally available (facultative frugivores). Approximately one-third of avian species in temperate and tropical communities consume fruits as an important component of their diet, making fruits the second most common food category following insects. Many birds are opportunistic and consume fruits as part of a diverse diet containing many other types of foods. These omnivorous birds (e.g., many pigeons, parrots, cardinals, grosbeaks) do not markedly change their digestive strategy to accommodate the fruits and often grind fruit seeds in their gizzard to give efficient digestion (Morse, 1975; Levey and Grajal, 1991).

FRUITS AS FOOD

Fruits vary widely in size, nutrient content, numbers and sizes of seeds, toughness of fruit coat, and presence of toxic compounds. Pomologists classify fruits grown for human consumption as either moist-fleshy fruits, dry-fleshy fruits, or dry fruits. In moist-fleshy fruits, the entire pericarp and its accessory parts are succulent (e.g., bananas, dates, grapes, tomatoes, papayas, citrus, melons, squashes). Dry-fleshy fruits are those in which some parts of the pericarp are dry and usually hard, but other parts are succulent. These include the drupes (e.g., peaches, plums, cherries, olives) and pomes (e.g., apples and pears). Dry fruits do not have a moist and fleshy pericarp (e.g., legumes, sunflowers, oaks, walnuts, grains) and most are included in the granivore but not the frugivore diet category. Figs, strawberries, blackberries, pineapples, and mulberries are actually aggregates of dry fruits on the surface or interior of a succulent seed receptacle and are considered with the fleshy fruits for nutritional purposes. Among the fleshy fruits, there is a negative correlation between the sugar and lipid content of the pulp. The energy and protein content of fruit is extremely variable and generalizations are difficult. In general, the nutritional value of fleshy fruits diverges into two categories: nutrient-dilute and nutrient-dense. Likewise, the digestive strategies of birds that specialize in these two types of fruit categories diverge.

Nutrient-dilute fruits usually have a relatively indigestible exocarp (skin) and a large proportion of seeds, although domesticated forms have been bred to have few or no seeds. The pulp contains large amounts of water and, on a dry-matter basis, is high in carbohydrates and low in fiber and amino acids. In general, the most striking nutritional characteristic of these fruits is a very high ratio of energy to amino acids and other nutrients. For example, bird-dispersed fleshy fruits from east Mediterranean habitats in Israel have an average protein content of only 3.9% of the dry matter, and those from the central USA average 5.1% of the dry matter (Johnson *et al.*, 1985; Izhaki, 1993).

Nutrient-dense fruits (e.g., many laurels, including avocados, palms, dates, bay, sassafras, camphor, cinnamon, mistletoe) contain relatively large amounts of lipid (10–70% of the dry matter) and sometimes protein (5–20% of the dry matter), but less water and sugars (Snow, 1981; Stiles and Rosselli, 1993; Bosque *et al.*, 1995). These fruits are sometimes referred to as 'dry fruits,' because of their low moisture content and soft, oily texture. Storage of protein and lipid in the fruit requires a large nutritional investment for the plant. Consequently, these fruits are often less common but they are nutritionally superior to nutrient-dilute fruits, containing about three times the energy density on a fresh-matter basis.

Nutrient-dense fruits may be consumed preferentially by some birds for pre-migratory fattening or, in the winter, when energy requirements of nonmigrating birds are high in temperate climates. Nutrient-dense fruits are commonly consumed by primary frugivores of the tropics (e.g., Oilbird, quetzals, toucans, bellbirds, manakins, birds of paradise). Depending on the type of fruit, the lipid may be primarily triglycerides or waxes.

The type of sugar in the pulp is variable among fruits. Most fruits dispersed by mammals contain high levels of sucrose, and a subset are cultivated for human consumption (e.g., apples, strawberries, blueberries). Most bird-dispersed fruits are rich in glucose and fructose and are more rarely cultivated commercially. Levey and Grajal (1991) argue that frugivores may select for nutrients in fruit pulp that can be quickly absorbed, such as monosaccharides and free amino acids, and against more complex nutrients, such as sucrose and polypeptides. For birds lacking sucrase activity, such as thrushes (Muscicapidae), starlings (Sturnidae), and thrashers (Mimidae), sucrose is not only a useless energy source but can cause osmotic diarrhea and produce a conditioned aversion. Consequently, sucrose-intolerant birds prefer monosaccharide-dominated fruits. The clear dichotomy in the kinds of sugars offered as rewards for seed dispersal by plants may be maintained by the enzymatic limitations and digestive strategies of their avian dispersers (Martinez del Rio and Restrepo, 1993).

PHYSICAL ADAPTATIONS

A diet of fruit with soft pulpy flesh requires fewer adaptations to the beak, gastrointestinal tract, and body morphology than any of the other consumption categories (Moermond and Denslow, 1985; Jordano, 1986; Remsen *et al.*, 1993). This is because fruit is abundant, easy to find – due to colorful advertisement – and sessile and easy to obtain, and has simple nutrients that require minimal digestive sophistication. The body morphology of frugivores, including wings, legs, and feet, often reflects the preferred method for 'capturing' fruits (e.g., perching, hanging, hovering, stalling, swooping, or snatching). Compared with less frugivorous relatives, frugivores typically are larger and have a wider beak and oral cavity, giving them a wider gape. Among highly frugivorous birds, the beak shape tends to reflect the preferred type of fruit or feeding method. Birds that have strong beaks are most likely to break up a fruit prior to consumption (e.g., toucans, tanagers, and some finches). A wide beak helps the handling of fruits prior to swallowing. Some birds mandibulate fruits by rapid movements of their beaks to mash the pulp and facilitate swallowing, or to remove the pulp and juices so that a tough fruit coat and large seeds can be discarded (e.g., tanagers). A long beak permits an extended reach for feeding from a perch (e.g., toucans, hornbills). Beaks of facultative frugivores typically show few adaptations specific for fruit consumption, but rather are adapted for procuring insects or hard seeds during seasons when fruit is not consumed.

The gizzards of frugivores are variable in size but are usually less muscular than those of more granivorous or insectivorous relatives (Ziswiler and Farner, 1972; Landolt, 1985; Walsberg and Thompson, 1990). This is particularly true for species that consume very soft fruits. In these species, the openings of the proventriculus and the pyloric region of the gizzard are closely aligned, so that fruit can bypass the lumen of the small gizzard. The gizzard size of a frugivore may vary seasonally as the type of fruit consumed changes or, in the case of facultative frugivores, when insects are consumed. Gizzard musculature is related to the amount of effort required to remove the outer skin from the fruit and to remove the pulp from the seeds. For example, the gizzard of the Phainopepla changes in size by more than twofold over the course of a year, depending on the amount of mistletoe consumed. Grit is usually less prevalent in the gizzard of frugivores compared with granivores and herbivores, but the gizzard may have adaptations that permit the internal manipulation of the fruit to remove the pulp from the seeds and exocarp, facilitating their differential processing. The gizzards of fruit-eating Columbidae are particularly well studied. For example, some species of pigeons that primarily eat stone fruits have cone-like projections arising from the cuticle. These dentitions alternate longitudinally on either side of the gizzard so that during contraction they interdigitate and rake the flesh off the central seed. In other fruit pigeons, the cuticles under the two thick gizzard muscles form round pads, which partially occlude the lumen. Fruit is extruded through this small opening, squeezing the flesh from the seeds. The size of the cuticle pads and the width of the lumen are species-dependent and determine the size of seeds that can be processed. Finally, some species of fruit pigeons have very large and muscular gizzards, which grind the stone or seeds in fruits and permit their partial digestion.

The small intestine of frugivores is usually relatively short compared with other florivores and they typically have ceca that are small or vestigial. Alloenzymatic digestion is minimal, due to the small size of the digestive tract and the fast rate of passage of the food.

NUTRITIONAL STRATEGIES

The digestive strategies of frugivores adapted to consume nutrient-dilute fruit diverge from those adapted to consume nutrient-dense fruits (Witmer, 1996). Complete digestion of the pulp and seeds of nutrient-dilute fruits requires a slow rate of passage, high weight of retained digesta, and slow rate of energy acquisition. Due to the high weight of seeds and low nutrient density of the pulp, this strategy is incompatible with flight. Further, the complete digestion of pulp, including fermentable fiber, results in a ratio of energy to amino acids and other essential nutrients that is too great to be a complete diet. Frugivorous birds consuming nutrient-dilute fruits have adopted a nutritional strategy that avoids these problems, including: (i) a quick rate of passage of food through the digestive tract, resulting in poor digestion of lipid, fiber, and other complex carbohydrates; (ii) relatively poor glucose absorption; and (iii) high levels of fruit consumption. Digestive strategies that forfeit complete digestion of nutrients for high rates of intake are called skimming.

Frugivores process as much nutrient-dilute fruit pulp as physically possible and are limited by the rate at which they can pass the fruit through their digestive system (Worthington, 1989; Levey and Duke, 1992). In fact, many frugivores spend long periods of time inactive after a meal, apparently waiting for sufficient clearance of digesta so that they can consume another meal. Examination of the digestive tract of feeding frugivores reveals that the organs are packed with digesta. In Cedar Waxwings, for example, the intestinal lumen is completely filled with seeds following a meal and additional digesta cannot move from their distensible esophagus and gizzard until defecation makes room. Even though some frugivores spend less than 10% of their day foraging, they often devour double their body weight in fruit each day, though much of this consumption is water. The very simple digestive tract of frugivores permits fast rates of passage of digesta (Fig. 4.3). In studies of passerines, the rate of passage and the amount of food consumed are highly correlated with the degree of frugivory. The high rate of passage of fruit accommodates high intakes but limits digestion of complex nutrients, such as starch and lipids. This poor digestive efficiency is reflected in low *MEC values, which average 51% for the whole fruit and 64% for the pulp and skin. When these fruits are fed to granivores that do not adopt a

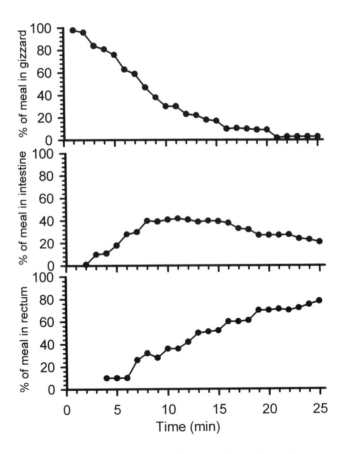

Fig. 4.3. Percent of a single meal remaining in the gizzard, intestine, and rectum over time since ingestion by Cedar Waxwings. The meal consisted of artificial fruit containing pulp, seeds, and a barium label. (Data from Levey and Duke, 1992.)

skimming strategy, the *MEC often exceeds 80% for the whole fruit, including seeds. Yet slow rates of passage of nonfrugivores preclude them from consuming enough nutrient-dilute fruits to meet their nutrient requirements (Herrera, 1984; Karasov, 1990; Karasov and Levey, 1990).

Low digestibility of nutrient-dilute fruits has been observed in a variety of frugivores. For example, the Yellow-rumped Warbler digests less than 20% of the lipid and starch when adapted to a nutrient-dilute fruit diet. Cedar Waxwings are one of the rare year-round frugivores of temperate North America. They have the highest intestinal sucrase activity measured among passerines, and yet sucrose digestibility is still only 61%, because of the very short time that digesta spends in the intestine and rectum. Likewise, frugivorous pigeons are very inefficient at the hydrolysis of disaccharides and the absorption of glucose is incomplete, resulting in considerable excretion (Ziswiler, 1990; Levey and Duke, 1992; Afik and Karasov, 1995).

Amino acid absorption is usually more efficient than monosaccharide absorption in frugivores. The poor digestibility of energy components of the fruit (fiber, lipid, sucrose and glucose) coupled with high digestibility of amino acids partially corrects the poor energy : amino acid balance of fruits. Yet many highly frugivorous birds are unable to survive on a diet of only fruit, due partly to an amino acid deficiency. Among fruits, the nutritional value varies directly with the proportion of amino acids to carbohydrate plus fat in the pulp. Electrolytes may also be deficient in some fruits. Even for adapted frugivores, the poor nutrient balance make most fruits a poor dietary choice when amino acid demands are especially high, such as in growing chicks. In fact, most birds that are highly frugivorous as adults feed their young more nutrient-dense foods, such as arthropods. As the chicks mature and their amino acid requirements decrease, fruit becomes a higher percentage of their diet. Fruits also have low levels of calcium, as do most insects, and female frugivores often ingest snails or grit to obtain calcium and they also feed these items to their young (Walsberg, 1975; Studier *et al.*, 1988; Izhaki, 1992).

There are several potential disadvantages of the skimming strategy adopted by many frugivores. First, skimming is wasteful. More than 50% of the possible dietary energy is excreted. Thus the biomass of fruit needed to support a given bird is high. Second, the high rate of intake increases the bird's exposure to toxins present in fruits. Fruits may contain tannins, alkaloids, terpenoids, saponins, steroids, and a variety of other toxins to prevent predation by nondispersing consumers, such as insects and microorganisms. Frugivorous birds may partially ameliorate this situation by eating a variety of different fruits. Because each fruit species has a different proportion of toxins, varied consumption keeps the total level of any one compound below a toxic level. Most frugivores consume dozens of different types of fruits on any given day and may be more tolerant to some toxins than other birds. Consumption of a wide variety of different fruits may also serve the purpose of nutritionally balancing the diet (Herrera, 1982; White and Stiles, 1990; Bairlein, 1996a,b).

Birds adapted to the consumption of nutrient-dense, high-fat fruits have a slower rate of passage and more complete digestion than those adapted to nutrient-dilute fruits (Holthuijzen and Adkisson, 1984; Place and Stiles, 1992; Afik and Karasov, 1995). For example, Yellow-rumped Warblers adapted to consumption of waxy bayberry fruit have relatively slow rates of passage and correspondingly high lipid digestibility. When these warblers are adapted to the consumption of nutrient-dilute fruits, they adopt a fast rate of passage and lower digestibility. Cedar Waxwings are capable of adjusting their rate of passage to the nutrient content of the fruit consumed in specific meals without long periods of adaptation. A variety of tropical birds appear to specialize in the consumption of energy-dense fruits. For example, Oilbirds eat high-fat (33–67% of the dry matter) tropical fruits, such as those of palm and laurel, which they efficiently digest, due to an exceptionally slow rate of passage of the pulp and regurgitation of the indigestible seeds. In fact, Oilbirds can raise their nestlings exclusively on these nutrient-dense fruits. However, the low protein/energy ratio of this diet results in very slow growth, with a high proportion of adipose tissue – hence their name. Many frugivores do not consume high-lipid fruits and prefer high-sugar fruits that are low in lipids. This may be because they lack the capacity to slow their rate of passage sufficiently to digest high-lipid foods. Alternatively, they may lack the specific lipases, esterases, bile storage, or bile-concentrating capacity required to digest high-lipid fruits (Bosque and Parra, 1992; Martinez del Rio and Restrepo, 1993; Thomas *et al.*, 1993; Witmer, 1996).

Seeds make up 30–50% of the dry weight of a fleshy fruit. When retained in the gastrointestinal tract, this bulk increases the energy expenditure needed for flight and physically impairs the digestibility of the pulp (Levey and Grajal, 1991). Frugivorous birds have adopted several methods for dealing with the problems associated with indigestible seeds.

1. Birds may selectively consume the pulp and not ingest the seed (pulp predators). Eating fruit piecemeal is not the specialty of any single species of frugivore but rather appears to be a response to fruits that are too large for a bird's gape. In some cases, birds (e.g., many tanagers and finches) mandibulate the fruit and drop the seeds.
2. Some species may have a sufficiently strong jaw or gizzard to crack the seed coat and digest the high-protein seed (seed predators). Species that have adopted this nutritional strategy are sometimes considered as granivores in the ecology literature, but not by nutritionists.
3. Intermediate- to large-sized seeds may be collected in the gizzard and regurgitated.
4. Small seeds may pass through the entire digestive system without digestion.

Species that regurgitate or defecate undamaged seeds are called seed dispersers, because they typically drop the seed beyond the tree in which they consume it. The last two categories are referred to as true frugivores in the ecological literature, because these are the species that have evolved mutualistic interactions with their host plants, but nutritionists consider all four categories as frugivores.

A large gape and a gizzard and intestine with a large diameter accommodate a seed defecation strategy. The mass of seeds in the gastrointestinal tract and the resulting negative effect on digestion and flight may contribute to the fast rate of passage of fruits. The weight of an actively feeding Mistletoebird increases 20% due to the bulk of indigestible seeds in the gastrointestinal tract. The mechanical separation of the pulp from the seed by the gizzard may be an important limitation to the processing rate of some fruits, limiting their consumption by birds that lack the capacity. The seeds may have a rate of passage that is faster than the pulp, minimizing physical hindrance of digestion. However, this is not always observed. Emus may retain seeds for over a week, probably due to selective retention as grit in the gizzard (Liddy, 1982; Willson, 1989).

OTHER CONSIDERATIONS

Most trees and shrubs are dependent on animals, including birds, for seed dispersal (Willson and Whelan, 1990; Herrera, 1995). Dispersing birds and host plants have a mutually beneficial arrangement. The pulp of the fruit is typically designed as the minimal investment by the plant that will ensure consumption by the dispersers. The nutrients in the plant benefit the bird. Plucking of fruits and consumption at distant locations or local consumption of the entire fruit and regurgitation or defecation of the seeds at distant locations provides the translocation that aids plant reproduction. Although there are a variety of colors of fruits, plants that rely on birds for dispersal usually produce conspicuous fruits, which are frequently red or black in color when ripe. Seeds of most fruits are impervious to digestive enzymes, due to a woody husk. The husk frequently contain toxins to further thwart digestive efforts. Often, the softening of the husk by the digestive process facilitates sprouting. A classic example of plants completely dependent on bird dispersal is mistletoes, parasites of trees through-out the world. Birds that consume mistletoe berries often excrete the seeds when perched in another tree. The voided seed may stick to a branch and germinate to form a new parasitic plant. Seeds that fall to the ground do not produce viable plants.

An unanswered question in ecology concerns the reason for the poor nutritional balance and presence of toxins found in many nondomestic fruits (Izhaki and Safriel, 1990). A suggestion, yet to be confirmed, is that the nutritional imbalance forces birds to leave the plant frequently in search of other types of fruits or insects in order to balance their diet. The plant benefits because the bird's increased movement distributes seeds more widely. According to this hypothesis, trees with nutritionally balanced, toxin-free fruits would be inhabited by birds that would not leave the area to disperse the seeds.

Not surprisingly, soft-bodied arthropods are often consumed as a supple-ment to a frugivorous diet (Moermond and Denslow, 1985; Remsen *et al.*, 1993; Fuentes, 1994). Large frugivores, such as toucans, may occasionally take small vertebrates, such as lizards and nestling chicks. The high protein content of these animal foods nutritionally balances the low protein levels of most fruits. The gastrointestinal morphology needed for digestion of insects is similar to that needed for adaptation to a fruit-based diet, although the rate of passage differs. Frugivores that supplement their diet with grains or foliage, instead of insects, are less common. This is not surprising, given the differing gastrointestinal morphol-ogy of frugivores and granivores. A large muscular gizzard and slow rate of passage permit the efficient digestion of grains but are inappropriate for digestion of most fleshy fruits. Further, most grains contain high levels of metabolizable energy and low levels of poor-quality protein, making them a poor nutritional complement to fruits. In temperate climates, many species switch from fruit in fall and winter to insects and other arthropods in the spring and summer. However, the digestive morphology required to assimilate an insect-based diet during part of the year usually precludes complete adaptation to a fruit-only diet in other parts of the year. Thus, facultative frugivores must consume a small portion of protein-rich foods (usually arthropods) year-round in order to meet their amino acid requirements (Levey and Karasov, 1989).

Granivores

Many birds eat grains and other hard, dry seeds as their primary dietary component during all or parts of the year. In temperate climates, seeds are most available in summer through early winter and facultative granivores switch to other foods when seeds become scarce. Some birds are able to find sufficient supplies year-round and are obligatory granivores (e.g., many sparrows, quail, doves, parrots, and finches). Representative species from about 20% of the avian families consume seeds as an important component of their diet, although only about 2% are primarily granivorous (Morse, 1975). Domestic granivores and a number of wild species are dependent on cultivated grains to support their large populations. Much of our detailed nutritional knowledge of granivores comes from domestic chickens, turkeys, quail, and pigeons and much of our under-standing of the nutritional value of grains is from domestic crops. Domestic poultry have been highly selected for productive efficiency on domesticated grains and may not always be representative of granivores in general. Their nutrition has been extensively reviewed elsewhere (Scott *et al.*, 1982; Leeson and Summers, 1991; Larbier and Leclercq, 1992) and will not be emphasized in this section. Most companion birds kept in captivity are granivores, including the Canary, Zebra Finch, Rock Dove, Budgerigar, and a variety of other finches and parrots.

GRAINS AND SEEDS AS FOODS

According to plant taxonomy, grains are the seeds of *Gramineae*, the grasses. This family includes a variety of species grown for animal and human food, including, wheat, corn, rice, rye, sorghum, sugar cane, oats, and barley. Grains are actually dry fruits, but nutritionally birds eating grains as well as a variety of other hard, dry fruits (e.g., beans, nuts, peanuts, sesame, rape, sunflowers, safflowers, and many wild varieties) are considered granivores, not frugivores. The term graminivore is reserved for birds eating the blades and rhizomes of grasses (see Chapter 2).

Seeds of most wild plants are small, usually ranging between 0.1 and 10 mg in weight. They are typically very abundant in open habitats, such as deserts, steppes, grasslands, and agricultural lands. The seeds of domesticated and wild plants are often protected externally by a hard seed coat. This coat may be tough and difficult to penetrate, such as that of nuts, or less well developed, such as that of grains. Seeds have the highest nutrient density of any part of the plant and are very rich in starch and low to moderate in protein content (Earl and Jones, 1962). They are low in fiber and have variable oil contents, ranging from low levels in grass seeds to high levels in oil seeds (e.g., soybeans, sunflowers, safflowers). Grains are usually very low in calcium. Most seeds have a moderate level of phosphorus, but much of this is a component of phytate, which is not efficiently digested by birds.

PHYSICAL ADAPTATIONS

The presence of a seed coat and the hard, dry nature of seeds require several morphological adaptations. The hull or shell must be cracked to access the nutritive interior. Some granivorous birds swallow seeds whole and use their gizzard to grind off the shell (e.g., doves, quail, chickens, turkey). More frequently, granivorous birds use their beak to remove and discard the shell. In some cases, a bird will ingest small or unripe seeds whole but remove the shell prior to consumption of larger or mature seeds. In species that dehull seeds prior to swallowing, the beak is typically short and stout, with pronounced ridges for seed processing. The coat of dicotyledonous seeds (e.g., legumes, sunflowers, and many woody trees and shrubs) is usually cut off, whereas a crushing action is used to remove the coat from monocotyledonous seeds (e.g., grains). In many granivorous finches (Fringillidae and Estrildidae), the upper palate has two lateral grooves, which normally accommodate the edges of the lower beak when closed (Newton, 1967). These grooves are wide and deep at the posterior end of the palate and shallow and narrow near the beak tip. Larger seeds are cradled in the broad ridge toward the posterior of the beak, and smaller seeds are cradled in the narrower ridge toward the tip. Seeds are positioned by the action of the tongue and by closing of the lower beak; they are repeatedly turned until weak spots are found in the shell (Fig, 4.4). Once wedged in place, the shell is cut by rapid anterior–posterior movements of the sharp-edged lower beak. The tongue supports the seed while it is cut and also transfers it to the other side of the beak so that the opposite side of the seed can be cut in turn. These processes are repeated until the seed is exposed and the shell can be removed by the tongue and discarded. Birds using this technique have beaks with very sharp edges (tomia) and unique articulation of the jaw. Birds (buntings, Cardinals, tanagers, weavers, and various other Emberizidae, Estrildidae, and Ploceidae) that feed on monocotyledons position elongated seeds against an enlarged buttress-like ridge in the roof of the mouth (hard palate) and crush them with an upward thrust of the lower mandible. In contrast, round seeds are placed into the lateral groove of the palate and crushed by the lower mandible. In both cases, the tongue is used to separate the seed from the crushed shell. Birds that crush seeds usually have a more robust beak, with very pronounced ridges and plates. Birds, such as large parrots (macaws, cockatoos), that feed on very hard seeds can develop tremendous pressure with their enlarged jaw muscles. These birds usually have a cartilaginous connection between their beak and the cranium to permit flexion for absorbing the shock of seed cracking.

The size and shape of the beak, the extent of the gape, and the strength of

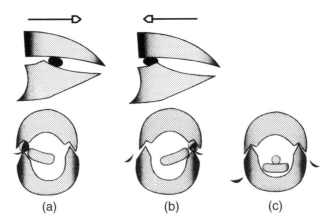

Fig. 4.4. The goldfinch removes the shell (shown in black) from an oval-shaped seed by wedging it into the grooves of its beak and cutting it with anterior–posterior movements of the mandibles (lateral views shown in the two top illustrations; cross-sectional views shown in bottom three illustrations). The tongue is used to position the seed in one side of the beak, where part of the shell is removed (a). The seed is transferred to the other side of the beak, where the rest of the shell is removed (b). The deshelled seed (light color) is then swallowed (c). (Redrawn from Dorst, 1974.)

the accompanying muscles of the jaw establish the speed and efficiency with which a species can consume a given type and size of seed (Bock, 1966; Newton, 1967; Diaz, 1990). Beak characteristics can be more important than body size in determining the size and hardness of seeds selected by a bird. Further, seed hardness is a more important limitation on seed utilization than seed size, and attempts to crack too hard a seed can result in skull damage, due to excessive compression of the skull. Species that regularly include insects in their diet have longer, thinner beaks than completely granivorous birds. For example, the diet of Horned Larks is composed almost entirely of seeds during the fall and winter and over one-third seeds in the breeding season, and yet the long narrow beak more closely resembles that of insectivores than that of more granivorous birds.

Many granivorous birds have an area for the storage of seeds collected during an intense foraging episode. Bullfinches, for example, have special gular pouches used for carrying food to the young, and many other cardueline finches carry seeds in large esophageal pouches. Many granivores have a large distinct crop, which typically holds sufficient seeds so that only two or three foraging bouts are required daily, permitting roosting and nesting at considerable distances from granaries. For example, the crop of a 26 g Greenfinch holds 3 g of seeds, and about 6 g of seeds are consumed each day (Gillespie, 1982). In addition to seed storage, the crop provides a moist environment for the softening of the seeds and hydration of the starch prior to digestion.

The proventriculus is usually moderate in size compared with more faunivorous relatives but may have extensive gastric glands to maximize protein digestion; however, the gizzard of granivorous birds is always very large and muscular (Guentert, 1981; Landolt, 1985). The gizzard is particularly large in granivorous pigeons and doves that swallow seeds whole and rely on it for removing the seed coat. The gizzard is somewhat smaller in the many parrots that remove the seed coat and crush the seed prior to consumption. The cuticle lining the lumen of the gizzard is extensive, especially under the thick muscles, where it forms grinding plates. Grit is commonly trapped in the longitudinal folds of the cuticle and aids in grinding seeds. Grit consumption is correlated with the coarseness of the seeds in the diet and is lowest when birds are fed ground grains. The small intestine of granivores is relatively long and their pancreas is usually large. These large organs permit extensive autoenzymatic digestion of seeds that are high in starch and lipids. Ceca of granivores are typically small or vestigial, with a few exceptions (e.g., Galliformes). However, the ceca of granivorous Galliformes is small relative to those of their more herbivorous relatives.

NUTRITIONAL STRATEGY

Seeds are considerably more nutritionally complete than nectar or fruits, the ratio of protein to energy and the levels of many vitamins and minerals being particularly superior. Many birds are sufficiently adapted to survive on a relatively strict granivorous diet with little or no animal foods. In fact, many adult passerines that are normally omnivorous can be sustained on a strict diet of good-quality seeds. The efficiency of digestion of seeds by granivores is high, due to the low fiber content of husked seeds, extensive autoenzymatic digestion, facilitated by a long small intestine, high enzyme production, and moderate rate of passage. Domestic chickens have been highly selected for efficiency of digestion of grains and legumes for over a century and have an *MEC of about 85% for ground corn. The digestibilities of starch, lipid, and protein in grains from domestic plants are usually greater than 80% in most wild birds. The digestive efficiency is independent of body size, as illustrated by the similar efficiencies for small granivores, such as Cardinals (45 g body weight) and Song Sparrows (25 g body weight), compared with larger granivores (Willson and Harmeson, 1973; Berthold, 1977; Shuman *et al.*, 1989).

Seeds from wild plants are usually less digestible than those from cultivated crops (Billingsley and Arner, 1970; Robel *et al.*, 1979). This is due to the higher levels of fiber, tannins, trypsin inhibitors, and other toxins and antimetabolites found in many wild seeds. Protein digestibility is particularly sensitive to the presence of tannins and other natural pesticides. On average, passerines are more efficient at digesting wild seeds than nonpasserines (*MEC of 75% and 59%, respectively; Karasov, 1990). This difference is mostly due to the greater number of passerine species that remove the seat coat prior to consumption.

An interesting divergence from the strategy of very efficient digestion is seen in a variety of waxbills (Estrildidae) and weavers (Ploceidae) that consume and raise their young exclusively on high-starch, low-protein grass seeds (Ziswiler, 1986, 1990). When fed seeds with only 4.8% protein and 82% starch, the Masked Finch excretes 61% of the starch consumed (Fig. 4.5). These birds have very extensive gastric glands for secretion of pepsin and presumably have highly efficient protein digestion and absorption. Thus, they improve the nutrient

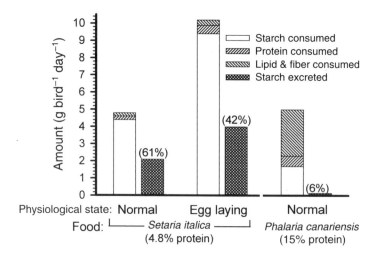

Fig. 4.5. Female Masked Finches adjust their starch digestion to balance the protein : energy ratio of the seeds that they consume in a way that meets their requirements for amino acids without metabolizing too much energy. Finches were fed either *Setaria italica* seeds (4.8% protein and 82% starch) or *Phalaria canariensis* seeds (15% protein, 42% starch). When birds were in a normal physiological state, they digested the starch in *S. italica* very poorly (39% digested, 61% excreted), but digested the protein efficiently. The high energy and protein demands of egg production resulted in increased food consumption and improved digestion of the starch in *S. italica* (58% digested, 42% excreted). When the finches consumed seeds (*P. canariensis*) with an appropriate protein : energy ratio, both the protein and the starch were completely digested (Ziswiler, 1986).

balance of their food by repressing energy digestion and maximizing protein digestion. When amino acid requirements are high, such as during egg laying, they increase their food intake as much as twofold to obtain the additional dietary protein.

A variety of factors may contribute to the selection of seed types for consumption, including ease and speed of husking, toxin content, amino acid balance, and caloric content (Willson and Harmeson, 1973; Shuman *et al.*, 1990; Murphy, 1994). The balance of amino acids in grains is poor relative to the balance required for growth or reproduction (see Chapter 6). The exact deficiency varies between types of seeds. For example, millet is proportionally low in methionine and lysine but hemp is proportionally high. Birds have some capacity to choose between diet components to balance their amino acid intake, although this adjustment is incomplete and an alternative protein source may be needed during periods of high amino acid needs. Soybean meal is often used to increase the protein level of modern grain-based diets fed to poultry and companion birds during growth or reproduction. This works well because the protein to energy ratio of soybean meal is remarkably high, due to the removal of lipid during processing. Legumes are also rich in lysine, which complements this deficiency in many grains. In the wild, insects are consumed by many species to provide the additional protein required for growth and reproduction, and the timing of reproduction corresponds with seasonal insect abundance. Granivorous pigeons use a different strategy. They supplement protein to their granivorous young through crop milk. Some specialized cardueline finches feed their young an exclusively grain diet, but seeds are selected prior to ripening and drying, resulting in a softer and higher-protein diet (Newton, 1967; Gluck, 1985).

The low calcium content of some seeds makes supplementation with food sources higher in calcium imperative during egg laying and chick growth. Another problem with seeds for young chicks is their digestibility. The immature gizzard is too weak to crack hard seeds and low enzyme levels limit the digestion of lipid and raw starch.

OTHER CONSIDERATIONS

Seed availability follows a distinct annual rhythm outside of the tropics. Seeds are usually produced at discrete times and surfeit amounts diminish to scarcity during specific times of the year. In temperate climates, two periods are especially critical for granivores: after seeds fall from the plants to the ground, but are difficult for the bird to find due to vegetative cover (early winter); and at the onset of the growing period, when old seeds germinate and the new year's seeds are not yet available (usually the spring). At these times, granivores may switch to other foods or they may migrate. Often, the scarcity of seeds at the beginning of the growing season coincides with the emergence of substantial numbers of insects, which are consumed by facultative granivores. Scarcity in the winter sometimes forces a switch to less nutritious foods, such as leaves, fruits, and buds. In the case of chickens and Wood Pigeons, the capacity to switch completely to a herbivorous diet is incomplete. Poor fermentation capacity, combined with an inability to sufficiently increase rate of passage and consequently consumption, results in an inadequate rate of nutrient absorption on high-fiber diets (Kenward and Sibly, 1977). When seeds are available in the winter, their high energy content makes them an excellent food. High rates of seed consumption due to winter's cold temperature result in higher levels of protein intake, ameliorating the protein inadequacy of seeds.

Many of the most successful avian species are granivores that have developed a close association with agriculture (Dyer and Ward, 1977). Due to their food preference, granivores frequently are uninvited participants in the grain harvest and in certain areas become a serious agricultural pest. In Argentina, depending on the month of the year, between 72 and 94% of the diet of the Eared Dove consists of cultivated grains, particularly sorghums, wheat, and millet. Although the total impact of birds on grain harvests is probably not substantial, the damage to isolated fields can be devastating to individual farmers. Locally huge concentrations of granivorous birds may occur, due to their flock-feeding habits. Exacerbating the situation are seasonal migrations, where large concentrations of birds pass through relatively small areas resulting in spectacular clouds of birds (e.g., American Blackbirds, Red-billed Quelea, and Common Starlings) and focused foraging areas. Research on breeding crops for resistance to bird predation and developing bird repellents dwarfs the effort involved in understanding the nutrition of nondomestic birds.

Herbivores

Vegetative parts of plants make up a significant portion of the diet of only about 3% of avian species, many of which are predominantly terrestrial (Morton, 1979). Birds that are primarily herbivorous include the Ostrich, Hoatzin, screamers, geese, swans, diving ducks, grouse, and plantcutters. Though relatively few avian species are herbivorous, they are well studied, due to their conspicuous nature and economic value for meat and sport.

Vegetation is usually very abundant, but its high fiber content presents a major nutritional challenge for birds (Dudley and Vermeij, 1991; Cork, 1994). This is because fiber digestion requires exposure to symbiotic microbes in a large gut chamber for long periods of time and yields energy slowly. The long retention time required for alloenzymatic digestion limits intake and increases body weight. Herbivorous birds that depend on flight are particularly constrained by these factors due to the high energy requirement of flight and the need to restrict the weight of digestive organs and their contents. Flight places a maximum on body weight of about 12 kg. Studies of mammalian herbivores suggest that a digestive strategy which relies entirely on fermentation to release digestible energy is not capable of supplying energy at a sufficient rate to meet the requirements for basal metabolism of an animal with a body mass less than 15 kg. These constraints prevent flying birds from occupying the niche filled by mammalian bulk-roughage feeders (e.g., cattle, deer, sheep). Thus, the capacity for flight, which makes ubiquitous vegetation readily obtainable, constrains its digestion. Herbivorous birds capable of flight must always be highly selective and consume plant parts that are relatively high in protein and low in fiber, such as rhizomes, buds, flowers, catkins, very young leaves, and young grasses. Mature leaves, stems, and grass are selected against. Nonflying birds are not restricted in body size and are often better at utilizing higher-fiber foods than flying birds. However, most nonflying birds rely on postgastric fermentation and have not evolved the efficiency of mammalian pregastric fermenters, such as ruminants and marsupials, which would permit primary consumption of mature foliage.

FOODS OF AVIAN HERBIVORES

The terms herbage and roughage are often used synonymously to refer to the components of leafy and woody plants consumed by herbivorous animals. In general, herbage is coarse-textured and high in cell walls and has a low weight for its volume. Division of roughages into specific nutritional groups is arbitrary, due to the extreme variability in composition of plants. Age, growing conditions, and species markedly affect the composition of a plant. Birds often select the most nutritious components of a specific plant, making the nutritional characteristics of the entire plant of little value. Plants commonly cultivated as roughage for animals include members of the *Leguminosae* (legumes), *Gramineae* (grasses), and *Brassica* (watercress, kale, rape) taxa. Selected components of plants from these taxa, as well as from a variety of different forbs, sedges, trees, and aquatic plants, are commonly consumed by wild birds. Alfalfa (lucerne) is the most common roughage fed to captive avian herbivores.

The predominant nutritional characteristic of roughages is their high content of cell-wall material (dietary fiber is explained in more detail in Chapter 8). The cell wall is high in cellulose, the most abundant organic compound on earth. Cellulose, together with lignin, provides the structural support for the plant. Young plant tissue has low amounts of lignin, but, when growth ceases, lignification occurs. Silica is also deposited in the cell wall of some plants, especially grasses. Silica and lignin are formidable impediments to digestion. The most digestible components are found inside the cell wall, including the cytoplasm, vacuole, chloroplasts, and mitochondria. The nutritional value of roughage depends largely on the relative proportions of cell contents and cell-wall constituents and on the degree of lignification of the cell walls.

The protein content of green plant tissue is variable, ranging from 5 to 35% of the dry matter. Although some types of roughage have moderately high levels of protein, the digestibility is usually poor, due to the cell walls. The balance of amino acids within young green plant tissue is fairly good, except for low levels of methionine (see Chapter 6). The balance is usually constant across plant types because most of the protein is found in chloroplasts, which have similar composition between species. Some plants store high amounts of starch and sugars within their vacuoles. These nonstructural carbohydrates are much more digestible than structural carbohydrates, such as cellulose. The moisture content of plants decreases with maturity and is highly correlated with digestibility. Many plants contain deleterious factors, such as alkaloids, glycosides, toxic amino acids, and mycotoxins, which limit intake and digestibility (Akin, 1989; Jakubas *et al.*, 1995).

PHYSICAL ADAPTATIONS

Morphological adaptations to a high-fiber diet are seen throughout the gastro-intestinal tract. The beak is usually modified to permit the harvesting of plant material. For example, the plantcutters and the Tooth-billed Bowerbird have fine serrations on the edges of their beaks, which facilitate cutting leaves and plucking buds and shoots. The beaks of Anatidae (ducks, geese, swans) are often modified to permit grazing or digging. For example, the edge of the posterior third of the beak of the Canada Goose is sharp and serrated. This adaptation permits the scissors-like cutting of grass stems by the forceful abduction of the lower jaw against the upper jaw. The anterior end of the upper mandible is modified to form a pincher-like nail for grasping. The nail permits grasses to be pulled up, so that the nutritious tubers and rhizomes can be obtained. The flat, shovel-like shape of the Anatidae beak permits rooting in soft mud for rhizomes and roots (Goodman and Fisher, 1962).

Herbivores have a large muscular gizzard, required for the extensive reduction of plant material, which is often physically tough, stringy, and difficult to fracture (McBee and West, 1969). The grinding of plant tissue is facilitated by grit. The gizzard may be adapted to separating more fibrous food components from more digestible small particles. For example, the large gizzard of the Alaskan

Willow Ptarmigan grinds and strips the bark and cambium layers from willow twigs so that they can be processed separate from the indigestible wood. The small intestine of herbivores does not show any major morphological adaptations. It is of average length and is important for the autoenzymatic digestion of protein, starch, and lipids found within the plant cells. Sorting of the digesta that remains following autoenzymatic digestion is accomplished by the interplay of cecal sphincters and the sieving action of long villi located at the entrance of the ceca. The ceca are usually large, contain anaerobic bacteria, and are the primary location for fermentation of dietary fiber in most avian herbivores. Spiral folds or sacculations increase the volume of the ceca and provide pockets where digesta may collect undisturbed by the flowing and mixing of digesta that occurs in the lumen. In the Sage Grouse, the total absorptive surface area in the ceca equals that of the small intestine. Among closely related species, cecal size is correlated with dietary fiber intake (Fenna and Baog, 1974; Kehoe and Ankney, 1985; Moss, 1989; Obst and Diamond, 1989).

NUTRITIONAL STRATEGIES

The strategies used by birds to deal with plant foods containing fiber diverge into the same four general digestive categories described in Chapter 3 (see Fig. 3.3):

1. Autoenzymatic digestion, followed by alloenzymatic digestion.
2. Autoenzymatic digestion, followed by alloenzymatic digestion and then coprophagy.
3. Primarily autoenzymatic digestion.
4. Alloenzymatic digestion, followed by autoenzymatic digestion.

The first strategy predominates among herbivorous birds. Individuals of many species shift between the first three categories, depending on their food source and nutritional demands. The fourth strategy is relatively rare among birds.

The digestive efficiencies of herbivores are the most variable of any consumption category, due to the high variability in the composition of foods of plant origin. The variety of digestive strategies that occur across species, as well as within an individual bird, depending on its diet, also contributes substantially to this variability. These factors make it inappropriate to apply an average *MEC value to all herbivorous birds eating a specific type of vegetation or even to a single animal eating different types of plants.

Strategy 1. The general strategy of most herbivores has two components. First, there is extensive use of autoenzymatic digestion in the proventriculus, gizzard, and small intestine. Second, components of the food that were not digested autoenzymatically may be fermented in the lower intestines, primarily the ceca (alloenzymatic digestion). Autoenzymatic digestion supplies more nutrients than alloenzymatic digestion in all species, with the exception of the Ostrich. Nevertheless, alloenzymatic digestion makes a nutritionally important contribution, particularly for energy, and is a dominant part of the digestive strategy of most herbivores. To accommodate the conflicting demands imposed by flight and alloenzymatic digestion, birds have made two principal adaptations. First, they

maintain high intakes and passage rates at the expense of lower fiber digestibilities. Second, they separate the digesta in the gizzard, rectum, and ceca, so that
large particles can be excreted quickly and small particles can be retained for
fermentation (see Fig. 3.2). These two adaptations are interrelated, because the
selective defecation of large particles that would otherwise require very long
fermentation times facilitates high food intake. As a result of these adaptations,
the weight of the gastrointestinal tract and digesta of flying avian herbivores is
considerably lower than in mammals consuming similar diets. However, these
adaptations limit the efficiency of fiber digestion in birds.

Most of the autoenzymatically digestible nutrients of plant tissue are found
in the cell contents, whereas the cell walls require microbial digestion. Access to
the cell contents requires considerable grinding in the gizzard to break the
structural barrier presented by the cell wall. Once exposed, autoenzymatic
digestion proceeds in a manner similar to that described for granivores. Those
autoenzymatically digestible components (protein, starch, lipids) that are not
exposed within the upper gastrointestinal tract may become available following
microbial attack on the cell walls in the ceca. However, alloenzymatic digestion
of protein, starch, and lipids results in considerably lower nutritional returns for
the bird than their autoenzymatic digestion. For example, fermentation of dietary
protein to volatile fatty acids is energetically inefficient and does not supply
essential amino acids to the bird. Thus, herbivorous birds invest considerably in
autoenzymatic digestion.

Following autoenzymatic digestion, the digesta arriving at the rectum may
be quite viscous and difficult to sort. In some species, urine refluxed from the
cloaca may aid in liquefying the digesta to permit sieving into the ceca. The
retrograde flow of urine may also wash electrolytes and other soluble nutrients
from the digesta in the rectum for transport into the ceca. The water-soluble
components and small particles that enter the ceca are only a small proportion
of the total dietary fiber (< 33% in the Blue Grouse) and are enriched in
hemicellulose, pectin, protein, starch, and other polysaccharides. In Rock Ptarmigan, only 18% of the solid digesta that passes through the ileocecal junction
enters the ceca and most of this consists of very small particles. In contrast, 96%
of the liquid phase enters the ceca. Due to their high ratio of surface area to
volume and low lignin content, the small particles and soluble components that
are retained in the ceca are quickly and completely (> 95%) fermented. Clearly,
this selective transport into the ceca facilitates the acquisition of energy from
only those food components that can be fermented without a long retention
time. Almost all of the volatile fatty acids produced during fermentation are
absorbed, and these fermentation products contribute 18% of the energy
required for basal metabolism and thermoregulation in Rock Ptarmigan. The
excluded larger particles that remain in the rectum are often high in lignin,
making them resistant to mechanical and microbial attack (see Fig. 3.2). These
relatively indigestible components are quickly excreted and include the bark,
epidermis, and vascular bundles of stems, leaves, and grasses. Many of the
mechanically resistant structures can be seen intact in the feces. Because dietary
fiber is only partially utilized, total food intake has to be correspondingly high

(Gasaway, 1976a,b; Björnhag and Sperber, 1977; Moss, 1989; Redig, 1989; Remmington, 1989).

Another important function of the separation mechanism at the entrance to the ceca is the retention of microorganisms. Because of their small size, microorganisms are suspended in the fluid phase of the digesta and selectively retained in the ceca. This prevents their excretion, maintaining high populations for continued fermentation. The transport of urine into the ceca permits nitrogen recycling via microbial use of uric acid and ammonia. Cecal ammonia absorption may play a role in the nitrogen economy of birds eating low-protein foods. The ceca of herbivorous birds can absorb amino acids by active transport. Presumably, these free amino acids would arise from microbial synthesis or from microbial hydrolysis of food or endogenous proteins; however, their nutritional contribution remains to be quantified (Björnhag,1989; Obst and Diamond, 1989; Foley and Cork, 1992; Karasawa and Maeda, 1994).

The Ostrich is among the most efficient birds at utilizing dietary fiber (Swart *et al.*, 1993a,b,c). This terrestrial bird is not constrained by weight limitations and has adapted a different digestive strategy from that of birds that fly. The rectum is by far the longest of any bird and accounts for 52% of the length of the gastrointestinal tract. Fermentation in the highly sacculated rectum and ceca produces high levels of volatile fatty acids, particularly acetate, and may supply as much as 75% of the metabolizable energy intake. Dietary fiber digestibility in the Ostrich approaches that in the horse (cell walls, 47%; hemicellulose, 66%; and cellulose, 38%). The digestibility of dietary fiber increases markedly with age and is also highest when forage quality is high and intake is low. With increasing dietary intake or increasing proportions of the diet as fiber, rate of passage increases and digestive efficiency falls progressively.

Microbial fermentation of fiber is not limited to terrestrial birds and the relatively poor-flying gallinaceous birds (e.g., grouse and ptarmigan). Ducks and geese are good fliers and, in some cases, acquire significant energy from fiber, despite possessing small ceca relative to herbivorous Galliformes (Buchsbaum *et al.*, 1986; Dawson *et al.*, 1989; Prop and Vulink, 1992; Sedinger *et al.*, 1995). Canada Geese, Atlantic Brant Geese, and Maned Ducks can obtain up to 30% of their resting energy requirements from the fiber of natural forage. In Maned Ducks consuming grass and clover, hemicellulose is the component of fiber that is most efficiently utilized. It has a digestive efficiency of 40%, compared with 11% and 4% for cellulose and lignin, respectively. Some components of hemicellulose may be partially hydrolyzed and solubilized in the gizzard, resulting in rapid fermentation in the lower small intestine and ceca. In general, geese are more efficient at fermenting cellulose than are ducks. Cellulose digestibilities of 30% and 45% have been measured in Canada Geese and Lesser Snow Geese, respectively. The large size of geese compared with that of most herbivorous ducks presumably permits longer retention times for fermentation. Fiber digestion of ducks and geese is dependent on the selective consumption of vegetation that is relatively low in cellulose and lignin; aquatic vegetation meets these criteria, because structural support comes from the dense water medium. When presented with choices, geese are remarkably capable of selecting the most nutritious species and components of aquatic vegetation, grasses, and sedges (Sedinger and Raveling, 1984; Prop and Deerenberg, 1991; Gadallah and Jefferies, 1995).

Strategy 2. The benefits of microbial fermentation in the ceca are magnified by the practice of coprophagy (consumption of feces). Cecal feces (cecotropes) contain considerable microbial protein, lipid, and vitamins. Preferential consumption of cecotropes over rectal feces, or cecotrophy, is common in several species of Galliformes and Ostriches, and is occasionally observed in a variety of other species. Cecotrophy permits the autoenzymatic digestion of microbial nutrients and their subsequent absorption in the small intestine. The cecal feces of Willow Ptarmigan contain four times higher concentrations of amino acids and only a fraction of the lignin, compared with rectal feces (Fig. 4.6). The quantitative significance of cecotrophy to the amino acid and vitamin requirements of birds awaits further investigation, but is probably critical in some species when diet quality is poor. This strategy markedly increases the digestive efficiency and lowers protein and mineral requirements in many mammalian species (e.g., many rodents) that have similar postgastric morphology. Maintaining birds that use coprophagy as part of their nutritional strategy in cages with wire bottoms may lower their digestive efficiency and increase their vitamin requirements.

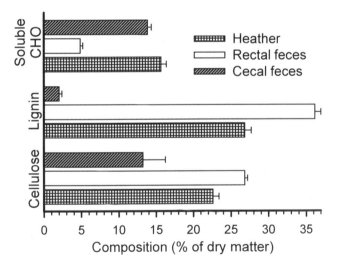

Fig. 4.6. Composition of the heather (*Calluna vulgaris*) consumed by Willow Ptarmigan, and in their rectal and cecal feces (cecotropes). Relative to rectal feces, cecotropes are low in lignin and cellulose, but high in soluble carbohydrates (CHO), reflecting the sorting action of the illeocecal junction. (Data calculated from Moss and Parkinson, 1972.)

Strategy 3.   Herbivorous birds have limits to which they can utilize dietary fiber and, in some circumstances, a strategy is adopted in which fiber is largely ignored, instead of fermented. For example, Blue Grouse eat conifer needles in the winter and ferment very little of the fiber in their ceca. On this high-fiber diet, the strategy appears to be one in which a very high rate of intake compensates for the low digestive efficiency, netting the required level of nutrients predominantly by autoenzymatic digestion. Many herbivores can switch between this 'skimming' strategy and a higher-efficiency, cecal fermentation strategy, depending on the fiber content of their diet or the quantity of diet available. In fact, geese in the Arctic summer can space their feeding bouts over a 24 h period and adopt a strategy of slow rate of passage and higher fermentation. On short days, when only 8 h are available for eating, their rate of passage increases, accompanied by a marked reduction in the efficiency of fiber digestion (Fig. 4.7). In birds adapted to a high-fiber diet, a skimming strategy markedly lowers the weight of digesta in the gastrointestinal tract and aids flight (Moss, 1989; Remmington, 1989; Prop and Vulink, 1992).

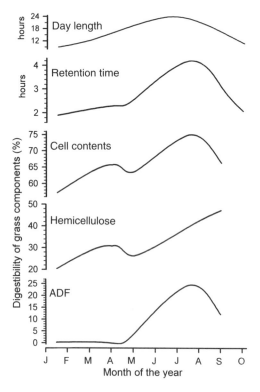

Fig. 4.7. The impact of time available for feeding (day length) on the retention time of digesta and the digestibility of lignin, hemicellulose, and acid detergent fiber (ADF) in Barnacle Geese consuming grasses in the Arctic. The spacing of meals increased with increasing day length. A greater interval between meals permitted longer retention of digesta and more efficient digestion. (Data from Prop and Vulink, 1992.)

When geese are adapted to a skimming strategy, they still autoenzymatically digest nonfiber carbohydrate and protein from grasses relatively efficiently. High rates of consumption compensate for the low level of autoenzymatically digestible nutrients in their food. The combination of high intake, fast rate of passage, and incomplete fermentation of fiber has made graminivorous ducks and geese famous for their defecation rates and volumes. For example, Barnacle Geese feeding on grasses defecate every 3–5 min, with each deposit averaging 5.5 cm in length (Buchsbaum *et al.*, 1986; Prop and Vulink, 1992).

In many species of Galliformes, the size of the gastrointestinal tract and its capacity to ferment dietary fiber change throughout the season, in conjunction with the quantity and quality of food consumed (Gasaway, 1976a,b; Redig, 1989; Remmington, 1989). For example, gizzard and cecal sizes are smallest in the summer, when the birds feed on tender shoots of newly growing herbaceous plants. The gizzard is largest and most muscular and the ceca are longest and yield the most volatile fatty acids in the winter, when birds feed primarily on woody stems, buds, and leaves. The time required for adaptive hypertrophy appears to be about 2–3 months.

Strategy 4. Hoatzins have adopted a unique pregastric alloenzymatic strategy that permits survival on a diet high in foliage (Dominguez-Bello *et al.*, 1993; Kornegay *et al.*, 1994; Grajal and Parra, 1995). This strategy is accommodated by radical anatomical modifications, including a voluminous crop and esophagus for fermentation (Fig. 4.8). The crop is lined by a cornified epithelium and is enveloped by several layers of thick muscles. The grinding action of the crop markedly reduces the particle size of the food. Numerous ridges and sacculations in the lower esophagus selectively delay the passage of smaller food particles. The retention time is exceptionally long for a small bird (> 24 h), giving a large digesta volume. A dense population of Gram-negative bacteria produces high levels of volatile fatty acids within the crop and esophagus. Because of its location, the microbial mass produced can be autoenzymatically digested in the proventriculus, gizzard, and small intestine. In fact, an intestinal lysozyme is expressed that facilitates the digestion of bacterial cell walls. Although the Hoatzin's strategy is similar to that of mammalian pregastric fermenters (e.g., kangaroo, deer), it is considerably less efficient at fiber digestion, due to a relatively smaller fermentation compartment. Yet the weight of its gastro-intestinal tract and digesta is sufficiently great to make the Hoatzin a very poor flier. Apparently, the Hoatzin has established an acceptable compromise between efficiency of fiber digestibility and flight. The benefits of this compromise include: the capability to obtain the most nutritious foliage provided by flight; the ability to take advantage of microbial detoxification of toxic components; and the synthesis of digestible vitamins and amino acids. The Kakapo has a similar digestive morphology and strategy as the Hoatzin, although it is probably much less efficient at fiber digestion. This flightless bird eats only the most nutritious young leaves and shoots. It masticates twigs and the central veins of leaves to extract juices and then regurgitates the fiber (Morton, 1979).

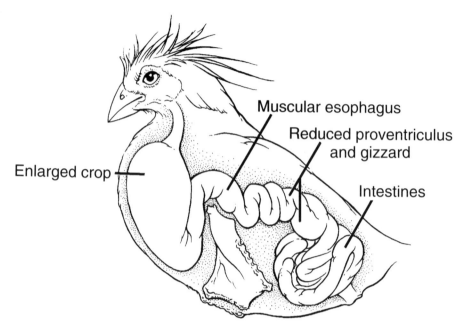

Muscular esophagus

Reduced proventriculus and gizzard

Enlarged crop

Intestines

Fig. 4.8. The Hoatzin alloenzymatically digests the fibrous leaves of its diet in an enlarged crop and long esophagus. Both of these organs are very muscular and serve to extensively grind the food. The large size of the digestive tract causes the Hoatzin to be a very poor flier. (Reproduced with permission from Proctor and Lynch, 1993.)

OTHER CONSIDERATIONS

The chicks of many herbivorous birds consume a more faunivorous or granivorous diet than their parents. The development of the ceca and other areas for alloenzymatic digestion requires several weeks to months; higher-fiber foods slowly become a greater proportion of the diet, in parallel with the maturation of the digestive tract. Also, young growing chicks have very high amino acid requirements, which are more easily met when arthropods or other readily digestible, high-protein foods are consumed. For example, Ruffed Grouse are highly herbivorous as adults, consuming less than 5% of their diet as insects in the summer when insect populations peak; yet insects and other invertebrates provide the majority of the young chick's diet during its first month. The availability of insects and highly nutritious forbs (dandelions, clover, milkvetch, etc.) influences the growth rate of Sage Grouse chicks and the size and density of populations. Often, the chicks of herbivorous birds are precocial, especially if they are herbivorous from hatching (e.g., geese). Herbivory in chicks is associated with comparatively slow growth rates, presumably due to the relatively low amino acid levels in their food (Savory, 1989; Johnson and Boyce, 1990; Drut *et al.*, 1994).

The smaller ceca and faster rates of passage in herbivorous birds relative to comparably sized mammals yield poorer efficiency of fiber utilization. For example, the apparent metabolizable energy of alfalfa for a goose is 7.4 kJ g^{-1} compared with 10.5 kJ g^{-1} in a rabbit. Compensatory high levels of food intake are associated with several disadvantages (Joyner *et al.*, 1987; Gauthier and Hughes, 1995; Guglielmo and Karasov, 1995).

1. High intake increases the predation on plants, more quickly exhausting the food supply and limiting sustainable population sizes. Many breeding and staging habitats of ducks and geese are at their carrying capacity and digestive inefficiencies translate into lower numbers of birds. Further, the grazing of geese and ducks on hay crops and lawns results in severe conflicts with farmers, landscapers, and golfers.
2. High levels of food intake are associated with increased effort and energy expenditure for food acquisition.
3. Higher food intake increases the exposure to toxins and antinutritional compounds and may preclude the consumption of otherwise nutritious vegetation. For example, the intake and preference of Ruffed Grouse for various components of winter browse (e.g., buds, twigs, catkins) are more dependent on the content of secondary metabolites than on nutrient levels. Geese apparently select among foods to maximize the consumption of protein and minimize the consumption of plant toxins.

On the positive side, the use of postgastric fermentation permits efficient autoenzymatic digestion of high-quality foods, such as grains, which approaches that of granivores. For example, Ostriches utilize corn with similar digestive efficiency as chickens, but are much more efficient at utilizing alfalfa. The pregastric fermentation strategy used by many mammals and a few avian species decreases the efficiency of utilization of high-quality foods and is disadvantageous when highly nutritious foods are periodically available (Cilliers *et al.*, 1994).

Omnivores

One-third of the avian families consume a wide variety of plant and animal foods on any given day and are omnivorous. These birds are generalists and have no primary specialization (Schoener, 1971; Morse, 1975). Few species regularly exploit the whole variety of foods available. Rather, they use foraging strategies adapted to finding food items that are appropriate relative to their body size and eat both highly nutritious and less nutritious food items that are discovered. Omnivores must be adaptable, because the proportions of various food types consumed change seasonally. Birds that primarily consume a single category of food and then switch to another category on a seasonal basis are not to be considered omnivorous. For example, a bird that primarily consumes insects in the spring and seeds in the fall and winter is a facultative insectivore-granivore, not an omnivore. Facultative consumers often have specific morphological adaptations (beak, gizzard) for their primary foods. The digestive tract of omnivores lacks accentuation of any organ and represents a composite of the various tracts described previously. The digestive adaptations required for omnivory and for facultative consumption patterns are considered together in the next section.

Digestive Adaptations to Changes in Diet

Adaptations of the digestive tract to various diets or physiological states occur on several time scales.

1. Gross morphological changes to the size and types of organs present in a bird's digestive tract can occur over many generations in an evolutionary time scale (thousands to millions of years) and require marked changes in gene frequencies or gene mutations. The study of the relationship between physical adaptations and environmental components, such as diet, is known as ecomorphology. Changes in the morphology of the beak due to genetic selection constitute one of the most celebrated areas of biology and are the subject of voluminous literature. Adaptations of other digestive structures and processes are less well characterized, but are often more important in determining a bird's diet.

2. Functional changes in the size or musculature of a digestive organ can occur within a bird's lifetime. For example, increase in the size of the cecum in response to dietary fiber level takes many weeks, requiring the proliferation and maturation of the cells in the cecal wall and epithelium.

3. Adaptations in the levels of enzymes and transport proteins involved in digestion or metabolism occur over a time frame of several days. For example, the adaptation of glucose-metabolizing enzymes to shifts in dietary energy or protein composition is complete within 72 h.

4. Adaptations can occur during the course of a single meal. Adjustments in the rate of passage of digesta in response to the rate of hydrolysis of fats and proteins occur over the course of several minutes and are an important fine-tuning of the digestive process.

5. Finally, the very fast nerve reflex networks result in actions within seconds. For example, salivary secretions begin immediately after the sight of food.

Excepting genetic and developmental changes, adaptations within the digestive system are often reversible over similar time frames and are collectively referred to as digestive plasticity.

Changes in the types of food consumed by a bird may be gradual over the changing seasons or may be abrupt due to sudden eruptions in the abundance of a food item, such as insects, fish, fruits, or flowers. Plasticity of the digestive tract in response to the type or quantity of food available includes changes in: the size and musculature of the crop, proventriculus, and gizzard; the length of the small intestine and ceca; the height of the villi in the small intestine; the production of pancreatic and brush-border enzymes; the numbers of nutrient-transport proteins on enterocytes; and the rate of passage.

The storage capacity of the esophagus or crop (if present) increases as the size of the meals consumed increases. Crop capacity also increases when larger food items or more fibrous food items are consumed. The musculature of the gizzard increases with the coarseness of the food consumed and the amount of effort required to crush and remove seed coats, fruit coats, shells, or exoskeletons. For example, the weight of the gizzard of the Wood Duck increases by 82% between the spring and the fall, due to a shift in feeding on relatively soft invertebrates to high-fiber plant foods and hard seeds (Drobney, 1984). Birds adapted to soft foods are typically unwilling to consume significant quantities of coarse, hard foods if they suddenly become available. As the gizzard strength and cuticle integrity increase, greater proportions of the food are consumed. These physical adaptations of the gizzard require several weeks for completion and are a primary reason why some types of dietary changes must be made gradually. Changes from hard foods to soft foods are more easily made by birds.

The length of the intestines increases with increased food intake or increased dietary fiber (Fenna and Boag, 1974; Savory and Gentle, 1976a,b; Brugger, 1991). For example, a 34% increase in dry-matter intake by Japanese Quail induces a 43% increase in the length of the small intestine. The ceca also respond to dietary changes, with cecal size and microbial activity being proportional to the dietary fiber content in a variety of omnivorous birds. Changes in intestinal and cecal size usually take 1–3 months to complete.

Intestinal surface area may change with diet quantity or quality, often by a change in the length of villi (Diamond, 1991). The length of villi in the upper jejunum is most flexible, presumably because it is the major site for digestion and absorption of most nutrients. Villi may have a critical length, which is determined by the balance between improved digestion and the cost of maintenance. According to this concept, the surface area of the small intestine should continue to increase by lengthening and elongation of villi in order to extract more nutrients until the cost of maintenance of the increased intestinal mass outweighs the diminishing increment in digestive efficiency. Because of their high metabolic activity and rate of turnover, the villi are among the most metabolically expensive tissues to maintain. A similar balance exists for digestive enzymes; the amount released following a meal is often modulated by the dietary composition of previous meals and, to some extent, the composition of the present meal. For example, in Yellow-rumped Warblers, aminopeptidase nitrogen activity doubles within a week of switching from a low-protein, fruit-based diet to a high-protein, insect-based diet. Similar changes have been observed in brush-border carbohydratases and lipases in chickens following increases in the carbohydrate and lipid content of the diet, respectively (Krogdahl and Sell, 1989; Biviano *et al.*, 1993; Afik *et al.*, 1995).

Changes in the size of digestive organs in order to optimize the efficiency of digestion are probably limited by two factors. First, the increase in maintenance costs described above for the small intestine also applies to other organs, such as the ceca and gizzard. Further, the extra weight associated with longer or larger digestive organs and their associated contents has to be carried around by its owner. This added weight decreases flying efficiency and proficiency, resulting in greater energy needs for mobility and increased vulnerability to predators (Sibly, 1981).

Rate of passage of the digesta is sensitive to diet composition (Levey and

Karasov, 1989; Karasov and Levy, 1990; Afik and Karasov, 1995). Adjustments in the rate of passage are most obvious in those birds that make dramatic changes in their diet on a seasonal basis. The three most common changes are between: insects and fruits; insects and seeds; and seeds and roughage. In American Robins and Yellow-rumped Warblers, digesta retention times are about 60% shorter when adapted to fruit than to insects (Fig. 4.9). Following a switch from fruit to insects, changes in rate of passage can be seen within 2 h, but require a few days for complete adjustment. The rapid change in rate of passage may be mediated by nutrient concentrations in the lumen of the intestine. High concentrations of amino acids and lipids in the duodenum inhibit the passage of digesta from the gizzard by neural and hormonal reflexes (see Chapter 3). The rate of passage of sunflower seeds (45% fat content) through Yellow-rumped Warblers is almost twice as slow as that of insects. The slower the rate of passage of digesta, the more efficient the extraction of dietary lipid (Fig. 4.9). Because fruit-adapted birds have high rates of passage, the occasional insect or seed consumed is digested at a lower efficiency than in birds adapted to these items. Switching between grain and roughage is also accompanied by changes in rates of passage. In birds that are not primarily herbivorous, the rate of passage usually increases with increasing dietary fiber. For example, Spur-winged Geese consuming corn have a slow rate of passage and relatively complete digestion, but, when they consume a high-fiber diet based on alfalfa, the rate of passage increases more than twofold

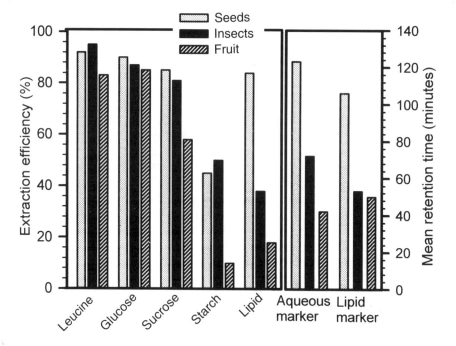

Fig. 4.9. The trade-offs between the rate of passage of digesta and the efficiency of nutrient digestion in Yellow-rumped Warblers (Afik and Karasov, 1995).

and digestive efficiency decreases by almost 30% (Halse, 1984).

The above discussion has focused primarily on adaptations to dietary type. During a bird's life cycle, marked changes in the amount of food consumed may also drive adaptations in the digestive-tract size and function. For example, food intake may increase by twofold due to the cold temperatures of winter, premigratory fattening, or laying large clutches of eggs. Molting and egg incubation are annual events that reduce food intake in many species. In general, high rates of food intake cause a generalized increase in the size of the gastrointestinal tract and the amount of digestive enzymes excreted, with little change in the rate of passage, resulting in little change in the digestive efficiency. Conversely, low rates of food intake cause a relatively rapid decrease in the size of the gastrointestinal tract and its digestive enzymes (Nir *et al.*, 1978; Karasov, 1996).

The amount and activity of enzymes responsible for metabolism of nutrients in tissues such as liver, muscle, kidney, and adipose tissue also adapt to dietary quantity and type. Many of these adaptations have been shown in controlled nutrition experiments, using quail, turkeys, chickens, and ducks, and will be discussed in Chapters 5–7. The capacity for digestive and metabolic adaptations has been mostly examined in species of Galliformes and Anseriformes whose natural diets are highly variable in nutrient content. Species that consume diets with relatively consistent nutrient content, such as faunivores, may have considerably less adaptability.

References

Adams, N.J. (1984) Utilization efficiency of a squid diet by adult king penguins (*Aptenodytes patagonicus*). *Auk* 101, 884–886.

Afik, D. and Karasov, W.H. (1995) The trade-offs between digestion rate and efficiency in warblers and their ecological implications. *Ecology* 76, 2247–2257.

Afik, D., Vidal, E.C., Delrio, C.M. and Karasov, W.H. (1995) Dietary modulation of intestinal hydrolytic enzymes in yellow-rumped warblers. *American Journal of Physiology – Regulatory Integrative and Comparative Physiology* 38, R413–R420.

Akin, D.E. (1989) Histological and physical factors affecting digestibility of forages. *Agronomy Journal* 81, 17–25.

Bairlein, F. (1996a) Food choice and nutrition in plant-feeding birds. *Comparative Biochemistry and Physiology A – Physiology* 113, 203–213.

Bairlein, F. (1996b) Fruit-eating in birds and its nutritional consequences. *Comparative Biochemistry and Physiology A – Physiology* 113, 215–224.

Baker, H.G. and Baker, I. (1990) The predictive value of nectar chemistry to the recognition of pollinator types. *Israel Journal of Botany* 39, 157–166.

Barton, N.W.H. and Houston, D.C. (1993a) A comparison of digestive efficiency in birds of prey. *Ibis* 135, 363–371.

Barton, N.W.H. and Houston, D.C. (1993b) The influence of gut morphology on digestion time in raptors. *Comparative Biochemistry and Physiology A – Comparative Physiology* 105, 571–578.

Barton, N.W.H. and Houston, D.C. (1994) Morphological adaptation of the digestive tract in relation to feeding ecology of raptors. *Journal of Zoology* 232, 133–150.

Berthold, P. (1977) The control and significance of animal and vegetable nutrition in omnivorous songbirds. *Ardea* 64, 140–154.

Beuchat, C.A., Calder, W.A. and Braun, E.J. (1990) The integration of osmoregulation and energy balance in hummingbirds. *Physiological Zoology* 63, 1059–1081.

Billingsley, B.B. and Arner, D.H. (1970) The nutritive value and digestibility of some winter foods of the eastern wild turkey. *Journal of Wildlife Management* 34, 176–182.

Biviano, A.B., Delrio, C.M. and Phillips, D.L. (1993) Ontogenesis of intestine morphology and intestinal disaccharidases in chickens (*Gallus gallus*) fed contrasting purified diets. *Journal of Comparative Physiology B – Biochemical Systemic and Environmental Physiology* 163, 508–518.

Björnhag, G. (1989) Sufficient fermentation and rapid passage of digesta – a problem of adaptation in the hindgut. *Acta Veterinaria Scandinavica*, 586, 204–211.

Björnhag, G. and Sperber, I. (1977) Transport of various food components through the digestive tract of turkeys, geese and guinea fowl. *Swedish Journal of Agriculture Research* 7, 57–66.

Bock, W. J. (1966) An approach to the functional analysis of bill shape. *Auk* 83, 10–51.

Bosque, C. and Parra, O. (1992) Digestive efficiency and rate of food passage in oilbird nestlings. *Condor* 94, 557–571.

Bosque, C., Ramirez, R. and Rodriguez, C. (1995) The diet of the oilbird in Venezuela. *Ornitologia Neotropical* 6, 67–80.

Brice, A.T. (1992) The essentiality of nectar and arthropods in the diet of the Anna hummingbird (*Calypte anna*). *Comparative Biochemistry and Physiology A – Comparative Physiology* 101, 151–155.

Brice, A.T. and Grau, C.R. (1989) Hummingbird nutrition – development of a purified diet for long-term maintenance. *Zoo Biology* 8, 233–237.

Brice, A.T. and Grau, C.R. (1991) Protein requirements of Costa's hummingbirds *Calypte costae*. *Physiological Zoology* 64, 611–626.

Brice, A.T., Dahl, K.H. and Grau, C.R. (1989) Pollen digestibility by hummingbirds and psittacines. *Condor* 91, 681–688.

Brugger, K.E. (1991) Anatomical adaptation of the gut to diet in red-winged blackbirds (*Agelaius phoeniceus*). *Auk* 108, 562–567.

Brugger, K.E. (1993) Digestibility of 3 fish species by double-crested cormorants. *Condor* 95, 25–32.

Bryant, D.M. and Bryant, V.M.T. (1988) Assimilation efficiency and growth of nestling insectivores. *Ibis* 130, 268–274.

Buchsbaum, R., Wilson, R. and Valiela, I. (1986) Digestibility of plant constituents by Canada geese and Atlantic brant. *Ecology* 67, 386–393.

Burton, P.J.K. (1974). *Feeding and the Feeding Apparatus in Waders*. British Museum, London.

Calder, W.A.I. and Hiebert, S.M. (1983) Nectar feeding, diuresis, and electrolyte replacement of hummingbirds. *Physiological Zoology* 56, 325–334.

Campbell, E.G. and Koplin, J.R. (1986) Food consumption, energy, nutrient and mineral balances in a Eurasian kestrel and screech owl. *Comparative Biochemistry and Physiology* 83A, 249–254.

Cannon, C.E. (1979) Observations on the food and energy requirements of rainbow lorikeets, *Trichoglossus haematodus*. *Australian Wildlife Research* 6, 337–346.

Castro, G., Stoyan, N. and Nyers, J.P. (1989) Assimilation efficiency in birds: a function of taxon or food type? *Comparative Biochemistry and Physiology* 92A, 271–278.

Chaplin, S.B. (1989) Effect of cecectomy on water and nutrient absorption of birds. *Journal of Experimental Zoology* S3, 81–86.

Chivers, D.J. and Langer, P. (1994) *The Digestive System in Mammals: Food, Form and Function.* Cambridge University Press, Cambridge.

Cilliers, S.C., Hayes, J.P., Maritz, J.S., Chwalibog, A. and Dupreez, J.J. (1994) True and apparent metabolizable energy values of lucerne and yellow maize in adult roosters and mature ostriches (*Struthio camelus*). *Animal Production* 59, 309–313.

Clark, A. (1980) The biochemical composition of krill, *Euphausia superba*, from South Georgia. *Journal of Experimental Marine Biology and Ecology* 43, 221–236.

Cork, S.J. (1994) Digestive constraints on dietary scope in small and moderately-small mammals: how much do we really understand? In: Chivers, D.J. and Langer, P. (eds) *The Digestive System in Mammals: Food, Form and Function.* Cambridge University Press, Cambridge, pp. 324–336.

Crome, F.H.J. (1985) An experimental investigation of filter-feeding on zooplankton by some specialized waterfowl. *Australian Journal of Zoology* 33, 849–862.

Croxall, J.P. (1987) *Seabirds: Feeding Ecology and Role in Marine Ecosystems.* Cambridge University Press, Cambridge.

Dawson, T.J., Johns, A.B. and Beal, A.M. (1989) Digestion in the Australian wood duck (*Chenonetta jubata*): a small avian herbivore showing selective digestion of the hemicellulose component of fiber. *Physiological Zoology* 62, 522–540.

des Lauriers, J.R. (1994) Hummingbirds eating ashes. *Auk* 111, 755–756.

Diamond, J.M. (1991) Evolutionary design of intestinal nutrient absorption: enough but not too much. *News Physiology Science* 6, 92–96.

Diamond, J.M. and Karasov, W.H. (1987) Adaptive regulation of intestinal nutrient transporters. *Proceedings of the National Academy of Sciences of the USA* 84, 2242–2245.

Diamond, J.M., Karasov, W.H., Phan, D. and Carpenter, F.L. (1986) Digestive physiology is a determinant of foraging bout frequency in hummingbirds. *Nature* 320, 62–63.

Diaz, M. (1990) Interspecific patterns of seed selection among granivorous passerines – effects of seed size, seed nutritive value and bird morphology. *Ibis* 132, 467–476.

Dominguez-Bello, M.G., Lovera, M., Suarez, P. and Michelangeli, F. (1993) Microbial digestive symbionts of the crop of the Hoatzin: an avian foregut fermenter. *Physiological Zoology* 66, 374–383.

Dorst, J. (1974) *The Life of Birds*, 1st edn. Columbia University Press, New York, USA.

Drobney, R.D. (1984) Effect of diet on visceral morphology of breeding wood ducks. *Auk* 101, 93–98.

Drut, M.S., Pyle, W.H. and Crawford, J.A. (1994) Technical note – diets and food selection of sage grouse chicks in Oregon. *Journal of Range Management* 47, 90–93.

Dudley, R. and Vermeij, G.J. (1991) Do the power requirements of flapping flight constrain folivory in flying animals? *Functional Ecology* 6, 101–104.

Duke, G.E. and Rhoades, D.D. (1977) Factors affecting meal to pellet intervals in great horned owls (*Bubo virginianus*). *Comparative Biochemistry and Physiology* 56A, 283–289.

Duke, G.E., Jegers, A.A., Loff, G. and Evanson, O.A. (1975) Gastric digestion in some raptors. *Comparative Biochemistry and Physiology* 50A, 649–656.

Duke, G.E., Evanson, O.A., Redig, P.T. and Rhoades, D.D. (1976) Mechanism of pellet egestion in great horned owls (*Bubo virginianus*). *American Journal of Physiology* 213, 1824–1828.

Duke, G.E., Bird, J.E., Daniels, K.A. and Bertoy, R.W. (1981) Food metabolizability and water balance in intact and cecectomized great-horned owls. *Comparative Biochemistry and Physiology* 68A, 237–240.

Duke, G.E., Place, A.R. and Jones, B. (1989) Gastric emptying and gastrointestinal motility in Leach's storm petrel chicks. *Auk* 106, 80–85.

Duke, G.E., Jackson, S. and Evanson, O.A. (1993) Great horned owls do not egest pellets prematurely when presented with a new meal. *Journal of Raptor Research* 27, 39–41.

Duke, G.E., Reynhout, J., Tereick, A.L., Place, A.E. and Bird, D.M. (1997) Gastrointestinal morphology and motility in American kestrels receiving high or low fat diets. *Condor* 99, 123–131.

Dyer, M.I. and Ward, P. (1977) Management of pest situations. In: Pinowski, J. and Kendeigh, S.C. (eds) *Granivorous Birds in Ecosystems*. Cambridge University Press, Cambridge, pp. 267–300.

Earl, F.R. and Jones, Q. (1962) Analyses of seed samples from 113 plant families. *Economic Botany*, 26, 221–250.

Fenna, L. and Boag, D.A. (1974) Adaptive significance of the caeca in Japanese quail and spruce grouse (Galliformes). *Canadian Journal of Zoology* 52, 1577–1584.

Foley, W.J. and Cork, S.J. (1992) Use of fibrous diets by small herbivores – how far can the rules be bent? *Trends in Ecology and Evolution* 7, 159–162.

Fuentes, M. (1994) Diets of fruit-eating birds – what are the causes of interspecific differences? *Oecologia* 97, 134–142.

Gadallah, F.L. and Jefferies, R.L. (1995) Comparison of the nutrient contents of the principal forage plants utilized by lesser snow geese on summer breeding grounds. *Journal of Applied Ecology* 32, 263–275.

Gasaway, W.C. (1976a) Seasonal variation in diet, volatile fatty acid production and size of the cecum of rock ptarmigan. *Comparative Biochemistry and Physiology* 53A, 109–114.

Gasaway, W.C. (1976b) Volatile fatty acids and metabolizable energy derived from cecal fermentation in the willow ptarmigan. *Comparative Biochemistry and Physiology* 53A, 115–121.

Gauthier, G. and Hughes, R.J. (1995) The palatability of arctic willow for greater snow geese – the role of nutrients and deterring factors. *Oecologia* 103, 390–392.

Gillespie, G.D. (1982) Factors affecting daily seed intake of the greenfinch, *Carduelis cholis*. *New Zealand Journal of Zoology* 9, 295–300.

Gluck, E.E. (1985) Seed preference and energy intake of gold finches *Carduelis carduelis* in the breeding season. *Ibis* 127, 421–429.

Goodman, D.C. and Fisher, H.I. (1962) *Functional Anatomy of the Feeding Apparatus in Waterfowl*. Southern Illinois University Press, Carbondale.

Grajal, A. and Parra, O. (1995) Passage rates of digesta markers in the gut of the hoatzin, a folivorous bird with foregut fermentation. *Condor* 97, 675–683.

Guentert, M. (1981) Morphologische Untersuchungen zur adaptiven Radiation des Verdauungstraktes bei Papegeien (Psittaci). *Zoologische Jahrbuecher Abteilung fuer Anatomie* 106, 471–526.

Guglielmo, C.G. and Karasov, W.H. (1995) Nutritional quality of winter browse for ruffed grouse. *Journal of Wildlife Management* 59, 427–436.

Guillemette, M. (1994) Digestive-rate constraint in wintering common eiders (*Somateria mollissima*): implications for flying capabilities. *Auk* 111, 900–909.

Hainsworth, L.R. and Wolf, L.L. (1976) Nectar characteristics and food selection by hummingbirds. *Oecologia* 25, 101–113.

Halse, S.A. (1984) Food intake, digestive efficiency and retention time in spur-winged geese *Plectropterus gambensis*. *South African Journal of Wildlife Research* 14, 106–110.

Heath, R.G.M. and Randall, R.M. (1985) Growth of jackass penguin chicks (*Spheniscus demersus*) hand reared on different diets. *Journal of Zoology London A* 205, 91–105.

Heinrich, B. and Bell, R. (1995) Winter food of a small insectivorous bird, the golden-crowned kinglet. *Wilson Bulletin* 107, 558–561.

Herpol, C. (1964) Activité protéolytique de l'appareil gastric d'oiseaux granivores et carnivores. *Annales de Biologie Animale Biochemie, Biophysique* 4, 239–246.

Herrera, C.M. (1982) Defense of ripe fruit from pests: its significance in relation to plant–disperser interactions. *American Naturalist* 120, 218–241.

Herrera, C.M. (1984) Adaptation to frugivory of Mediterranean avian seed dispersers. *Ecology* 65, 609–617.

Herrera, C.M. (1995) Plant–vertebrate seed dispersal systems in the Mediterranean – ecological, evolutionary, and historical determinants. *Annual Review of Ecology and Systematics* 26, 705–727.

Holmes, R.T. (1990) Ecological and evolutionary impacts of bird predation on forest insects: an overview. *Studies in Avian Biology* 13, 6–13.

Holthuijzen, A.M. and Adkisson, C.S. (1984) Passage rate, energetics, and utilization efficiency of the cedar waxwing. *Wilson Bulletin* 96, 680–684.

Izhaki, I. (1992) A comparative analysis of the nutritional quality of mixed and exclusive fruit diets for yellow-vented bulbuls. *Condor* 94, 912–923.

Izhaki, I. (1993) Influence of nonprotein nitrogen on estimation of protein from total nitrogen in fleshy fruits. *Journal of Chemical Ecology* 19, 2605–2615.

Izhaki, I. and Safriel, U.N. (1990) Weight losses due to exclusive fruit diet – interpretation and evolutionary implications – reply. *Oikos* 57, 140–142.

Jackson, S. (1986) Assimilation efficiencies of white-chinned petrels (*Procellaria aequinoctialis*) fed different prey. *Comparative Biochemistry and Physiology* 85A, 301–303.

Jackson, S. and Place, A.R. (1990) Gastrointestinal transit and lipid assimilation efficiencies in 3 species of sub-antarctic seabird. *Journal of Experimental Zoology* 255, 141–154.

Jakubas, W.J., Guglielmo, C.G., Vispo, C. and Karasov, W.H. (1995) Sodium balance in ruffed grouse as influenced by sodium levels and plant secondary metabolites in quaking aspen. *Canadian Journal of Zoology – Revue Canadienne de Zoologie* 73, 1106–1114.

Jenkin, P.M. (1957) The filter-feeding and food of flamingoes (Phoenicopteri). *Transactions of the Royal Society of London Series B* 240, 401–493.

Johnson, G.D. and Boyce, M.S. (1990) Feeding trials with insects in the diet of sage grouse chicks. *Journal of Wildlife Management* 54, 89–91.

Johnson, R.A., Wilson, M.F., Thompson, J.N. and Bertin, R.I. (1985) Nutritional values of wild fruits and consumption by migrant frugivorous birds. *Ecology* 66, 819–827.

Johnston, R.D. (1993a) Effects of diet quality on the nestling growth of a wild insectivorous passerine, the house martin *Delichon urbica*. *Functional Ecology* 7, 255–266.

Johnston, R.D. (1993b) The effect of direct supplementary feeding of nestlings on weight loss in female great tits *Parus major*. *Ibis* 135, 311–314.

Jordano, P. (1986) Frugivory, external morphology and digestive system in Mediterranean sylfiid warblers *Sylvia* spp. *Ibis* 129, 175–189.

Joyner, D.E., Jacobson, B.N. and Arthur, R.D. (1987) Nutritional characteristics of grains fed to Canada geese. *Waterfowl* 38, 89–93.

Karasawa, Y. and Maeda, M. (1994) Role of caeca in the nitrogen nutrition of the chicken fed on a moderate protein diet or a low protein diet plus urea. *British Poultry Science* 35, 383–391.

Karasov, W.H. (1990) Digestion in birds: chemical and physiological determinants and ecological implications. *Studies in Avian Biology* 13, 391–415.

Karasov, W.H. (1996) Digestive plasticity in avian energetics and feeding ecology. In: Carey, C. (ed.) *Avian Energetics and Nutritional Ecology*. Chapman & Hall, New York, pp. 61–84.

Karasov, W.H. and Cork, S.J. (1994) Glucose absorption by a nectarivorous bird – the passive pathway is paramount. *American Journal of Physiology* 267, G18–G26.

Karasov, W.H. and Cork, S.J. (1996) Test of a reactor-based digestion optimization model for nectar-eating rainbow lorikeets. *Physiological Zoology* 69, 117–138.

Karasov, W.H. and Levey, D.J. (1990) Digestive system trade-offs and adaptations of frugivorous passerine birds. *Physiological Zoology* 63, 1248–1270.

Kaspari, M. (1991) Prey preparation as a way that grasshopper sparrows (*Amodramus savannarum*) increase the nutrient concentration of their prey. *Behavioral Ecology* 2, 234–241.

Kaspari, M. and Joern, A. (1993) Prey choice by three insectivorous grassland birds – reevaluating opportunism. *Oikos* 68, 414–430.

Kehoe, F.P. and Ankney, C.D. (1985) Variation in digesta organ size among five species of diving ducks (*Aythya* spp.). *Canadian Journal of Zoology* 63, 2339–2342.

Kenward, R.E. and Sibly, R.M. (1977) A woodpigeon (*Columba palumbus*) feeding preference explained by a digestive bottle-neck. *Journal of Applied Ecology* 14, 815–826.

Kirkwood, J.K. (1979) The partitioning of food energy for existence in the kestrel (*Falco tinnunculus*) and the barn owl (*Tyto alba*). *Comparative Biochemistry and Physiology* 63A, 495–498.

Kornegay, J.R., Schilling, J.W. and Wilson, A.C. (1994) Molecular adaptation of a leaf-eating bird: stomach lysozyme of the hoatzin. *Molecular Biology and Evolution* 11, 921–928.

Krebs, J.R. and Avery, M.I. (1984) Chick growth and prey quality in the European bee-eater (*Merops apiaster*). *Oecologia* 64, 363–368.

Krogdahl, A. and Sell, J.L. (1989) Influence of age on lipase, amylase and protease activities in pancreatic tissue and intestinal contents of young turkeys. *Poultry Science* 68, 1561–1568.

Landolt, R. (1985) A comparative morphological study of the gizzard in the Columbidae. *Fortschritte der Zoologie* 30, 225–268.

Landry, S.W., Defoliart, G.R. and Sunde, M.L. (1986) Larval protein quality of six species of Lepidoptera. *Journal of Economic Entomology* 79, 600–604.

Larbier, M. and Leclercq, B. (1992) *Nutrition and Feeding of Poultry*. Nottingham University Press, Loughborough.

Laugksch, R.C. and Duffy, D.C. (1986) Food transit rates in cape gannets and jackass penguins. *Condor* 88, 117–119.

Leeson, S. and Summers, J.D. (1991) *Commercial Poultry Nutrition*. University Books, Guelph.

Levey, D.J. and Duke, G.E. (1992) How do frugivores process fruit? Gastrointestinal transit and glucose absorption in cedar waxwings (*Bombycilla cedrorum*). *Auk* 109, 722–730.

Levey, D.J. and Grajal, A. (1991) Evolutionary implications of fruit-processing limitations in cedar waxwings. *American Naturalist* 138, 171–189.

Levey, D.J. and Karasov, W.H. (1989) Digestive responses of temperate birds switched to fruit or insect diets. *Auk* 106, 675–686.

Liddy, J. (1982) Food of the mistletoebird near Pumicetown Passage, south-eastern Queensland. *Corella* 6, 11–15.

McBee, R.H. and West, G.C. (1969) Cecal fermentation in the willow ptarmigan. *Condor* 71, 54–58.

Martinez del Rio, C.M. (1990a) Dietary, phylogenetic, and ecological correlates of intestinal sucrase and maltase activity in birds. *Physiological Zoology* 63, 987–1011.

Martinez del Rio, C.M. (1990b) Sugar preferences in hummingbirds – the influence of subtle chemical differences on food choice. *Condor* 92, 1022–1030.

Martinez del Rio, C.M. and Restrepo, C. (1993) Ecological and behavioral consequences of digestion in frugivorous animals. *Vegetatio* 108, 205–216.

Meyers, M.R. (1995) Comparative tolerance and metabolic adaptations to glucose of the barn owl (*Tyto alba*) and chicken (*Gallus domesticus*). PhD thesis, University of California, Davis, California.

Moermond, T.C. and Denslow, J.S. (1985) Neotropical avian frugivores: patterns of behavior, morphology, and nutrition, with consequences for fruit selection. In: Buckley, P.A., Foster, M.S., Morton, E.S., Ridgely, R.S., and Smith, N.G. (eds) *Neotropical Ornithology*. Monograph 36, American Ornithologists' Union, Ames, Iowa, pp. 865–897.

Morse, D.H. (1975) Ecological aspects of adaptive radiation in birds. *Biological Reviews* 50, 167–214.

Morton, E. (1979) Avian arboreal folivores: why not? In: Montgomery, G.G. (ed.) *The Ecology of Arboreal Folivores*. Smithsonian Institution Press, Washington, DC, pp. 123–130.

Moss, R. (1989) Gut size and the digestion of fibrous diets by tetraonid birds. *Journal of Experimental Zoology*, Suppl. 3, 61–65.

Moss, R. and Parkinson, J.A. (1972) The digestion of heather (*Calluna vulgaris*) by red grouse (*Lagopus lagopus scoticus*). *British Journal of Nutrition* 27, 285–298.

Murphy, M.E. (1994) Dietary complementation by wild birds – considerations for field studies. *Journal of Biosciences* 19, 355–368.

Newton, I. (1967) The adaptive radiation and feeding ecology of some British finches. *Ibis* 109, 33–98.

Nir, I., Nitsan, Z., Dror, Y. and Shapira, N. (1978) Influence of overfeeding on growth, obesity and intestinal tract in young chicks of light and heavy breeds. *British Journal of Nutrition* 39, 27–35.

Obst, B.S. and Diamond, J.M. (1989) Interspecific variation in sugar and amino acid transport by the avian cecum. *Journal of Experimental Zoology*, Suppl. 3, 117–126.

Paton, D.C. (1980) The importance of manna, honeydew and lerp in the diets of honeyeaters. *Emu* 80, 213–226.

Paton, D.C. (1981) The significance of pollen in the diet of the New Holland honeyeater, *Phylidonyris novaehollandiae*. *Australian Journal of Zoology* 29, 217–224.

Paton, D.C. (1982) The diet of the New Holland honeyeater, *Phylidonyris novaehollandiea*. *Australian Journal of Ecology* 7, 279–298.

Paton, D.C. and Collins, B.G. (1989) Bills and tongues of nectar-feeding birds – a review of morphology, function and performance, with intercontinental comparisons. *Australian Journal of Ecology* 14, 473–506.

Piersma, T., Koolhaas, A. and Dekinga, A. (1993) Interactions between stomach structure and diet choice in shorebirds. *Auk* 110, 552–564.

Pimm, S.L., Rosensweig, M.L. and Mitchell, W. (1985) Competition and food selection: field tests of a theory. *Ecology* 66, 798–807.

Place, A.R. and Stiles, E.W. (1992) Living off the wax of the land – bayberries and yellow-rumped warblers. *Auk* 109, 334–345.

Prince, P.A. (1980) The food and feeding ecology of blue petrel *Jalobaena caerulea* and dove prion *Pachyptila desolata*. *Journal of Zoology London* 190, 59–76.

Proctor, N.S. and Lynch, P.J. (1993) *Manual of Ornithology*. Yale University Press, New Haven, 340 pp.

Prop, J. and Deerenberg, C. (1991) Spring staging in brant geese *Branta bernicla* – feeding constraints and the impact of diet on the accumulation of body reserves. *Oecologia* 87, 19–28.

Prop, J. and Vulink, T. (1992) Digestion by barnacle geese in the annual cycle – the interplay between retention time and food quality. *Functional Ecology* 6, 180–189.

Redford, K.H. and Dorea, J.G. (1984) The nutritional value of invertebrates with emphasis on ants and termites as food for mammals. *Journal of Zoology London* 203, 385–395.

Redig, P.T. (1989) The avian ceca – obligate combustion chambers or facultative afterburners – the conditioning influence of diet. *Journal of Experimental Zoology,* Suppl. 3, 66–69.

Remmington, T.E. (1989) Why do grouse have ceca? A test of the fiber digestion theory. *Journal of Experimental Zoology* 3, 87–94.

Remsen, J.V., Hyde, M.A. and Chapman, A. (1993) The diets of neotropical trongons, motmots, barbets and toucans. *Condor* 95, 178–192.

Rhoades, D.D. and Duke, G.E. (1975) Gastric function in captive American bittern. *Auk* 92, 786–792.

Richardson, K.C. and Wooller, R.D. (1986) The structures of the gastrointestinal tracts of honeyeaters and other small birds in relation to their diets. *Australian Journal of Zoology* 34, 119–124.

Richardson, K.C. and Wooller, R.D. (1990) Adaptations of the alimentary tracts of some Australian lorikeets to a diet of pollen and nectar. *Australian Journal of Zoology* 38, 581–586.

Robel, R.J., Bisset, A.R., Clement, T.M. and Dayton, A.D. (1979) Metabolizable energy of important foods of bobwhites in Kansas. *Journal of Wildlife Management* 43, 982–986.

Robel, R.J., Press, B.M., Henning, B.L., Johnson, K.W., Blocker, H.D. and Kemp, K.E. (1995) Nutrient and energetic characteristics of sweepnet-collected invertebrates. *Journal of Field Ornithology* 66, 44–53.

Roby, D.D. (1991) Diet and postnatal energetics in convergent taxa of plankton-feeding seabirds. *Auk* 108, 131–146.

Rubega, M.A. and Obst, B.S. (1993) Surface-tension feeding in phalaropes – discovery of a novel feeding mechanism. *Auk* 110, 169–178.

Savory, C.J. (1989) The importance of invertebrate food to chicks of gallinaceous species. *Proceedings of the Nutrition Society* 48, 113–133.

Savory, C.J. and Gentle, M.J. (1976a) Effects of dietary dilution with fibre on the food intake and gut dimensions of Japanese quail. *British Poultry Science* 17, 561–570.

Savory, C.J. and Gentle, M.J. (1976b) Changes in food intake and gut size in Japanese quail in response to manipulation of dietary fibre content. *British Poultry Science* 17, 571–580.

Schoener, T.W. (1971) The theory of foraging strategies. *Annual Review of Ecology and Systematics* 2, 369–404.

Scott, M.L., Nesheim, M.C. and Young, R.J. (1982) *Nutrition of the Chicken*, 3rd edn. M.L. Scott & Associates, Ithaca.

Sedinger, J.S. and Raveling, D.G. (1984) Dietary selectivity in relation to availability and quality of food for goslings of cackling geese. *Auk* 101, 295–306.

Sedinger, J.S., White, R.G. and Hupp, J. (1995) Metabolizability and partitioning of energy and protein in green plants by yearling lesser snow geese. *Condor* 97, 116–122.

Shuman, T.W., Robel, R.J., Zimmerman, J.L. and Kem, K.E. (1989) Variance in digestive efficiencies of four sympatric avian granivores. *Auk* 106, 324–326.

Shuman, T.W., Robel, R.J., Zimmerman, J.I. and Dayton, A.D. (1990) Influence of handling time and metabolizable energy on seed selection by 4 emberizids. *Southwestern Naturalist* 35, 466–468.

Sibly, R.M. (1981) Strategies of digestion and defecation. In: Townsend, C.R. and Calow, P. (eds) *Physiological Ecology: an Evolutionary Approach to Resource Use.* Blackwell, Oxford, pp. 109–139.

Snow, D.W. (1981) Coevolution of birds and plants. In: Foley, P.L. (ed.) *The Evolving Biosphere.* Cambridge University Press, Cambridge, pp. 169–178.

Speakman, J.R. (1987) Apparent absorption efficiencies for redshank (*Tringa totanus* L.)

and oystercatcher (*Haemoatopus ostralegus* L.): implications for the predictions of optimal foraging models. *American Naturalist* 130, 677–691.

Stiles, F.G. (1995) Behavioral, ecological and morphological correlates of foraging for arthropods by the hummingbirds of a tropical wet forest. *Condor* 97, 853–878.

Stiles, F.G. and Rosselli, L. (1993) Consumption of fruits of the melastomataceae by birds – how diffuse is coevolution? *Vegetatio* 108, 57–73.

Studier, E.H., Szuch, E.J., Tompkins, T.M. and Cope, V.W. (1988) Nutritional budgets in free flying birds: cedar waxwings (*Bombycilla cedrorum*) feeding on Washington hawthorn fruit (*Crataegus phaenopyrum*). *Comparative Biochemistry and Physiology* 89A, 471–474.

Swart, D., Mackie, R.I. and Hayes, J.P. (1993a) Influence of live mass, rate of passage and site of digestion on energy metabolism and fibre digestion in the ostrich (*Struthio camelus* var. *domesticus*). *South African Journal of Animal Science* 23, 119–126.

Swart, D., Mackie, R.I. and Hayes, J.P. (1993b) Fermentative digestion in the ostrich (*Struthio camelus* var. *domesticus*), a large avian species that utilizes cellulose. *South African Journal of Animal Science* 23, 127–135.

Swart, D., Siebrits, F.K. and Hayes, J.P. (1993c) Utilization of metabolizable energy by ostrich (*Struthio camelus*) chicks at 2 different concentrations of dietary energy and crude fibre originating from lucerne. *South African Journal of Animal Science* 23, 136–141.

Tabaka, C.S., Ullrey, D.E., Sikarskie, J.G., Debar, S.R. and Ku, P.K. (1996) Diet, cast composition, and energy and nutrient intake of red-tailed hawks (*Buteo jamaicensis*), great horned owls (*Bubo virginianus*), and turkey vultures (*Cathartes aura*). *Journal of Zoo and Wildlife Medicine* 27, 187–196.

Taylor, J.M. and Kamp, J.W. (1985) Feeding activities of the Anna's hummingbird at subfreezing temperatures. *Condor* 87, 292–293.

Thomas, D.W., Bosque, C. and Arends, A. (1993) Development of thermoregulation and the energetics of nestling oilbirds (*Steatornis caripensis*). *Physiological Zoology* 66, 322–348.

Walsberg, G.E. (1975) Digestive adaptations of *Phainopepla nitens* associated with the eating of mistletoe berries. *Condor* 77, 169–174.

Walsberg, G.E. and Thompson, C.W. (1990) Annual changes in gizzard size and function in a frugivorous bird. *Condor* 92, 794–795.

White, D.W. and Stiles, E.W. (1990) Co-occurrences of foods in stomachs and feces of fruit-eating birds. *Condor* 92, 291–303.

Wijnandts, H. (1984) Ecological energetics of the long-eared owl (*Aiso otus*). *Ardea* 72, 1–92.

Willson, M.F. (1989) Gut retention times of experimental pseudoseeds by emus. *Biotropica* 21, 210–213.

Willson, M.F. and Harmeson, J.C. (1973) Seed preference and digestive efficiency of cardinals and song sparrows. *Condor* 75, 225–234.

Willson, M.F. and Whelan, C.J. (1990) The evolution of fruit color in fleshy-fruited plants. *American Naturalist* 136, 790–809.

Wing, S.L. and Tiffney, B.H. (1987) Interactions of angiosperms and herbivorous tetrapods through time. In: Friis, E.M., Challoner, W.G. and Crane, P.R. (eds) *The Origins of Angiosperms and Their Biological Consequences.* Cambridge University Press, Cambridge, pp. 203–224.

Witmer, M.C. (1996) Annual diet of cedar waxwings based on US biological survey records (1885–1950) compared to diet of American robins: contrasts in dietary patterns and natural history. *Auk* 113, 414–430.

Wooller, R.D., Richardson, K.C. and Wells, D.R. (1990) Allometric relationships of the gastrointestinal tracts of insectivorous passerine birds from Malaysia, New-Guinea and Australia. *Australian Journal of Zoology* 38, 665–671.

Worthington, A.H. (1989) Adaptations for avian frugivory – assimilation efficiency and gut transit time of *Manacus vitellinus* and *Pipra mentalis. Oecologia* 80, 381–389.

Ziswiler, V. (1986) Comparative morphology of the avian digestive tract. *Acta XIX Congress of International Ornithology* 19, 2436–2444.

Ziswiler, V. (1990) Specialisation in extremely unbalanced food: possibilities and limits of its investigation exclusively by functional morphology. *Netherlands Journal of Zoology* 40, 299–311.

Ziswiler, V. and Farner, D.S. (1972) Digestion and the digestive system. In: Farner, D.S., King, J.R., and Parkes, K.C. (eds) *Avian Biology*, Vol. II. Academic Press, New York, pp. 343–430.

Zweers, G.A., Gerritsen, A.F.C. and van Kranenburg-Voddgd, P.J. (1977) *Mechanisms of Feeding of the Mallard* (Anas platyrhynchos *L., Aves, Anseriformes*). Karger, Basle.

Zweers, G., Dejong, F., Berkhoudt, H. and Vandenberge, J.C. (1995) Filter feeding in flamingos (*Phoenicopterus ruber*). *Condor* 97, 267–324.

CHAPTER 5
Nutrient Requirements

Nutrients in the diet supply energy to fuel metabolism and provide the precursors for synthesis of structural and functional macromolecules. The macronutrients make up the bulk of the diet and include water, protein, lipid, and carbohydrates. The micronutrients are less prevalent dietary constituents and include the vitamins and minerals. Energy itself is not a nutrient; rather, it is a property that some nutrients possess. Nevertheless, birds have well-defined requirements for the amount of energy that must be supplied by dietary nutrients and energy is typically included in discussions on nutrients. The nutrients that must be in the diet for optimum health and productivity are referred to as essential nutrients and the amounts needed are the requirements. Some nutrients are essential because they are needed in metabolism but cannot be synthesized by the bird; others are essential because the rate of their synthesis is insufficient to meet metabolic demands. The quantitative and qualitative aspects of nutrient requirements are well understood for Galliformes, such as chickens, turkeys, and Japanese Quail (Table 5.1), and for domestic ducks. It appears that these domesticated species require similar essential nutrients in roughly the same proportions, and further work is needed to extend this knowledge to species in other orders.

Some nutrients are considered conditionally essential in that they become essential under specific dietary situations. For example, tyrosine is essential if an adequate amount of its precursor, phenylalanine, is not supplied. Other conditionally essential nutrients for growing chickens include glucose (if there are insufficient gluconeogenic precursors in the diet), niacin (if insufficient dietary tryptophan), arachidonic acid (if insufficient linoleic acid), and cystine (if insufficient dietary methionine). The specific nutrients that are conditionally essential vary across species and within a species, according to age and physiological state. In general, there are more nutrients that can be considered conditionally essential in birds than in mammals.

A bird's physiological state is a major determinant of its nutrient requirements. Requirements are usually determined for three physiological states: basal, maintenance, and total. The basal requirement is that needed by a bird that is not engaged in any important function. In general, the basal requirement is the minimal amount of nutrient that a bird must consume to replace losses inherent in being alive, but not growing, reproducing, or engaging in activities. The maintenance requirement is equal to the basal requirement plus the additional amount of nutrient needed for typical daily functions, including the activity of finding and consuming food, interacting with other animals, and defending body

Table 5.1. Essential nutrients for growing Japanese Quail.

Nutrient	Amount required for growth[*]
Oxygen	600 l kg^{-1} diet
Water[†]	≈ 500 ml kg^{-1} diet
Energy[‡]	> 10.0 MJ kg^{-1} diet
Amino acids (total)	24.0%
Arginine	1.25%
Glycine + serine	1.15%
Histidine	0.36%
Isoleucine	0.98%
Leucine	1.69%
Lysine	1.30%
Methionine	0.50%
Methionine + cystine	0.75%
Phenylalanine	0.96%
Phenylalanine + tyrosine	1.80%
Threonine	1.02%
Tryptophan	0.22%
Valine	0.95%
Fatty acids	
Linoleic acid	1.0%
α-Linolenic acid	Not known
Macrominerals	
Calcium	0.80%
Chlorine	0.14%
Magnesium	0.03%
Nonphytate phosphorus	0.30%
Potassium	0.40%
Sodium	0.15%
Trace minerals	
Copper	0.0005%
Iodine	0.00003%
Iron	0.012%
Manganese	0.006%
Selenium	0.00002%
Zinc	0.0025%
Fat-soluble vitamins	
A (retinol)	0.00005%
D$_3$ (cholecalciferol)	0.000002%
E (D-α-tocopherol)	0.0008%
K (menadione)	0.0001%
Water-soluble vitamins	
B$_{12}$	0.0000003%
Biotin	0.00003%
Choline	0.20%
Folacin	0.0001%
Niacin	0.004%
Pantothenic acid	0.001%
Pyridoxine	0.0003%
Riboflavin	0.0004%
Thiamin	0.0002%

[*]From Shim and Vohra (1984) and National Research Council (1994a), for growing Japanese Quail chicks. Assumes a diet of 90% dry matter and 12.1 MJ of metabolizable energy kg^{-1} and an intake typical of thermoneutral conditions.
[†]Water needs are extremely variable and depend upon environmental temperature.
[‡]Actual dietary energy levels can vary over a wide range. At energy densities below 10 MJ kg^{-1} diet, quail chicks are unable to eat sufficient diet to grow at a maximal rate.

temperature during hot or cold weather. For some nutrients, such as energy and water, the difference between the basal and the maintenance requirement may be quite large. For others, such as amino acids, vitamins, and minerals, the differences between basal and maintenance requirements are minimal. Functions, such as growth, molting, and reproduction, increase nutrient needs above maintenance levels, and pathological states, such as injuries or infections, modify a bird's nutrient requirements. The sum total of nutrient needs for all purposes is the bird's total requirement. Differences between maintenance requirements and total requirements are not uniform across all nutrients. Consequently, the optimum proportion of nutrients in the diet changes throughout a bird's life. The greatest nutritional demands during a bird's life occur during the rapid growth that occurs immediately after hatching. Egg production is also a period of heightened nutrient demands.

Determining Nutrient Requirements

Several different methods have been used to estimate nutrient requirements. Often they are determined by empirical methods, in which experimental diets containing graded nutrient levels are fed and the minimal level that optimizes the bird's health and performance is set as the requirement. For example, when groups of young chicks are fed diets containing increasing levels of methionine, their growth rates increase linearly with dietary methionine level until the requirement is met. Above the requirement, growth rate is not further augmented by further additions of methionine. Similarly, blood hemoglobin levels increase with increasing dietary iron levels until the dietary requirement is reached, after which higher dietary iron levels do not result in an additional response. For many nutrients, the transition between deficient levels and the requirement is not sharp, but follows a pattern of diminishing returns. In other words, incremental additions well below the requirement give large increases in the response, whereas the same increment slightly below the requirement results in a much smaller increase in the response. A good example of this principle of diminishing returns is the use of calcium for bone calcification by growing quail. At low levels of dietary calcium (0.4%), an additional 0.1% calcium results in a 20% increase in bone calcification, whereas at levels just below the requirement (0.7%), a 0.1% addition results in only a 5% increase in calcification.

A second method to determine requirements is by calculations based on factorial summation of needs for specific functions. For example, the methionine requirement for egg production can be calculated by summing the needs for normal maintenance plus the additional amount for the accretion of the reproductive tract, plus that for egg yolk, plus that for egg albumen. A factorial-summation approach is commonly used to determine energy requirements and sometimes for amino acids and calcium. In theory, the factorial approach should be very accurate, but a lack of knowledge of the efficiency of use of absorbed nutrients for various purposes often results in errors.

A combination of empirical and factorial-summation methods has been used to estimate the requirement of chickens, turkeys, quail, ducks, geese, and pheasants in several physiological states and published by the US National Academy of Sciences (NRC, 1994a). It should be noted that these requirements are for an average bird in a population and do not make allowances for genetic variation or environmental influences. Because nutrient requirements are minimal levels established under optimal conditions, nutritionists that formulate diets for captive birds usually add a margin of safety to provide for uncertainties. A margin of safety permits the building of nutrient storage pools that buffer periods of infectious diseases, environmental stresses, etc. For some nutrients (e.g., many vitamins and some trace minerals), digestibility is extremely variable across foods and the danger of toxicity is relatively low, so large margins of safety are employed when diets are formulated for captive birds. For other nutrients, cost or toxicity considerations may limit the amount of nutrient included in formulated diets to levels very near the bird's minimum requirement. High levels of some nutrients have pharmacological actions that are not related to their normal nutritional function. For example, very high levels of tryptophan may modify behavior and decrease aggression.

Expression of Nutrient Requirements

Nutrient requirements can be expressed in two ways: on an intake basis (e.g., mg day^{-1} or mg kg^{-1} body weight day^{-1}), referred to as the daily requirement; or on a concentration basis (e.g., % of the diet or g kg^{-1} diet), referred to as the dietary requirement. These two forms of expression are easily interconvertible if the food intake of the bird is known. In practice, requirements are typically expressed as concentrations, because it is more convenient, for three reasons. First, expressing requirements as a concentration, such as % of the diet, permits their direct comparison with the nutrient levels in foods, which are typically expressed as a concentration (usually % or mg kg^{-1}). Second, this form of expression is necessary to formulate novel diets from a collection of individual foods or specific ingredients. Third, requirements expressed on a percentage basis change slowly and linearly throughout a bird's life cycle (Fig. 5.1). This is because a bird's food intake usually changes proportionally with changes in its nutrient needs, minimizing the change in the concentration of the nutrient that is required in the diet.

Birds actually require a specific quantity of a nutrient each day and the daily requirement (mg day^{-1}) is the most accurate form of expression because it is not subjected to variability in food intake or composition. In practice, this form of expression is useful if a bird's daily requirements are relatively constant, as in adult birds that are not reproducing. But in growing birds the daily requirements change very rapidly, making this form of expression severely cumbersome in application (Fig. 5.1). Correction of the daily requirement for metabolic body size (BW$^{0.75}$) eliminates part of this volatility and is especially useful for comparison of maintenance energy needs across species.

Further refinements in nutrient requirements include expression on a digestible or metabolizable nutrient basis. This is particularly important for energy, where the difference between the gross energy in the diet and the amount that is actually digestible and available for metabolic purposes is extremely variable and the requirement must be met precisely. Amino acid requirements are often expressed on a digestible basis and vitamin and mineral requirements are expressed on a bioavailable basis (see Chapter 4).

Energy-related Interrelationships

A bird's energy needs are considerably more variable than the other nutrient requirements (see Chapter 9). Because birds usually eat a quantity of food necessary to satisfy their energy needs, their food intake fluctuates with environmental temperature, their activity level, and the energy concentration of the diet. Two important nutritional concepts can be gained by teasing apart the relationships between a bird's energy requirement and its intake of other nutrients.

1. If a bird decreases its intake because of lower energy needs, its dietary requirement for other nutrients, expressed on a % basis, increases proportionally (Fig. 5.2a). Conversely, if a bird's food intake increases, the required concentration

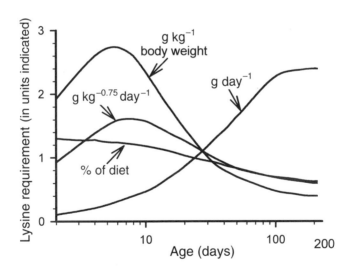

Fig. 5.1. The expression of a dietary requirement on a g 100 g^{-1} basis (%) gives the most consistent values over the period of growth of broiler chickens. Expressed as %, the requirement gradually declines by about half with age; when expressed as mg lysine day^{-1}, the requirement increases by about 25-fold with age. Chickens of this type weigh 45, 205, 580, and 2100 g on days 1, 10, 21, and 56, respectively. They reach an adult body weight of 6000 g at 168 days of age. The requirement at various ages was determined using prediction equations in NRC (1994b).

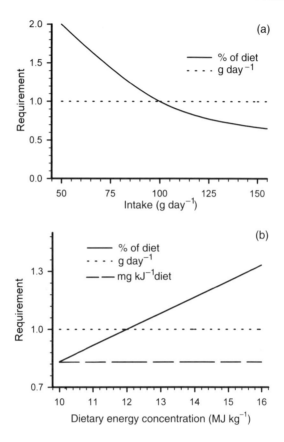

Fig. 5.2. A major disadvantage of expressing requirements on a % basis is the dependency on the level of food intake (a) and consequently the energy density of the diet (b). As intake declines, the requirement (%) increases proportionally. High dietary energy concentrations are among the factors that cause intake to decline and, in general, intake decreases proportionally with increasing dietary energy density. Expression of the requirement on the basis of g day^{-1} or mg kJ^{-1} of diet eliminates these problems.

of nutrients in the diet decreases proportionally. In each case, the bird's daily requirement, expressed as mg day^{-1}, for nutrients other than energy is relatively constant. This relationship between food intake and dietary requirements has many important implications for both wild and captive birds. For example, an adult finch living in the temperate zone may meet its amino acid requirements by consuming low-protein seeds in the winter, when food intake is high to support thermogenesis, but not in the summer, when intake is lower. Also, parrots in the wild must consume high amounts of food to support flight, thermoregulation, etc. and have a much lower requirement for other nutrients (expressed as % of the diet) than sedentary parrots in captivity. As a corollary, foods that are nutritionally adequate in the wild may often cause deficiencies in captivity.

2. As the energy concentration of a bird's diet changes, the requirement for other nutrients (expressed as % of the diet) changes proportionally (Fig. 5.2b). Thus, the concentration of essential nutrients in an energy-dense diet, for example 15 kJ g^{-1} diet, must be 25% higher than in a low-energy diet (12 kJ g^{-1}) to correct for the difference in food intake. When dietary requirements are expressed as mg kJ^{-1} of dietary energy, these problems are mostly eliminated. Because it is usually more practical to express requirements on a % of the diet basis, the energy level of the diet and the energy requirement of the bird must be accurately specified.

Nutrient Deficiencies, Imbalances, and Toxicities

A nutrient is considered to be deficient in the diet when the addition of more of that nutrient improves growth, reproduction, or general fitness, or relieves some pathological condition. Birds are very sensitive to acute deficiencies (several days) of some nutrients, such as water, energy, and amino acids, whereas chronic deficiencies (several months) of some vitamins and minerals are tolerated with minimal impact. Nutrient deficiencies commonly observed in captive birds are described in Table 5.2. For captive birds, the proportions of dietary components can be mixed in such a way as to meet or exceed the requirement as it changes throughout the life cycle. Wild birds must change their foraging habits in order to procure the appropriate proportions of available foods. For omnivores and florivores, energy is often the most consequential (first limiting) nutrient in birds that are not growing or reproducing. However, at the most nutritionally demanding periods of a bird's life (growth and egg production), amino acids and

Table 5.2. Most likely nutrient deficiencies of birds raised in captivity and fed common diets.

Nutritional strategy	Likely deficiencies*	Comments
Faunivore		
Insectivore	Calcium	
Carnivore	Calcium, vitamin E, thiamin	Assuming prey are fed whole
Piscivore	Calcium, vitamin E, thiamin	Assuming prey are fed whole
Florivore		
Herbivore	Energy	
Nectarivore	Thiamin, methionine	
Frugivore	Methionine, thiamin	
Granivore	Vitamins A, E, thiamin, niacin, biotin, methionine, lysine, zinc	Using diets based on domestic grains
Omnivore	Vitamins A, E, methionine	

*Listings are based on deficiency syndromes reported for captive and free-living birds in the literature. The amount of species-to-species variation in critical vitamins is not known, but is probably large. For all nutritional strategies, captive birds raised without normal levels of sunlight are susceptible to a vitamin D deficiency.

macrominerals may be most limiting in many natural foods. Growth, unlike egg production, cannot be stopped before completion without the potential for permanent damage. Stunting is a common outcome of early nutrient deficiencies. Even a slightly delayed growth curve can have devastating effects on the fitness of wild birds.

A nutrient imbalance can occur when high levels of one nutrient increase the requirement for another nutrient. Usually, the imbalance is due to the excessive nutrient impairing the metabolism of another structurally or functionally similar nutrient, by decreasing its absorption or increasing its catabolism or excretion. An imbalanced diet may appear to have levels of all nutrients that are above the bird's requirement, but is nutritionally inadequate due to the interaction of specific nutrients. Amino acids and minerals are very susceptible to imbalances, as will be discussed in more detail in later chapters.

A nutrient toxicity exists if the level of the imbalancing nutrient is sufficiently high that the addition of other nutrients does not correct the problem. Toxic levels of a nutrient induce specific pathologies, which are commonly useful for diagnosis.

References

National Research Council (NRC) (1994a) *Nutrient Requirements of Poultry.* National Academy Press, Washington, DC.

National Research Council (NRC) (1994b) *Metabolic Modifiers: Effects on the Nutrient Requirements of Food-producing Animals.* National Academy Press, Washington, DC.

Shim, K.F. and Vohra, P. (1984) A review of the nutrition of Japanese quail. *Worlds Poultry Science* 40, 261–280.

CHAPTER 6
Amino Acids

Birds synthesize proteins that contain 20 L-amino acids and utilize free amino acids to fulfill a variety of functions. Birds are unable to synthesize nine of these amino acids due to the lack of specific enzymes. These essential amino acids are also sometimes referred to as indispensable amino acids and are: arginine, isoleucine, leucine, lysine, methionine, phenylalanine, threonine, tryptophan, and valine. Histidine, glycine, and proline can be synthesized, but the rate is insufficient to meet the bird's metabolic demand in some situations and they are also considered essential in growing birds. Those amino acids that can be synthesized by the bird in sufficient amounts at all times during its life are termed nonessential or dispensable. The diet must supply enough nitrogen to synthesize these nonessential amino acids. Tyrosine and cysteine can be synthesized when the essential amino acid precursors phenylalanine and methionine are adequate, and are referred to as conditionally essential or semiessential. Birds do not specifically require protein in their diet and can live and mature on a diet with free amino acids as the only nitrogen source (Baker *et al.*, 1979; Murphy, 1993). Nevertheless, most foods consumed by birds supply the essential amino acids as intact protein. Dietary protein must supply sufficient levels of amino acids to meet essential amino acid requirements with enough excess amino acids to supply the nitrogen needed to synthesize the nonessential amino acids. Thus, stating both protein and amino acid requirements is a practical way of ensuring that all amino acid needs are provided.

Amino Acid Requirements

Amino acids are used predominantly for the synthesis of proteins – the principal structural and catalytic components of all tissues. Free amino acids and small peptides act to maintain osmolarity of body fluids, buffer pH, and serve as neurotransmitters, antioxidants, and shuttles in intermediary metabolism. Further, amino acids can be metabolized to a wide variety of metabolically important molecules. Quantitatively, the requirement for an amino acid is driven by the sum of three processes: (i) the rate at which amino acids are needed for functional purposes such as protein accretion, or as precursors for other metabolites; (ii) the rate of *endogenous losses* of protein and amino acids – these occur mostly from the digestive tract; and (iii) the rate at which amino acids are lost to *oxidation* or other metabolic pathways.

Adults at maintenance have minimal protein accretion, so the requirements are driven by endogenous losses and oxidation. These two types of losses are difficult to separate in birds because of their excretory anatomy, so they are considered together as obligatory losses. As this name suggests, these are losses that will occur throughout life and cannot be further minimized without impinging on important functions. These losses are at their minimum rate during a state of amino acid deficiency.

In the young growing chick and in egg-laying females, the amino acid requirement is largely for protein accretion and is proportional to the rate of growth or egg production. The specific balance of essential amino acids needed for protein accretion differs from that needed to replace obligatory losses. Thus, the amount and ratio of amino acids required by a bird change during its life cycle. In almost all species of birds, the requirements are highest at hatching and gradually decline until adult body weight and composition are achieved. The requirements increase again during reproduction, decline following reproduction, and increase once more during molt.

Scientists interested in domestic birds have long considered knowledge of amino acid requirements to be of premier nutritional importance. This importance is based on both metabolic and economic factors. Dietary protein is metabolically important because it can supply essential amino acids and, through intermediary metabolism, it can give rise to glucose, fat, energy, and several vitamins. The economic importance of protein derives from the high cost of high-protein foodstuffs. Not surprisingly, considerable research on amino acid requirements and metabolism has been conducted with domestic species in an effort to minimize feed costs. In fact, the number of studies on amino acids in poultry exceeds that on any other nutrient category, including energy. The requirement for each specific amino acid needed by chickens for growth and reproduction is known very accurately, to an even greater precision than that of laboratory rodents or humans. The requirements of other commercially important birds (e.g., turkeys, pheasants, ducks, quail) are known for several amino acids that are most commonly deficient in many diets (Table 6.1).

In contrast to our excellent state of knowledge of amino acid requirements and metabolism in domestic birds, very little is known about wild or companion birds. This is unfortunate because dietary amino acid supply is of paramount importance at the most critical stages of a bird's life. Clearly, acquisition of sufficient essential amino acids determines reproductive success, including clutch size and the growth rate and health of the young. Reproduction of most avian species is timed to coincide with the availability of high-protein foods. Despite the critical importance of dietary amino acids, research on nutritional energetics vastly exceeds that on amino acids. This state of affairs may be driven by the technical difficulty of research on amino acids relative to energy, especially in free-living birds. Fortunately, the critical importance of amino acid nutrition is beginning to be recognized by nutritional ecologists (Thomas *et al.*, 1993; White, 1993; Williams, 1996).

Table 6.1. Protein and amino acid requirements for birds in various physiological states.*

Species	Protein	Amino acid requirement as % of the diet						
		Methionine	Methionine + cysteine	Lysine	Arginine	Threonine	Tryptophan	Glycine + serine
White-crowned Sparrow, maintenance[†]	8.7	0.14	0.27	0.20	0.27	0.15	<0.05	—
Budgerigar, maintenance[‡]	6.7	—	0.25	0.15	0.26	—	—	—
Chicken, maintenance[§]	5.3	0.33	0.42	0.14	0.56	0.35	0.09	—
Chicken, growth	18	0.29	0.60	0.82	0.96	0.65	0.17	0.67
Chicken, laying	15	0.37	0.72	0.84	0.86	0.58	0.20	—
White Pekin Duck, growth	22	0.40	0.69	0.86	1.1	—	0.23	—
White Pekin Duck, laying	15	0.27	0.49	0.59	—	—	0.14	—
Goose, growth	20	—	0.59	1.0	—	—	—	—
Turkey, growth	26	0.52	1.00	1.5	1.5	0.95	0.25	1.1
Turkey, laying	14	0.2	0.4	0.6	0.6	0.44	0.13	0.5
Japanese Quail, growth	24	0.49	0.74	1.27	1.23	1.00	0.22	1.18
Japanese Quail, laying	21	0.44	0.69	0.98	1.23	0.73	0.19	1.21

*Requirement values are adjusted to a common metabolizable energy density of 12.5 kJ g⁻¹ diet and are based on highly digestible protein. Chicken values are for white-egg-laying strains.

[†]From Murphy (1993).

[‡]From Drepper et al. (1988).

[§]From Leveille et al. (1960). For further information and references to relevant research for all other species, see NRC (1994).

Semiessential amino acids

Essential:semiessential pairs include: methionine and cysteine; phenylalanine and tyrosine; and glycine and serine. The requirement for a semiessential amino acid can be met by either that amino acid or its essential amino acid precursor. Because of this relationship, the requirement for the essential amino acid needs to be expressed in two ways. The first is the amount of essential amino acid needed if the diet has adequate amounts of the semiessential amino acid. For example, the methionine requirement for growing chickens in the presence of adequate dietary cysteine is 0.50% of the diet. Second, the requirement for the essential amino acid must be considered in regard to situations where the diet has low levels of the conditionally essential amino acid and some of the essential amino acid must be used to synthesize the conditionally essential amino acid. For example, the methionine requirement in the absence of any dietary cysteine is equal to the total amount needed to supply the requirement of both nutrients (e.g., the requirement for methionine to supply methionine + cysteine is 0.93% in growing chickens). It is not uncommon for a diet to have adequate levels of methionine to meet the requirement for methionine alone, but inadequate levels to meet the requirement for both methionine and cysteine.

Protein quality

The quality of protein present in a food is a function of its quantity, its digestibility, and the balance of amino acids. The amount and balance of amino acids present in a food rarely, if ever, match the exact requirement of a bird. The amino acid in a food that is most deficient relative to a bird's requirement is referred to as the first limiting amino acid (Figs 6.1 and 6.2). The next most deficient amino acid is referred to as the second limiting amino acid, etc. All of the amino acids present in excess relative to the first limiting amino acid are of no value, except as an energy source, and may even have a negative effect (see imbalances and antagonisms below). Indexes of protein quality are occasionally used to score the balance of amino acids relative to the requirement of a bird. The protein efficiency ratio, biological value, and chemical score are examples of such indexes. These indexes are of limited value for evaluating individual protein sources but are useful for evaluating diets. This is because birds rarely eat a single food and the quality of a combination of foods has little relationship to the quality of the components. The limiting amino acids in one food are often compensated by surpluses in another food in the diet. The degree to which two foods correct their respective deficiencies when combined in a diet is referred to as protein complementation. A common example of this is found with corn and soybean meal. These two foods are the most prevalent components of diets formulated for captive birds, excepting faunivores. For young growing birds, corn is first limiting in lysine but moderately rich in methionine and cysteine. Soybean is first limiting in methionine and rich in lysine and therefore these proteins are highly complementary, justifying their common use for avian diets.

Dietary amino acids that are not used for anabolic purposes are quickly catabolized and used for energy. Thus, the correct amount and ratio of essential amino acids must be consumed within a relatively short time frame. The consumption of

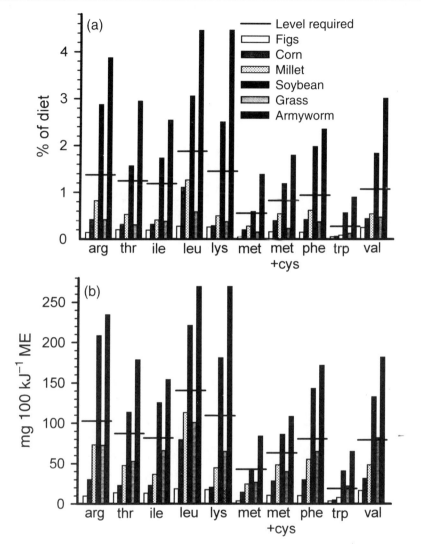

Fig. 6.1. Amino acid composition of several avian foods relative to the requirement of growing Japanese Quail chicks (NRC, 1994): (a) the amino acid levels as % of the diet (dry weight); (b) the amino acid concentration in each food is corrected for that food's metabolizable energy (ME) content and is expressed as mg amino acid 100 kJ^{-1} of metabolizable energy. The amino acid levels of young grasses commonly consumed by geese (Sedinger, 1984) are poorly digestible and have been corrected for their digestibility. The composition of figs is taken from FAO (1970) and the composition of the larval fall armyworm (*Spodopotera frugiperda*) is taken from Landry *et al.*, (1986). arg, Arginine; thr, threonine; ile, isoleucine; leu, leucine; lys, lysine; met, methionine; cys, cysteine; phe, phenylalanine; trp, tryptophan; val, valine.

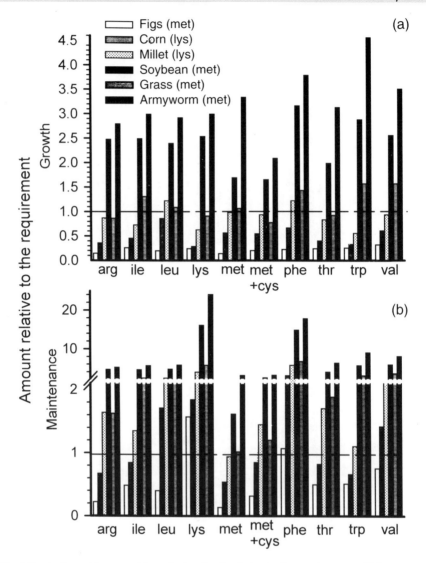

Fig. 6.2. Amino acid composition of avian foods relative to the requirement of (a) growing chickens and (b) adult chickens at maintenance. Data (mg amino acid kJ^{-1} metabolizable energy) are expressed as: concentration of amino acids in the food/concentration required. The most limiting amino acid for growth for each food is shown in parenthesis. For abbreviations, see caption to Fig. 6.1.

foods that have complementary amino acid patterns must occur during the same meal or foraging bout to be most effective (Murphy and Pearcy, 1993).

The ideal protein concept

A protein that has a balance of essential amino acids that exactly matches a bird's requirements, along with sufficient nonessential amino acid nitrogen to permit the synthesis of all of the nonessential amino acids, is referred to as an ideal protein (Baker and Han, 1994; Cole and Van Lunen, 1994). For convenience, the proportion of each amino acid is expressed relative to the amount of lysine (Fig. 6.3). Lysine was chosen as the standard because it is particularly well studied and because it is not used extensively for purposes other than protein synthesis. The amount of ideal protein needed to meet all of a bird's amino acid requirements is equal to that bird's minimum protein requirement.

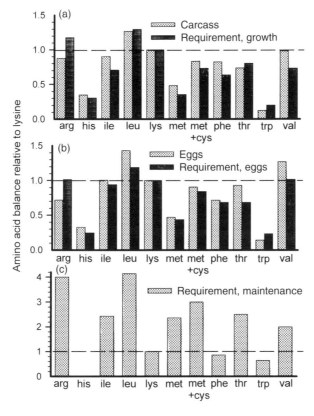

Fig. 6.3. The requirements for essential amino acids relative to the lysine requirement represent the balance of amino acids in an ideal protein. The ideal balance of amino acids for growth (a), egg production (b), and maintenance (c) is compared with the balance of amino acids in the carcass and eggs of chickens. The ideal balance of amino acids required for growth and for egg production are similar, due to the similarity in the balance of amino acids in tissue and eggs. The balance of amino acids required for maintenance is very different from that required for growth or egg production.

Real-world foods deviate from an ideal balance of amino acids, resulting in excesses of certain essential and nonessential amino acids, which are lost to catabolism. The greater the deviation in amino acid balance of dietary protein from that of an ideal protein, the higher the level of protein needed to meet the bird's amino acid requirements. For example, a diet that has an ideal balance of amino acids meets the amino acid requirement of young growing chickens at a level of only 18% protein, but a diet consisting of poorly balanced protein may need in excess of 25% protein to meet the chick's requirement for essential amino acids.

Requirements for maintenance

In amino acid nutrition, maintenance refers to a state in which protein is not being deposited for any purpose, including growth, reproduction, and feather replacement. Amino acid requirements are largely independent of activity and environmental conditions, so the term 'maintenance' is used over a much wider range of contexts when referring to amino acid nutrition than when referring to energy nutrition. Maintenance amino acid requirements are due to obligatory losses and are low relative to those needed by growing birds (Table 6.1). The balance of amino acids needed for maintenance is not proportional to the balance of amino acids in a bird's tissues, but rather reflects the relative rate of obligatory loss of each individual amino acid. For this reason, the balance needed for maintenance is considerably different from that needed for growth or egg production (Fig. 6.3). At maintenance, the requirement for lysine relative to other amino acids is very low and glycine and histidine are not needed. Methionine, arginine, and threonine are required at proportionally high levels. Dietary amino acid levels slightly below maintenance can sustain life, but muscle mass and function are impaired.

The amount of high-quality protein required at maintenance has been approximated for a few florivorous species (Table 6.2). The variability in maintenance protein requirements reflects the different rates of obligatory losses among species. Additionally, species vary in their capacity to digest dietary protein due to their diverse digestive strategies. In general, the dietary protein requirement (%) increases with the level of protein in a species' customary diet. For this reason, there are no valid equations for predicting maintenance protein requirements from body weight that can be used across all avian species. However, the protein requirement of birds with similar dietary strategies can be predicted empirically. Among granivorous and omnivorous species that have been studied, daily protein requirements follow the relationship: mg protein day^{-1} = $3489 \times$ kg body weight$^{0.58}$ (Fig. 6.4). This relationship might be used as a rough approximation of the maintenance requirement for high-quality protein of other omnivores and granivores, but certainly this predictive equation is not applicable for all florivorous species. For example, frugivores and herbivores, typified by the Oilbird and Emu, have exceptionally low rates of obligatory losses and consequently have low maintenance needs for their body size. Faunivores have high rates of endogenous losses, due mostly to high rates of amino acid degradation, and would be predicted to have correspondingly high maintenance protein

Table 6.2. Protein requirement for maintenance.[*]

Species	Body weight (BW) (g)	Protein (mg day^{-1})	Protein (g kg^{-1} BW)	Protein[†] (g kg^{-1} BW$^{-0.75}$)	Protein (% diet)	Diets (Reference)
Costa's Hummingbird	3.8	28	7.4	1.8	1.3	Sucrose + soymeal (Brice and Grau, 1991)
White-crowned Sparrow	27.7	436	15.7	6.4	8.7	Purified diet (Murphy, 1993)
Tree Sparrow	17.7	352	19.9	7.3	8.4	Corn + soymeal (Martin, 1968)
Dark-eyed Junco	17.8	304	17.1	6.2	9.2	Wheat + cornstarch (Parish and Martin, 1977)
Budgerigar	40	441	11.3	4.9	6.6	Seeds + cornstarch (Drepper et al., 1988)
Small frugivores[§]	Variable	–	–	4.8	6.6	Variety of fruits (Reanalysis of data from Izhaki, 1992)
Japanese Quail	130	715	5.5	3.3	5.5	Corn + soymeal + fishmeal (Yamane et al., 1979)
Oilbird	378	151	0.4	0.3	0.4	Avocado + casein (Bosque and Parra, 1992)
Ruffed Grouse	530	930	1.8	1.5	6.5	Corn + soymeal (Beckerton and Middleton, 1982)
Chicken	2,500	4,375	1.8	2.2	5.3	Purified diet (Leveille and Fisher, 1960)
Turkey	8,290	13,316	1.6	2.7	9.3	Corn + soymeal (Moran et al., 1983)
Emu	38,000	8,609	0.23	0.6	2.5	Grains (Dawson and Herd, 1983)

[*]Minimal level of protein required to prevent weight loss or to maintain nitrogen balance. These requirements were determined using diets with highly digestible protein, and levels should be adjusted upwards for foods that have less digestible protein.

[†]Regression analysis of all species, excluding the Emu, gives the following relationship: mg protein required day^{-1} = 2619 × kg body weight$^{-0.75}$ (r^2 = 0.92).

[‡]Adjusted to a common metabolizable energy content of 12.5 kJ g^{-1} diet.

[§]Summary of a variety of studies, most of which utilized inferior-quality protein sources and were not specifically designed to determine protein requirement.

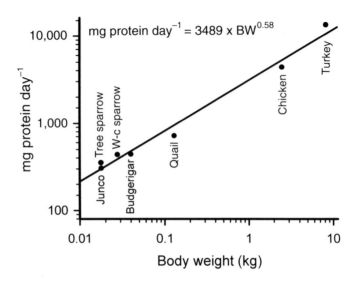

Fig. 6.4. The protein requirement for maintenance of granivorous and omnivorous birds listed in Table 6.2 scales with an allometric constant of body weight$^{0.58}$ ($r^2 = 0.98$). W-c, White-crowned.

requirements (Dawson and Herd; 1983; Brice and Grau, 1991; Bosque and Parra, 1992; Meyers and Klasing, 1997).

Expressing the maintenance protein requirement as % of the diet gives an indication of the relative proportions of protein and energy that must be provided by the diet. A variety of factors affect this relationship, including digestive strategy, body size, activity, and endogenous losses. Some of these relationships have not been experimentally tested in detail; however, two fundamental associations are evident. First, with decreasing adult body size, energy needs increase proportionally more than protein needs, resulting in a decline in dietary protein requirement (Fig. 6.5). For example, the % dietary protein required by a hummingbird is low compared with that of other birds that have been studied. Presumably, this is because its ravenous demand for energy drives high rates of food intake and permits its daily protein requirement to be met at low dietary protein concentrations. This allows the hummingbird to consume foods with very low protein and very high energy. Second, dietary protein requirements decrease with increasing activity level. A bird in the wild requires a lower % dietary protein than it does in captivity, because high energy needs associated with foraging and thermoregulation drive consumption of greater amounts of food (Fig. 6.5). Although birds in captivity must consume foods with relatively high levels of protein, their daily protein requirement on a mg day^{-1} basis has not been shown to differ between captive and wild environments.

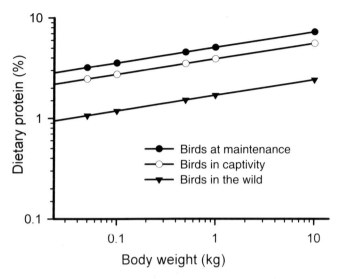

Fig. 6.5. The dietary protein requirement for nonreproducing adults is dependent upon their body weight and daily energy requirement. The following assumptions were made to calculate the dietary protein requirement: a dietary metabolizable energy content of 12.5 kJ g^{-1}; daily protein requirement = 3.5 × body weight$^{0.58}$; daily maintenance energy needs = 342 × body weight$^{0.73}$; energy requirement in captivity = 1.3 × maintenance; energy requirement in the wild = 3 × maintenance.

Requirements for growth

The balance of amino acids required for growth closely reflects the pattern of amino acids incorporated into tissue proteins. This is because needs for protein accretion are considerably greater than needs for maintenance. For example, in young growing chickens, 94% of the valine requirement is used to support growth and only 6% is required to replace obligatory losses (Baker *et al.*, 1996). In other words, the amount of essential amino acids deposited in tissues is 15 times greater than the amount needed for maintenance. It is instructive to compare this large demand for amino acids with that for energy; energy required to support growth in the young chicken is similar to the energy needed for maintenance functions. This large difference in the increased demand in amino acids relative to energy to support growth translates into a much greater protein requirement (% of the diet) for the growing chick compared with the adult (see Table 6.1).

Species vary widely in their developmental pattern and growth rate. However, the fractional rate of growth (% increase day^{-1}) of chicks in most species is highest after hatching and decreases steadily until an adult lean body mass is achieved. The requirement at any given age varies directly with a bird's fractional growth rate. Thus, the amino acid requirements (% of the diet) decrease with age and, at the same time, the ideal balance of amino acids changes gradually to reflect those of maintenance. The amino acid composition of the tissues of different species is relatively similar (Table 6.3) and differences in requirements are primarily determined by their relative growth rates.

Requirements for reproduction

Birds laying eggs need dietary amino acids for normal maintenance, growth of the oviduct, and accretion of egg proteins (Murphy, 1994). The diverse reproductive strategies among avian species make it difficult to generalize about the amino acid requirement to support egg production. The most important variables influencing requirements are the number of eggs laid in a clutch, the number of days between eggs, the size and composition of the eggs, and the timing of yolk and albumen accretion relative to egg laying. Reproduction has a marked impact on amino acid requirements in species that lay an egg every day and have large clutches (e.g., many Galliformes, Anseriformes). In species that lay a single egg or that skip several days between eggs in a clutch, the amino acid requirements are only slightly increased.

The growth of the oviduct and the synthesis of several yolks are mostly complete before the first egg is laid (Grau, 1984). Consequently, the female's requirement increases at least a week prior to her first oviposition. In some birds, such as petrels and penguins, yolk accretion is extended over several weeks and daily requirements increase very little. In most species, egg albumen

Table 6.3. Amino acid composition of the mixed proteins of avian eggs, carcasses, and muscles.

	Amino acid concentration (μmol g^{-1} protein)							
	Arg	His	Ile	Leu	Lys	Met	Cys	Phe
Eggs								
Zebra Finch	362	155	409	783	632	167	370	351
Pigeon	303	164	411	793	671	234	324	388
Chicken	379	172	531	755	528	248	231	377
Budgerigar	268	159	466	826	593	192	–	403
Finch	415	215	439	790	605	344	355	428
Carcass								
White-crowned								
Sparrow	455	182	380	857	679	168	245	306
Budgerigar	429	182	389	628	632	191	–	299
Gosling	461	204	327	672	554	145	108	285
Muscle								
White-crowned								
Sparrow	461	175	407	867	718	198	176	313
Canvasback	442	226	451	805	695	229	117	326

Source: Murphy (1994).
Arg, arginine; His, histidine; Ile, isoleucine; Leu, leucine; Lys, lysine; Met, methionine; Cys, Cysteine; Phe, phenylalanine.

is synthesized in the oviduct during a 24 h period before ovulation. Thus, dietary amino acid requirements are especially high on the day preceding each oviposition. Exceptions include the Emu and many sea birds, which fully form an egg several days prior to oviposition and have an earlier peak in their requirement.

The characteristic proportion of albumen to yolk varies widely among avian species and this variability influences the timing and total quantity of dietary amino acids needed for egg production. In general, species that have altricial young lay eggs with proportionally greater albumen content compared with precocial species. Because albumen is considerably higher in protein than yolk, altricial species would be expected to have a greater increase in amino acid requirements compared with precocial species laying eggs of similar size and number. The ratio of yolk to albumen has little effect on the balance of amino acids required because the amino acid compositions of these components are almost identical (Table 6.3).

The amino acid requirements to support egg production by domestic species are well characterized (see Table 6.1). These species have been bred to lay an egg each day for a year or more, using amino acids supplied daily by the diet, with very little from tissue stores. The ideal amino acid balance for continuous egg production differs from that needed for maintenance (see Fig. 6.3), but is generally similar to that needed for growth, although slightly higher proportions of methionine, valine, and isoleucine are needed. In nondomestic species, fewer eggs are laid in a clutch (1–30 eggs). For species that lay only a few eggs, such as Costa's Hummingbirds and Shags, the amino acid requirements for reproduction are probably less than twice maintenance for about a week's duration. For others that lay many large eggs, such as Pintail Ducks, the requirement for oviduct growth and yolk synthesis may increase by more than six times maintenance for critical amino acids, such as methionine and lysine. Energy requirements also increase during egg production, in order to deposit lipid in the yolk and to synthesize the protein and other molecules of the egg. In many species, food intake increases sufficiently to meet the heightened energy needs. Because the requirement for energy does not increase as much as that for amino acids, higher-protein foods are needed (Fig. 6.6). For example, the requirement of Japanese Quail for well-balanced protein increases from 5.5% of the diet at maintenance to 23.7% for daily egg laying. This increase in requirements during egg production is often accompanied by a change in food preferences. For example, many female waterfowl greatly increase their consumption of invertebrates prior to and during reproduction. This phenomenon is known as a protein shift and is essential for reproductive success (Yamane *et al.*, 1979; Brice, 1992; Krapu and Reinecke, 1992).

In many species, a portion of the amino acids needed for reproduction is mobilized from body tissues (Alisauskas and Ankney, 1992). For this reason, lean body tissue is often referred to as a protein reserve. The size of these reserves may influence the number and size of eggs laid. In Snow Geese, flight muscles increase markedly in size at staging areas prior to migration. These birds arrive at breeding grounds in the Arctic before the emergence of

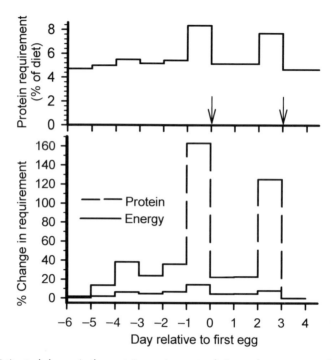

Fig. 6.6. Estimated change in the protein requirement, relative to the energy requirement of a small (15 g), free-living granivorous bird laying a clutch of two eggs, 3 days apart. The arrows indicate the time of egg laying. The protein and energy requirements begin to increase about 5 days prior to ovulation to support the growth of the two follicles. The dietary protein requirement (%) is based on a dietary energy density of 12.5 kJ g^{-1} and the assumption that food intake increases to match the increased energy needs.

vegetation, and flight muscles are preferentially used to supply amino acids for egg production. The balance of amino acids in skeletal muscle is sufficiently different from the balance in egg protein that 1.4 (chicken) to 2.1 (Pintail Duck) g of muscle protein are required for each 1 g of egg synthesized. This is because eggs are enriched in methionine and cysteine relative to skeletal muscle (Murphy, 1994).

Requirements for molt

Most adult birds molt several times a year (Turcek, 1966). This periodic event is associated with increased amino acid needs for the synthesis of replacement feathers and, to a lesser degree, for the synthesis of new feather follicles, feather sheaths, and epidermal blood vessels. Feathers lost and replaced during a full molt amount to about 25% of the protein mass of a bird. Feathers and sheaths contain more than 90% of their dry mass as protein, primarily keratins. The amino acid composition of feathers is considerably different from that of other body proteins or egg proteins (Table 6.3). Feathers are enriched in cysteine and, to a lesser extent, the branched-chain amino acids valine and leucine. The nutritional requirements for molt have been most thoroughly examined in White-crowned Sparrows (Murphy and King, 1990, 1991, 1992). In this small (18 g) bird, the ratio of amino acid needs for feathers to those for feather sheaths to those for other epidermal structures is 20:4:1. A peak protein deposition of about 75 mg day^{-1} occurs midmolt and represents a theoretical increase of about 19% in the protein requirement. However, this calculated increase in the protein requirement is considerably less than the actual increase measured by experimentation. The difference between the calculated and the observed protein requirements is due to very inefficient use of dietary amino acids for feather accretion. Feathers grow continuously throughout the day and night but dietary amino acids are supplied only following meals. In the postabsorptive state, most of the amino acids needed for keratin synthesis are mobilized from tissue proteins, although tissue (especially liver) glutathione may supply some of the cysteine needed. The mismatch in the amino acid composition of body protein and that of feather protein results in a bottleneck in the supply of cysteine and branched-chain amino acids to the feather follicles. It has been calculated that between 2.9 and 3.6 g tissue protein must be mobilized for each gram of feather synthesized in penguins. Most of the liberated methionine, leucine, and valine is used for feather synthesis, but a sizable portion of the other essential amino acids (e.g., lysine, arginine, histidine) are not needed and are oxidized. This diurnal deposition and mobilization of large amounts of tissue proteins markedly decreases the efficiency of amino acid and energy use for molt. It also moderates the balance of dietary amino acids required for molt to more closely resemble the pattern in tissue protein than in feather protein (Cherel *et al.*, 1994; Murphy, 1996).

Molt is energetically expensive, due to the loss of feather insulation, the cost of synthesizing feather protein, and increased body protein synthesis and degradation. The percent increase in the energy expenditure associated with molt often exceeds the percent increase in protein needed for the molt. Consequently, molt may not be associated with a change in the dietary amino acid requirement when expressed as percentage of the diet or as mg kJ^{-1}. This is because increased food consumption to meet energy requirements results in sufficiently increased amino acid intake to meet requirements for molting. Further, some birds, such as White-crowned Sparrows, can select among foods to maximize their protein and sulfur amino acid intake during molt (Murphy and King, 1987).

In birds living in the wild, a decrease in food intake may accompany a full molt, because birds are unable to fly to find food (Bailey, 1985; Cherel *et al.*, 1994). Nevertheless, feather production has a very high priority and continues at the expense of other body tissues. Not surprisingly, flight muscles are preferentially catabolized to supply amino acids for molt. Penguins swim with difficulty when they have shed their feathers and they undergo a complete fast during much of the molt. In the King Penguin, 2.9 g of body protein are required for each gram of feathers synthesized and an additional 1 g is needed to support other body functions due to the fast. Thus, the King Penguin pays a price of almost 4 g of body protein for each gram of feathers produced.

Predicting amino acid requirements

The requirement for amino acids during the life cycle of nondomestic species is almost completely unknown, with very few exceptions. Most notably, almost nothing is known about the requirements of faunivorous and frugivorous species. Even among granivores and omnivores, the validity of using amino acid requirements determined for domestic species as an estimate for nondomestic species is questionable. Certainly domestic chickens, Turkeys, and ducks have been bred for efficient growth and long periods of egg production. Further, these precocial birds have fractional growth rates that are considerably lower than those of most altricial species and many other precocial species. For these reasons, requirements determined for domestic birds should not be used directly to estimate those of nondomestic species, but this robust source of information can provide a starting point for various prediction approaches. For example, the ideal ratio of amino acids required by poultry is useful for predicting the requirement for all the essential amino acids, once the requirement for one amino acid has been determined experimentally. For example, growing Cockatiel chicks require 0.8% lysine in their diet (Roudybush and Grau, 1991). Using the ideal amino acid balance of growing chickens (see Fig. 6.3), the ratio of the methionine requirement to the lysine requirement is 0.35, giving a calculated methionine requirement of 0.28% for the Cockatiel. Similarly, the ratio of the arginine requirement to that of lysine is 1.17, giving a calculated arginine requirement of 0.93% of the diet for the growing Cockatiel chick.

Prediction of amino acid requirements using an ideal amino acid balance appears to be valid among the well-studied domestic species. It may also be a useful approach for approximating the requirement for other florivores that have low maintenance needs. This approach is probably inappropriate for determining the amino acid requirement of faunivorous birds, as their high maintenance requirements change the ideal balance of amino acids needed.

Another approach to estimating amino acid requirements is to sum the needs for various amino acid-consuming processes, such as growth, molting, reproduction, and maintenance (Emmans and Fisher, 1986; Baker *et al.*, 1996; Cilleirs and Hayes, 1996; Grau, 1996; O'Malley, 1996). This sum must then be corrected for the efficiency of protein digestion and for the efficiency of use of absorbed amino acids for protein accretion. In chickens, the efficiency (above maintenance) of using absorbed valine for growth is 73% and the efficiency for lysine is 80%. The efficiency of utilizing essential amino acids for accretion of egg protein ranges between 80 and 85%. This factorial summation method has been used successfully in poultry, Emus, and Ostriches. This method should be useful for all species but awaits an accurate estimation of obligatory losses before it can be applied to faunivores.

Amino acid deficiencies

In young chicks, a severe dietary deficiency of a single amino acid manifests as slowed growth rates and decreased food intake. The decrease in intake occurs within hours of consumption of the deficient meal and is due to a distortion in the plasma and tissue amino acid levels. The anorexia serves to establish learned avoidance of severely amino acid-deficient foods. A moderate deficiency of an essential amino acid results in decreased food intake and also causes disproportional deposition of adipose tissue relative to skeletal muscle. Very marginal deficiencies may not cause a decrease in body weight but can be identified by an increased proportion of adipose tissue. This is because a marginal deficiency may be compensated by an increase in food intake. The increases in intake provide for the needs of the limiting essential amino acid, but they also result in excess energy intake. For this reason, marginal amino acid deficiencies can also be viewed as excess dietary energy density. Obesity due to a marginal deficiency in amino acids has been observed in a variety of species, including chickens, turkeys, quail, ducks, and Budgerigars (Underwood *et al.*, 1991; D'Mello, 1994; Kirkpinar and Oguz, 1995).

An amino acid deficiency of a few days' duration in the growing bird can usually be compensated for during subsequent days of dietary adequacy with no permanent effect on adult body size. A more chronic amino acid deficiency during growth results in a delay in reaching adult weight and often in permanent stunting. Chronically deficient Cockatiel chicks experience difficulty in weaning or, in severe cases, fail to wean, due to an inability to balance food and water intake (Roudybush, 1986).

In adult birds, an amino acid deficiency results in net catabolism of body proteins, particularly those of skeletal muscle, and causes negative nitrogen balance (Cherel and Le Maho, 1985; Swain, 1992). Moderate and severe deficiencies may delay the onset of egg production or prevent it completely. An amino acid deficiency in laying females can result in fewer or smaller eggs. For example, Ruffed Grouse lay fewer and smaller eggs in their clutch as dietary methionine levels decrease. The decrease in egg size due to an amino acid deficiency is largely due to a decrease in albumen content. Chicks that hatch from small eggs are smaller, grow more slowly, have retarded behavioral development, and have higher mortality. In chickens, a marginal methionine deficiency decreases egg size without affecting the number of eggs laid. Other species may mobilize body protein to supply the amino acids necessary for egg formation, and egg size and numbers do not decline if dietary amino acids are inadequate (Beckerton and Middleton, 1982; Hill, 1993; Fisher, 1994).

In some species, such as White-crowned Sparrows, Japanese Quail, chickens, and turkeys, a deficiency of lysine may impair feather pigmentation (Grau *et al.*, 1989; Murphy and King, 1991). In other species, such as Cockatiels and pigeons, a lysine deficiency doesn't influence pigmentation. Amino acid deficiencies may also cause defects, such as fraying or curvature, in feathers replaced following accidental loss or molt. No other specific pathological indications result from an amino acid deficiency and diagnosis is difficult.

Commercial diets are typically formulated using a 'least cost' approach and meeting amino acid requirements greatly impacts the cost of the diet. For this reason, the most limiting amino acids are typically supplied with little or no margin of safety. Most commercial diets utilize plant protein sources, such as soybeans, and methionine is the first limiting amino acid, followed by lysine. Consequently, methionine and lysine are the most likely amino acids to be deficient in commercially prepared diets. Similarly, the diets of florivorous wild species are usually first limiting in methionine, followed by lysine.

Metabolism and Storage of Amino Acids

Transport

Amino acids are absorbed from the intestinal lumen as free amino acids and small peptides. The peptides are hydrolyzed in the enterocyte and free amino acids are transported across the basal membrane. A sizable portion of the free amino acids in blood are present inside of the red blood cells, with the remainder found in the plasma as free amino acids. Portal blood carries dietary amino acids to the liver, where some are removed to support hepatic functions, including synthesis of secretory and hepatocyte proteins, gluconeogenesis, and oxidation. The remaining amino acids pass through the liver and are delivered to tissues for their anabolic needs.

Storage

The concept of storage pools of amino acids is somewhat controversial. Cells need amino acids on a relatively continuous basis, but amino acids are provided only after a meal. Thus, birds apparently have a mechanism to buffer dietary amino acid influx for use by cells during times of deficit. Free amino acid concentrations do not increase sufficiently in the plasma or tissues following a meal to account for this buffering. The labile pool of amino acids is apparently present in body proteins. No single type or class of protein has been clearly identified as a storage pool and this function is probably served by a wide variety of individual proteins in most tissues. The amount of labile protein available to buffer dietary shortages before important tissue functions decline is not known, but certainly does not approach the amount of stored energy in triglycerides.

Very small amounts of amino acids may become available from small peptides present in tissues. For example, the tripeptide glutathione may buffer some of the demand for cysteine for synthesis of keratin during molt and synthesis of egg proteins during reproduction. Similarly, the dipeptide carnosine may supply histidine during periods of depletion. As both of these peptides have important functions and their pool sizes are small relative to synthetic needs, storage capacity is modest (Robbins *et al.*, 1977; Murphy and King, 1985, 1990).

Synthesis of amino acids

The carbon skeletons of nonessential amino acids synthesized by birds come from intermediates of carbohydrate metabolism. Oxaloacetate and α-keto-glutarate are intermediates of the citric acid cycle and can be transaminated or undergo reductive amination to give aspartic acid and glutamic acid, respectively. Aspartic acid and glutamic acid are precursors for asparagine and glutamine. Glutamic acid and arginine are precursors for the synthesis of proline, although this pathway is insufficient to completely meet the metabolic needs for proline in fast-growing chicks (Austic, 1976).

Pyruvate and 3-phosphoglycerate are intermediates of glycolysis and are transaminated to alanine and serine. Glycine can be synthesized from serine, although not at a sufficient rate to meet all metabolic requirements. A glycine requirement has been demonstrated in growing chickens, turkeys, quail, and Budgerigars (Featherston, 1976; NRC, 1994; Taylor *et al.*, 1994).

Histidine can be synthesized from 5-phosphoribosyl-1-pyrophosphate, an intermediate in purine synthesis from ribose. The rate is sufficiently slow that histidine is normally required by those birds that have been tested. Birds do not have a functioning urea cycle and lack the enzymatic capacity to synthesize ornithine, the metabolic precursor for arginine in mammals (Wu *et al.*, 1995). Thus, birds must obtain all of the arginine required for protein synthesis and other metabolic purposes from their diets. Some birds use ornithine to conjugate xenobiotics for excretion and it must be supplied by dietary arginine.

All natural diets that meet a bird's requirement for essential amino acids also provide sufficient amounts of nonessential amino acids, so that synthesis is normally minimal. It is possible to formulate a diet, using purified ingredients, that contains adequate essential amino acids but is deficient in nitrogen needed for nonessential amino acids. This diet can be made nutritionally complete by the addition of one or more nonessential amino acids or by the addition of ammonia (Allen and Baker, 1974). Ammonia can be 'fixed' into amino nitrogen by the synthesis of glutamic acid and glutamine. Once glutamic acid is synthesized, transamination permits the distribution of amino nitrogen to other nonessential amino acids.

Specific amino acid relationships

METHIONINE AND CYSTEINE

Methionine degradation occurs by the transsulfuration pathway and involves the net transfer of its sulfur to serine, resulting in cysteine biosynthesis. On a molar basis, L-methionine is 100% efficient as a precursor of L-cysteine in chickens. Thus, all dietary methionine in excess of its requirement can be completely used to satisfy the bird's metabolic need for cysteine. About equal amounts of methionine and cysteine are needed for metabolism and protein synthesis, and dietary cysteine can 'spare' 50% of the dietary methionine requirement. For this reason, the total sulfur amino acid requirement (methionine + cysteine) is about twice the methionine requirement in the young growing chick (see Table 6.1). Following feather growth, the methionine-sparing value of cysteine declines (Graber and Baker, 1971; Wheeler and Latshaw, 1981; Murphy, 1993).

PHENYLALANINE AND TYROSINE

The first step in the disposal of excess phenylalanine is hydroxylation via phenylalanine hydroxylase to give tyrosine. This simple pathway has a conversion efficiency of 100%. The metabolic requirement for these two amino acids is similar. Consequently, the requirement for phenylalanine plus tyrosine is about twice that for phenylalanine alone. In other words, dietary tyrosine can spare half of the dietary phenylalanine requirement (Sasse and Baker, 1972; Murphy, 1993).

GLYCINE AND SERINE

Excess dietary glycine can be converted to serine on an equimolar basis (Sugahara and Kandatsu, 1976). This reaction is reversible and glycine can be used to synthesize serine. Because of these efficient interconversions, the dietary requirement for glycine can be fulfilled by serine and vice versa. Thus, the requirement is expressed as glycine plus serine. Serine, and thus glycine, can also be synthesized during threonine catabolism.

IMBALANCES AND ANTAGONISMS

A dietary excess of certain amino acids may lead to reduced utilization of the first limiting amino acid in a diet. When this relationship is relatively nonspecific and manifests largely as an impairment in appetite, it is called an amino acid imbalance. A diet in which the amount of the first limiting amino acid is very low relative to the next most limiting amino acid is usually imbalanced. The quality of the diet could be improved either by providing more of the first limiting amino acid or by lowering the level of the next most limiting amino acids. Amino acid imbalances are most obvious at low levels of dietary protein (D'Mello, 1994).

When excesses of a specific amino acid increase the requirement for a metabolically similar amino acid, the relationship is called an amino acid antagonism. For example, high dietary levels of leucine or valine reduce the utilization of isoleucine (Calvert *et al.*, 1982). Decreased appetite mediates part of the antagonism, but there are also specific metabolic complications. Surplus leucine or valine induces the enzymes that catabolize all three branched-chain amino acids, causing greater loss of isoleucine and exacerbating its deficiency. The result of an amino acid antagonism created by high dietary levels of one branched-chain amino acid is an increase in the requirement for the other two. Birds are also susceptible to a lysine–arginine antagonism, in which excess levels of lysine increase renal arginase activity and increase the oxidation of arginine. Thus, high levels of dietary lysine increase the requirement for arginine.

AMINO ACID CONVERSIONS TO VITAMINS

During the course of its degradation, dietary tryptophan can be used to synthesize niacin, although the efficiency of this conversion is variable among species (Chapter 11). Only the dietary tryptophan that is in excess of the requirement is used for niacin synthesis. Consequently, high levels of niacin do not decrease the bird's requirement for tryptophan (Chen *et al.*, 1996).

Choline can be synthesized in the liver from ethanolamine and methyl groups from methionine (Chapter 11). Three moles of methionine are required for each mole of choline synthesized and this conversion is relatively slow and inefficient in chickens. Adequate dietary choline slightly reduces the bird's methionine requirement (Jukes, 1947).

OTHER PRODUCTS OF ESSENTIAL AMINO ACIDS

Methionine is the most metabolically active of the essential amino acids. In addition to choline, its methyl groups can be used to synthesize a very wide variety of important molecules, including carnitine, creatine, niacin, polyamines, and purines. These methyl groups are made available by the synthesis of S-adenosyl methionine. Methionine, via cysteine, is also used to synthesize taurine. This particular interconversion is inefficient in mammalian carnivores, leading to a taurine requirement; it is not currently known if avian carnivores require taurine. The presence of choline, niacin, or creatine in the diet decreases the requirement for methionine.

Several other amino acids are used to synthesize metabolically important compounds, although the quantitative impact of these interconversions is not generally known in birds. Tryptophan is used to synthesize the neurotransmitters, serotonin and melatonin. Phenylalanine, via tyrosine, is used to synthesize the melanin pigments and the hormones, epinephrine, norepinephrine, and thyroxine. Glycine is utilized to synthesize pyrimidines and purines, such as uric acid (see below), creatine, glutathione, and porphyrins, and in some avian species it is the predominant amino acid used for conjugation of bile salts and of xenobiotics to facilitate their solubility and excretion. Histidine is used for the synthesis of histamine and of carnosine, which acts as a pH buffer in many tissues. Arginine is utilized to synthesize creatine and polyamines. Lysine is used to synthesize carnitine.

Protein turnover and accretion

Proteins are a major constituent of both the structural and functional components of tissues. On a fat-free basis, the protein content of birds is relatively constant and approximately 85% of the dry matter. Variation in fat deposition is the primary contributor to the variability in carcass protein content between individuals within a species and between species. The contribution of individual tissues to the total body protein of quail is shown in Table 6.4. Each protein in the body turns over at its own characteristic rate. Confusion often exists as to the difference between the terms 'turnover' and 'degradation'. Protein degradation refers to the hydrolysis of protein to free amino acids. Protein turnover is defined as the continual synthesis and degradation of protein, resulting in no net change in the amount of protein. The rate of protein turnover equals the rate of protein degradation in birds that are growing or at maintenance, but the rate of turnover equals the rate of synthesis in the catabolic state (Fig. 6.7). In general, regulatory enzymes have very fast rates of turnover and structural proteins turn over slowly. The aggregate turnover rate of all of the proteins within a tissue determines the turnover rate of that tissue. Similarly, the aggregate rate of turnover of all the tissues determines the whole-body turnover rate. Relationships between adult body size and rates of protein turnover across avian species have yet to be characterized.

When the rate of protein synthesis exceeds the rate of protein degradation, protein accretion occurs. It is important to note that the rate of protein accumulation is independent of the rate of protein turnover. In other words, a given rate of protein accretion can occur at an infinite number of combinations of rates of protein synthesis and degradation (Fig. 6.7). The efficiency of protein synthesis for deposition is defined as the proportion of synthesized proteins that accumulates (accretion rate/synthesis rate). In young growing Japanese Quail,

Table 6.4. The contribution of individual tissues to
whole-body protein content of mature male
Japanese Quail (data from K.C. Klasing, unpublished).*

Tissue	% of whole body protein
Muscle	
Myofibrillar	29
Sarcoplasmic	12
Stroma	7
	48
Connective tissue	
Collagen	9
Elastin	5
Others	11
	24
Liver	6
Plasma	3
Gastrointestinal	6
Others	13

*Feathers are not included, but typically account for
an additional 25% of protein mass.

the efficiency of protein synthesis is around 70% and this declines until it reaches
0% at mature body weight, when none of the synthesized protein results in net
accretion. Rates of both protein synthesis and protein degradation are high in
young chicks and decline with time (Maeda *et al.*, 1986).

PROTEIN SYNTHESIS
Protein synthesis has priority over most other metabolic uses for amino acids. The
biochemical details of protein synthesis, including the role of messenger ribonucleic acid (mRNA), ribosomes and transcription factors, have been known for a long
time and are similar for birds and other vertebrates. Recent progress in molecular
biology has provided an understanding of the cellular and molecular events that
regulate the rate at which a given protein is synthesized, including the role of
promoters, inducers, enhancers, etc. Rates of protein synthesis are often expressed
in one of two different ways: as absolute rates, e.g., mg of protein synthesized day^{-1};
or as fractional rates, e.g. mg of protein synthesized 100 mg^{-1} of original protein
day^{-1} (% day^{-1}). Expression as a fractional rate permits easy comparison across

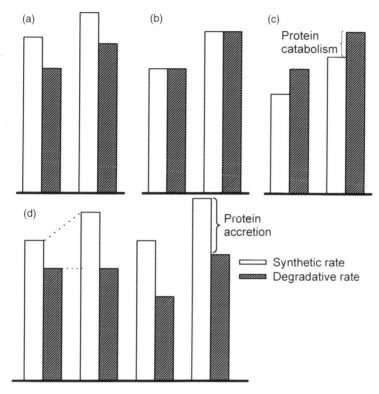

Fig. 6.7. Different combinations of protein synthetic and degradation rates can result in equal rates of: (a) protein accretion; (b) maintenance levels of protein turnover; and (c) protein catabolism. (d) The rate of protein accretion may increase by increasing synthesis, decreasing degradation, or both.

individual proteins, tissues, and species. The average fractional synthetic rates for proteins in various tissues of growing quail are shown in Table 6.5. In general, the greater the metabolic activity of a tissue, the greater its rate of protein synthesis. When a tissue becomes more metabolically active, its rate of protein turnover usually increases. A good example of this is molting in White-crowned Sparrows. During molt, metabolic rate increases by 20% and whole-body fractional protein synthesis rates increase by 40%. Rates of protein degradation increase similarly (Murphy and Taruscio, 1995; Taruscio and Murphy, 1995).

PROTEIN DEGRADATION
Following synthesis, most proteins are degraded within a few hours to days. This process is energetically expensive but vital for homeostatic and homeorhetic processes. Protein degradation serves the following functions: (i) it permits a rapid change in the amount of a protein (metabolic plasticity); (ii) it allows the release of amino acids during periods of nutritional deprivation; (iii) it permits the

Table 6.5. Fractional rate of protein synthesis in various tissues of 7-day-old Japanese Quail chicks.[*]

Tissue	% day^{-1}
Skin	39.1
Pectoralis muscle	33.4
Thigh muscles	39.8
Liver	96.2
Gastrointestinal tract	117.3
Heart	42.2
Kidney	71.5
Thymus	123.4
Bursa	77.2

[*]Rate determined according to the method of Barnes *et al.* (1995).

removal of faulty or damaged proteins; (iv) it permits the limited removal of signal sequences or sections of the protein necessary to establish tertiary structure; and (v) it allows for the restructuring or repair of cells and tissues. From this list of functional properties, it is easy to understand why fast-growing, metabolically active birds have very high rates of protein degradation compared with slower-growing, older birds.

The mechanisms responsible for protein degradation are not nearly as well understood as those for protein synthesis (Fagan *et al.*, 1992; Stevens, 1996). Some proteins are degraded in lysosomes by lysosomal proteases, such as the cathepsins. This is especially true for extracellular proteins that must be internalized before degradation and for proteins that are integral parts of organelles or membranes. The lysosome is also particularly active during pathological states, such as starvation or remodeling following injury. The lysosome is probably not solely responsible for normal, basal levels of proteolysis. Some cells do not have lysosomes and many structural proteins are too large to be internalized intact by lysosomes. Thus, proteolytic enzymes in the cytosol, such as the calpains and the proteosomal proteases, play a central role in degrading many proteins.

The processes of protein degradation and amino acid oxidation are independently regulated, and high rates of protein degradation are not necessarily indicative of high rates of amino acid loss. Thus, rates of protein degradation do not determine amino acid requirements. The amino acids liberated by protein degradation are largely reutilized for protein synthesis by a process that channels them directly to transfer RNA (tRNA) acylation. This coupling of protein degradation and protein synthesis prevents the loss of amino acids during protein turnover. Even in fast-growing chicks consuming high-protein diets, more than 75% of the amino acids used for protein synthesis arrive from protein degradation and not from the diet (Barnes *et al.*, 1992, 1994).

REGULATION OF PROTEIN ACCRETION BY DIET

The role of nutrition in regulating growth and protein accretion is well known. However, the influence of nutrition on the relative rates of protein synthetic and degradative processes is less well appreciated. Protein synthesis requires the presence of all of the essential, semiessential, and nonessential amino acids to be present in the cells of a tissue at adequate concentrations and at the same time. Protein synthesis is among the most energetically expensive of any metabolic process. Thus, it is not surprising that inadequate consumption of a single amino acid, total protein, or energy decreases the rate of protein synthesis and degradation in skeletal muscle and many other tissues. Because the decrease in the rate of protein synthesis is greater than the decrease in protein degradation, protein accretion is impaired in growing birds and protein catabolism occurs in adult birds. Nutrition can influence protein synthetic and degradative rates directly, by the supply of amino acids and energy to the tissues, or indirectly, through secondary changes in hormone concentrations. Glucose and branched-chain amino acids arriving from the diet are anabolic for many tissues, especially skeletal muscle. The exact mechanism through which an insufficient supply of amino acids decreases the rate of protein synthesis is not known. In young growing chickens, a severe methionine deficiency impairs protein synthesis but does not result in a decrease in methionine tRNA (met-tRNA) or cysteine tRNA (cys-tRNA) (Klasing and Calvert, 1987, Hiramoto *et al.*, 1990a,b; Barnes *et al.*, 1995).

High levels of dietary protein slightly increase protein deposition at levels well above that needed to meet essential amino acid requirements (Yeh and Leveille, 1969; Rosebrough *et al.*, 1988). Part of this induction in protein accretion may be due to higher insulin-like growth factor-1 (IGF-1) levels induced by high rates of amino acid absorption.

Protein synthesis requires about 6 kJ g^{-1} protein (not including the energy in the protein itself) and accounts for 28% of the heat production in 1-week-old chicks (Muramatsu and Okumura, 1985; Muramatsu *et al.*, 1987; Akoyagi *et al.*, 1988). One might expect protein degradation to liberate usable energy to offset part of this cost. However, no energy is captured as adenosine triphosphate (ATP) and, in fact, protein degradation consumes ATP. Part of this energy is required to maintain fidelity of degradation and part is required to maintain a hydrogen ion (H$^+$) concentration gradient across the lysosome. The exact amount of ATP required for degradation has not been measured or calculated, because the biochemical events are largely unknown. Interestingly, strains of chickens that are energetically efficient at protein accretion have similar rates of protein synthesis and lower rates of protein degradation than less efficient strains. Diets that maximize the efficiency of energy utilization for growth also minimize the rate of protein degradation (Klasing *et al.*, 1987).

Protein metabolism and egg formation

Following ovulation, the yolk is coated with egg-white proteins as it traverses the magnum section of the oviduct (Etches, 1996). The mucosa of the magnum contains tubular glands,which synthesize ovalbumin, ovotransferrin, lysozyme, and ovomucoid. These proteins are synthesized on the rough endoplasmic reticulum near the base of the secretory cells and are packaged by the Golgi apparatus into granules surrounded by a lipid membrane. The nascent granules coalesce into large storage granules, which collect in the apical region of the secretory cells. As the yolk tranverses the magnum, signals derived from the distention of the lumen induce the storage granules to fuse with the apical cell membrane and they are secreted. Avidin is synthesized in the cells lining the lumen of the magnum and is also added. The albumen proteins are synthesized continuously throughout the day preceding ovulation; however, synthesis increases by threefold as the egg tranverses the magnum. In many avian species, ovulation and oviposition occur in the morning. Thus, the period of maximum amino acid demand for synthesis of albumen for an egg also occurs in the morning. In chickens, increasing dietary protein during this period increases albumen accretion (Hiramoto *et al.*, 1990b; Penz and Jensen, 1991).

Yolk proteins are synthesized in the liver, largely in conjunction with lipid accretion and transport. Very-low-density lipoprotein, lipovitellin, phosvitin and plasma albumin make up greater than 95% of the yolk proteins. Unlike albumen, the synthesis of yolk proteins is continuous throughout the day. The peak daily need for dietary amino acids for yolk proteins depends upon the amount of overlap in growth cycles of individual ova and the number of follicles growing simultaneously. Specific aspects of yolk lipoprotein synthesis are covered in Chapter 7.

Amino acid losses

Excess dietary amino acids beyond the need for protein synthesis may be used for the synthesis of metabolites or they may be degraded. The first step in the degradation of most amino acids is deamination. The amino group can be channeled to uric acid synthesis, either through transamination or through glutamine synthesis. Depending on the bird's diet and physiological state, the carbon skeletons may be used for glucose synthesis or fat synthesis or oxidized to provide energy. Some amino acid degradation occurs at all times, even when dietary amino acid supplies are low. This constitutive rate of amino acid loss, or obligatory loss, accounts for a major part of the maintenance amino acid requirement. The rate of obligatory loss differs for each amino acid. For example, in chickens consuming a protein-free diet, lysine is lost much more slowly than methionine or isoleucine (Baker, 1991; Muramatsu *et al.*, 1991). Thus, a perfect pattern of dietary amino acids for maintenance (ideal protein) deviates from the pattern of amino acid present in body tissues (see Fig. 6.3). This relationship is manifest in the low proportion of lysine and high proportion of methionine required for maintenance relative to the proportions required to support growth.

Endogenous losses of amino acids also occur through the sloughing of skin, in the replacement of damaged feathers, and in protein in the feces. The endogenous fecal losses include unrecovered digestive enzymes, mucus, and sloughed cells. The amount of fecal loss is related to the level of food intake, and the amount of skin and feather loss is environment-dependent. The rates of these endogenous losses have not been accurately quantified in birds.

OXIDATION

Excess amino acids in the diet or prolonged periods of fasting result in the use of amino acids as an energy source. Most amino acids are oxidized in the liver or kidney. The branched-chain amino acids can also be transaminated and oxidized in skeletal muscle.

When dietary levels of amino acids are low, rates of oxidation are minimal due to an inhibition in the synthesis and activity of the catabolic enzymes. Further, the Michaelis constants (K_m) for the amino acid transaminases, which initiate oxidation, are considerably higher than the K_m of the tRNA acylating enzymes, which direct free amino acids to protein synthesis. These enzymatic relationships result in the priority use of amino acids for anabolic purposes. The activity of amino acid catabolic enzymes is normally much higher in faunivores than in florivores, reflecting their respective dietary protein levels. For example, hepatic arginase activity is about 100-fold higher in Great Blue Herons or Belted Kingfishers than in pigeons or chickens and is intermediate in omnivores such as Crows or Boat-tailed Grackles (Brown, 1966).

GLUCONEOGENESIS

When a bird is fasted or consumes a very low-carbohydrate diet, it must synthesize glucose from amino acids. The deamination and catabolism of most of the amino acids provides intermediates of glycolysis or the citric acid cycle, which can be converted to glucose. The catabolism of lysine and leucine results in intermediates that cannot give the net synthesis of glucose. Consequently, these two amino acids are not gluconeogenic but are ketogenic. Gluconeogenesis is particularly active in faunivores and occurs in both the fasted and fed state. The Barn Owl has lost some of its capacity to downregulate gluconeogenesis from amino acids, presumably because of the paucity of glucose in its diet of mice and other small vertebrates (Migliorini *et al.*, 1973; Meyers, 1995).

Excretion and Toxicities

Birds excrete most of their waste nitrogen as uric acid rather than urea or ammonia. This unique excretion strategy is considered to be an adaptation to the incubation requirements of the egg. During embryonic development waste uric acid precipitates out of solution, preventing the osmotic imbalance that would occur if urea or ammonia were excreted. This osmotic advantage may also aid in water conservation in adult birds.

Uric acid is a purine and is synthesized in the liver and, to a lesser extent, by the kidney (Campbell and Vorhaben, 1976; Wiggins *et al.*, 1982). Its synthesis utilizes the same metabolic pathway as the synthesis of adenine and guanine, found in deoxyribonucleic acid (DNA), and constitutes a cycle in which ribose-5-phosphate is utilized and regenerated. The origin of the carbon and nitrogen atoms of uric acid is shown in Fig. 6.8. Amino nitrogen is funneled to uric acid by transamination, eventually giving glutamic acid, or by the glutamine synthetase reaction. Birds lack carbamyl phosphate synthetase, which fixes free ammonia in

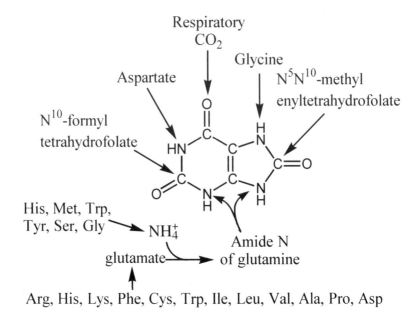

Fig. 6.8. The metabolic origins of the atoms in uric acid synthesized in birds. The two carbons of glycine contribute the shared carbons of the two rings. CO_2, carbon dioxide; NH_4^+, ammonium ion; His, histidine; Met, methionine; Trp, tryptophan; Tyr, tyrosine; Ser, serine; Gly, glycine; Arg, arginine; Lys, lysine; Phe, phenylalanine; Cys, cysteine; Ile, isoleucine; Leu, leucine; Val, valine; Ala, alanine; Pro, proline.

mammals. Instead, birds utilize mitochondrial glutamine synthetase to detoxify ammonia arising from amino acid catabolism. The glutamine synthesized by this reaction freely diffuses into the cytosol, where uric acid is synthesized. Amido-phosphoribosyltransferase is the primary regulatory enzyme in uric acid production and its activity changes directly with the dietary protein level.

   The disposal of excess amino acid nitrogen by the uric acid synthetic pathway by birds results in greater requirements for arginine, methionine, and glycine relative to mammals. High amounts of arginine are required by birds, even at maintenance, because it is not synthesized, due to the lack of a urea cycle. One mole of glycine is consumed in the synthesis of each mole of uric acid. Thus, the metabolic demand for glycine is great and cannot always be met by endogenous synthesis, resulting in a dietary requirement. Formyltetrahydrofolate donates methyl groups for the synthesis of uric acid and these methyl groups can originate from methionine, increasing its requirement. The consumption of glycine and methionine during uric acid synthesis causes the requirement for these two amino acids to increase directly with increasing dietary protein. Further, uric acid is a relatively reduced molecule and its excretion is accompanied by the loss of more energy than in urea excretion. The excretion of uric acid requires 3.75 ATP mol^{-1} of nitrogen, compared with 2 ATP mol^{-1} for urea excretion by mammals (Mapes and Krebs, 1978).

   The concentration of uric acid in the urine of adult birds ranges between 0.1

and 1.0 mol l^{-1}, which exceeds its solubility in water (1 mmol l^{-1}); but uric acid does not crystallize out of solution. Instead, it forms a supersaturated colloidal suspension of small spheres, ranging from 1 to 15 μm in diameter. These spheres consist of plasma protein, urate, calcium, and potassium. The excretion of colloidal uric acid complexed with electrolytes reduces the osmotic potential of the urine, because it is not osmotically active. Synthesis and excretion of colloidal spheres may prevent the blockage of nephrons that would occur with uric acid crystals (Skadhauge, 1983; Casotti and Braun, 1996).

Although uric acid is the most prevalent nitrogenous excretory product, ammonia excretion is relatively high in birds (Stewart *et al.*, 1968; McNabb *et al.*, 1980). Chickens fed high-protein diets excrete about 25% of their nitrogen as ammonia and this proportion decreases with the dietary protein level. The percentage of ammonia excreted by the carnivorous Turkey Vulture is slightly lower, and the excretion of large amounts of colloidal uric acid following a high-protein meal is an osmotic necessity, which permits this bird to survive only with the water found in its food. Conversely, freshwater ducks, which normally have a surfeit of water available, excrete larger amounts of ammonia. Small amounts of urea found in avian urine arise from the catabolism of excess dietary arginine and not from a functional urea cycle.

Toxicity

Moderately high dietary levels of well-balanced dietary protein are generally well tolerated by domestic birds, though this may not hold for all species (Parish and Martin, 1977). Very high protein levels result in decreased growth, decreased body fat, and increased uric acid levels in the blood. In small granivores, such as Tree Swallows and Dark-eyed Juncos, high intakes due to cold weather decrease the level of protein that can be tolerated. High dietary levels of a single amino acid are more debilitating than high levels of total protein. In addition to metabolic imbalances and antagonisms, discussed above, frank toxicity can occur. Methionine is the most toxic of the amino acids, depressing growth and food intake at levels of about three to four times the requirement. As methionine is the most common amino acid supplemented to commercial diets, it is also the amino acid that is most likely to be accidentally added in excess. The relative order of toxicity of amino acids for growing chickens is methionine > phenylalanine > tryptophan = histidine = lysine > tyrosine = threonine > isoleucine > arginine > valine = leucine. Nonessential amino acids are considerably less toxic than essential amino acids (Ueda *et al.*, 1981; Edmonds and Baker, 1987; Han and Baker, 1993).

The oxidation of excess dietary methionine and cysteine results in the production of sulfate. In the process, 2 mol of acid are produced per mole of amino acid oxidized. The oxidation of phosphorylated amino acids and dibasic amino acids also contributes to metabolic acidosis. Thus, high levels of dietary protein cause a metabolic acidosis and may contribute to a variety of problems in birds, including poor bone mineralization, thinning of egg shells, and poor growth.

Other Nutritional Considerations

Nutritional factors that effect amino acid requirements

Amino acid requirements are not markedly influenced by many of the same factors that impact energy requirements such as environmental temperature or activity. However, several dietary factors impact the utilization of essential amino acids and consequently a bird's requirement. As discussed above, dietary imbalances and antagonisms increase the requirement for essential amino acids. Amino acid requirements also increase with increasing levels of dietary protein. It appears that, in the disposal of excess dietary amino acids, the bird makes less efficient use of the limiting amino acids (Grau and Kamei, 1950; Morris *et al.,* 1987; Mendonca and Jensen, 1989).

A variety of secondary plant products affect amino acid requirements. Some, such as tannins and protease inhibitors, interfere with digestion of protein. Others, such as amino acid analogues, interfere with the metabolism of structurally similar amino acids. Some require the use of glycine, arginine, or taurine for conjugation and excretion.

Digestibility of dietary protein

Studies conducted to determine amino acid requirements have utilized diets in which the protein source was highly digestible. Thus, published amino acid requirement values, such as NRC (1994), are only valid for diets with highly digestible protein. In application, the amino acid content of foods should be corrected for its digestibility. Determining amino acid digestibility is technically difficult and subject to a variety of errors (Johnson, 1992; McNab, 1994). Nevertheless, the digestibility of amino acids in a wide variety of feedstuffs has been determined for the chicken. In general, the protein digestibility of unprocessed foodstuffs is positively correlated with energy digestibility. Fat-extracted foodstuffs, such as soymeal and other meals prepared from seeds, have highly available protein relative to their energy digestibility. Whether digestibility values determined in chickens are applicable to other species is unknown, although there is little difference between protein digestibility by chickens and Muscovy Ducks (Mohamed *et al.,* 1986). As the amino acid digestibility of foods consumed by wild birds is unknown, it may be useful to make estimates based on energy digestibility.

Activity and environment

The environmental temperature or the amount of activity that a bird engages in has not been shown to markedly affect daily amino acid requirements of birds consuming adequate energy. Free-living birds have much higher energy requirements than captive birds. The high rate of diet consumption to fulfill the energy requirements of thermoregulation, flight, and other activities results in a proportional decrease in amino acid requirements, expressed as % of the diet. Consequently, foods that may be a poor nutritional choice for captive birds, due to low levels of essential amino acids, may be adequate for wild birds, especially in the winter.

Supplementing amino acids

Supplementing diets with free amino acids permits the amino acid requirements to be met at lower levels of protein. This results in less excretion of uric acid and permits the commercial production of cheaper feeds. Consequently, most commercial feeds for captive birds are supplemented with methionine and often lysine. Amino acids supplemented to the diet are approximately 100% digestible, but their utilization may suffer if their absorption follows a faster time course than that of amino acids arriving from intact protein. If there is a transient surplus of an amino acid, it is oxidized and unavailable for protein synthesis. The inefficiency associated with supplementing free dietary amino acids is probably small and only apparent when birds are fed one large meal per day. It is not a problem when birds eat many meals throughout the day because transient surpluses are avoided (Baker and Izquierdo, 1985; Otto *et al.*, 1989).

Utilization of amino acid isomers and precursors

All amino acids used for protein synthesis by birds must be in the L form. D-Amino acids are present naturally in bacteria and some commercial sources of amino acids are chemically synthesized as racemic mixtures of the D and L isoforms. The D isoform must be converted to the L isoform to be nutritionally active. Conversion requires deamination by amino acid oxidase to give a keto acid. Some keto acids can then be transaminated to give L-amino acids. The efficiency of conversion differs between the amino acids (Table 6.6). Birds lack the transaminases to synthesize lysine and threonine from their keto acids and the D isoforms of these amino acids do not have nutritional efficacy. The keto acids of many other amino acids are diverted to other metabolic functions and are not efficiently transaminated. Some D-amino acids are inefficiently absorbed by the intestine or reabsorbed in the renal tubules and are excreted prior to conversion to the L isoform. When diets contain D-amino acids, it is important to correct the amount present by the efficiency of conversion to the L form. DL-Methionine is commonly supplemented to commercial bird diets. Conversion of the D isoform is about 90% in chickens. As the conversion of the D isoform is extremely variable across mammalian and fish species, it should not be assumed that all bird species can efficiently utilize this isomer.

In some cases, a precursor to an amino acid can be converted by the bird into nutritionally relevant amounts of an essential amino acid. For example, DL-2-hydroxy-4-methylmercaptobutyric acid (HMB) is often used commercially to provide methionine activity. Both isomers are converted to methionine and used for protein synthesis in growing chicks (Barnes *et al.*, 1995).

Lability of dietary amino acids

It is not uncommon for food items to be consumed following weeks or months of storage. In general, dietary protein is relatively stable in composition during storage at room temperature. Lysine is the most labile of the amino acids in intact protein. The epsilon amino group of lysine can react with reducing sugars, such as glucose, by the Maillard reaction. This reaction is accelerated by heat and moisture and renders lysine nutritionally unavailable (Robbins and Baker, 1980).

Table 6.6. Relative bioavailability of amino acid isomers, analogues, and precursors for the growing chicken. Values are expressed on a molar basis relative to the L isomer, which is assumed to be 100% bioavailable. (Data and references are from the summary of Baker, 1994.)

Amino acid	Bioavailability	Amino acid	Bioavailability
Arginine		Methionine	
L-Arginine	100	L-Methionine	100
D-Arginine	0	D-Methionine	90
L-Ornithine	0	DL-HMB*	80
L-Citrulline	90	L-Homocysteine	65
Histidine		Cysteine	
L-Histidine	100	L-Cysteine	100
D-Histidine	10	D-Cysteine	100
Carnosine	100	DL-Lanthionine	35
Anserine	0	Glutathionine	100
Isoleucine		L-Homocysteine	100
L-Isoleucine	100	Phenylalanine	
D-Isoleucine	0	L-Phenylalanine	100
L-Keto-isoleucine	85	D-Phenylalanine	75
D-Keto-isoleucine	0	Threonine	
Leucine		L-threonine	100
L-Leucine	100	D-Threonine	0
D-Leucine	100	Tryptophan	
Keto-leucine	100	L-Tryptophan	100
L-OH-Leucine	100	D-Tryptophan	20
D-OH-Leucine	100	Valine	
Lysine		D-Valine	100
L-Lysine	100	L-Valine	70
D-Lysine	0	Keto-valine	80
Lysinoalanine	0	L-OH-Valine	80
γ-Glutamyl-L-lysine	0	D-OH-Valine	70
L-Homoarginine	0		

*The bioavailability of DL-2-hydroxy-4-methylmercaptobutyric acid (DL-HMB) is higher when it is supplemented at low levels.

References

Alisauskas, R.T. and Ankney, C.D. (1992) The cost of egg laying and its relationship to nutrient reserves in waterfowl. In: Batt, B.D.J., Afton, A.D., Anderson, M.G., Ankney, C.D., Johnson, D.H., Kadlec, J.A. and Krapu, G.L. (eds) *Ecology and Management of Breeding Waterfowl.* University of Minnesota Press, Minneapolis, pp. 30–61.

Allen, N.K. and Baker, D.H. (1974) Quantitative evaluation of nonspecific nitrogen sources for the growing chick. *Poultry Science* 53, 258–264.

Aoyagi, Y., Tasaki, I., Okumura, J. and Muramatsu, T. (1988) Energy cost of whole-body protein synthesis measured *in vivo* in chicks. *Comparative Biochemistry and Physiology* 91A, 765–768.

Austic, R.E. (1976) Nutritional and metabolic interrelationships of arginine, glutamic acid and proline in the chick. *Federation Proceedings* 35, 1914–1916.

Bailey, R.O. (1985) Protein reserve dynamics in postbreeding adult male redheads. *Condor* 87, 23–32.

Baker, D.H. (1991) Partitioning of nutrients for growth and other metabolic functions – efficiency and priority considerations. *Poultry Science* 70, 1797–1805.

Baker, D.H. (1994) Utilization of precursors for L-amino acids. In: D'Mello, J.P.F. (ed.) *Amino Acids in Farm Animal Nutrition.* CAB International, Wallingford, UK, pp. 37–62.

Baker, D.H. and Han, Y.M. (1994) Ideal amino acid profile for chicks during the first three weeks posthatching. *Poultry Science* 73, 1441–1447.

Baker, D.H. and Izquierdo, M.S. (1985) Effect of meal frequency and spaced crystalline lysine ingestion on the utilization of dietary lysine by chickens. *Nutrition Research* 5, 1103–1112.

Baker, D.H., Robbins, K.R. and Buck, J.S. (1979) Modification of the level of histidine and sodium bicarbonate in the Illinois crystalline amino acid diet. *Poultry Science* 58, 749–755.

Baker, D.H., Fernandez, S.R., Parsons, C.M., Edwards, H.M., Emmert, J.L. and Webel, D.M. (1996) Maintenance requirement for valine and efficiency of its use above maintenance for accretion of whole body valine and protein in young chicks. *Journal of Nutrition* 126, 1844–1851.

Barnes, D.M., Calvert, C.C. and Klasing, K.C. (1992) Source of amino acids for transfer RNA acylation – implications for measurement of protein synthesis. *Biochemical Journal* 283, 583–589.

Barnes, D.M., Calvert, C.C. and Klasing, K.C. (1994) Source of amino acids for tRNA acylation in growing chicks. *Amino Acids* 7, 267–278.

Barnes, D.M., Calvert, C.C. and Klasing, K.C. (1995) Methionine deficiency decreases protein accretion and synthesis but not tRNA acylation in muscles of chicks. *Journal of Nutrition* 125, 2623–2630.

Beckerton, P.R. and Middleton, A.L. (1982) Effects of dietary protein levels on ruffed grouse reproduction. *Journal of Wildlife Management* 46, 569–579.

Bosque, C. and Parra, O. (1992) Digestive efficiency and rate of food passage in oilbird nestlings. *Condor* 94, 557–571.

Brice, A.T. (1992) The essentiality of nectar and arthropods in the diet of the Anna hummingbird (*Calypte anna*). *Comparative Biochemistry and Physiology A – Comparative Physiology* 101, 151–155.

Brice, A.T. and Grau, C.R. (1991) Protein requirements of Costa's hummingbirds *Calypte costae. Physiological Zoology* 64, 611–626.

Brown, G.W. (1966) Studies in comparative biochemistry and evolution. 1. Avian liver arginase. *Archives of Biochemistry and Biophysics* 114, 184–194.

Calvert, C.C., Klasing, K.C. and Austic, R.E. (1982) Involvement of food intake and amino acid catabolism in the branched-chain amino acid antagonism in chicks. *Journal of Nutrition* 112, 627–635.

Campell, J.W. and Vorhaben, J.E. (1976) Avian mitochondrial glutamine metabolism. *Journal of Biological Chemistry* 247, 781–786.

Casotti, G. and Braun, E.J. (1996) Functional morphology of the glomerular filtration barrier of *Gallus gallus. Journal of Morphology* 228, 327–334.

Chen, B.J., Shen, T.F. and Austic, R.E. (1996) Efficiency of tryptophan–niacin conversion in chickens and ducks. *Nutrition Research* 16, 91–104.

Cherel, Y. and Le Maho, Y. (1985) Five months of fasting in king penguin chicks: body mass loss and fuel metabolism. *American Journal of Physiology* 249, R387– R392.

Cherel, Y., Charrassin, J.B. and Challet, E. (1994) Energy and protein requirements for molt in the king penguin *Aptenodytes patagonicus*. *American Journal of Physiology* 266, R1182–R1188.

Cilleirs, S.C. and Hayes, J.P. (1996) Feedstuff evaluation and metabolisable energy and amino acid requirements for maintenance and growth in ostriches. In: Deeming, D.C. (ed.) *Improving our Understanding of Ratites in a Farming Environment.* Ratite Conference, Oxford, pp. 85–91.

Cole, D.J.A. and Van Lunen, T.A. (1994) Ideal amino acid patterns. In: D'Mello, J.P.F. (ed.) *Amino Acids in Farm Animal Nutrition.* CAB International, Wallingford, UK, pp. 99–112.

Dawson, T.J. and Herd, R.M. (1983) Digestion in the emu: low energy and nitrogen requirements of this large ratite bird. *Comparative Biochemistry and Physiology* 75, 41–45.

D'Mello, J.P.F. (1994) Responses of growing poultry to amino acids. In: D'Mello, J.P.F. (ed.) *Amino Acids in Farm Animal Nutrition,* CAB International, Wallingford, UK, pp. 37–62.

Drepper, K., Menke, K., Schulze, G. and Wachter-Vormann, U. (1988) Untersuchungen zum Protein- und Energiebedarf adulter Wellensittche (*Melopsittacus undulatus*). *Kleinteierpraxis* 33, 57–62.

Edmonds, M.S. and Baker, D.H. (1987) Comparative effects of individual amino acid excesses when added to a corn–soybean meal diet: effects on growth and dietary choice in the chick. *Journal of Animal Science* 65, 699–705.

Emmans, G.C. and Fisher, C. (1986) Problems in nutritional theory. In: Fisher, C. and Boorman, K.N. (eds) *Nutrient Requirements of Poultry and Nutritional Research.* Butterworths, London, pp. 9–39.

Etches, R.J. (1996) *Reproduction in Poultry.* CAB International, Wallingford, UK.

Fagan, J.M., Wajnberg, E.F., Culbert, L. and Waxman, L. (1992) ATP depletion stimulates calcium-dependent protein break-down in chick skeletal muscle. *American Journal of Physiology* 262, E637–E643.

FAO (1970) *Amino Acid Content of Foods.* Food Policy and Food Science Service, Rome.

Featherston, W.R. (1976) Glycine–serine interrelationships in the chick. *FASEB* 35, 1910–1915.

Fisher, C. (1994) Responses of laying hens to amino acids. In: D'Mello, J.P.F. (ed.) *Amino Acids in Farm Animal Nutrition.* CAB International, Wallingford, UK, pp. 245–280.

Graber, G. and Baker, D.H. (1971) Sulfur amino acid nutrition of the growing chick: quantitative aspects concerning the efficacy of dietary methionine, cysteine and cystine. *Journal of Animal Science* 33, 1005–1011.

Grau, C.R. (1984) Egg formation. In: Whittow, G.C. and Rahn, H. (eds) *Seabird Energetics.* Plenum Press, New York, pp. 33–57.

Grau, C.R. (1996) Nutritional needs for egg formation in the shag *Phalacrocorax aristotelis. Ibis* 138, 756–764.

Grau, C.R. and Kamei, M. (1950) Amino acid imbalance and the growth requirements for lysine and methionine. *Journal of Nutrition* 41, 89–101.

Grau, C.R., Roudybush, T.E., Vohra, P., Kratzer, F.H., Yang, M. and Nearenberg, D. (1989) Obscure relations of feather melanization and avian nutrition. *World's Poultry Science Journal* 45, 241–246.

Han, Y.M. and Baker, D.H. (1993) Effects of excess methionine or lysine for broilers fed a corn–soybean meal diet. *Poultry Science* 72, 1070–1074.

Hill, W.L. (1993) Importance of prenatal nutrition to the development of a precocial chick. *Developmental Psychobiology* 26, 237–249.

Hiramoto, K., Muramatsu, T. and Okumura, J. (1990a) Effect of methionine and lysine deficiencies on protein synthesis in the liver and oviduct and in the whole body of laying hens. *Poultry Science* 69, 84–89.

Hiramoto, K., Muramatsu, T. and Okumura, J. (1990b) Protein synthesis in tissues and in the whole body of laying hens during egg formation. *Poultry Science* 69, 264–269.

Izhaki (1992) A comparative analysis of the nutritional quality of mixed and exclusive fruit diets for yellow-vented bulbuls. *Condor* 94, 912–923.

Johnson, R.J. (1992) Principles, problems and application of amino acid digestibility in poultry. *World's Poultry Science Journal* 48, 232–246.

Jukes, T.H. (1947) Choline. *Annual Review of Biochemistry* 1947, 194–211.

Kirkpinar, F. and Oguz, I. (1995) Influence of various dietary protein levels on carcase composition in the male Japanese quail. *British Journal of Poultry Science* 36, 605–610.

Klasing, K.C. and Calvert, C.C. (1987) Control of fractional rates of protein synthesis and degradation in chick skeletal muscle by hormones. *Journal of Nutrition* Supplement 1987, 141–149.

Klasing, K.C., Calvert, V.J. and Jarrell, V.L. (1987) Growth characteristics, protein synthesis, and protein degradation in muscles from fast and slow growing chickens. *Poultry Science* 66, 1189–1196.

Krapu, G.L. and Reinecke, K.J. (1992) Foraging ecology and nutrition. In: Batt, B.D.J., Afton, A.D., Anderson, M.G., Ankney, C.D., Johnson, D.H., Kadlec, J.A. and Krapu, G.L. (eds) *Ecology and Management of Breeding Waterfowl.* University of Minnesota Press, Minneapolis, pp. 1–29.

Landry, S.W., Defoliart, G.R. and Sunde, M.L. (1986) Larval protein quality of six species of Lepidoptera. *Journal of Economic Entomology* 79, 600–604.

Leveille and Fisher (1960) Amino acid requirements for maintenance in the adult rooster. I. *Journal of Nutrition* 66, 441–453.

Leveille, G.A., Shapiro, R. and Fisher, H. (1960) Amino acid requirements for maintenance in the adult rooster. IV. *Journal of Nutrition* 72, 8–15.

McNab, J.M. (1994) Amino acid digestibility and availability studies with poultry. In: D'Mello, J.P.F. (ed.) *Amino Acids in Farm Animal Nutrition.* CAB International, Wallingford, UK, pp. 185–204.

McNabb, F.M.A., McNabb, R.A., Prather, I.D., Conner, R.N. and Adkisoon, C.S. (1980) Nitrogen excretion by turkey vultures. *Condor* 82, 219–223.

Maeda, Y., Hayashi, K. and Okamoto, S. (1986) Genetic studies on the muscle protein turnover rate of Coturnix quail. *Biochemical Genetics* 24, 207–216.

Mapes, J.P. and Krebs, H.A. (1978) Rate limiting factors in urate synthesis and gluconeogenesis in avian liver. *Biochemistry Journal* 172, 193–203.

Martin (1968) The effects of dietary protein on the energy and nitrogen balance of the tree sparrow (*Spizella arborea arborea*). *Physiological Zoology* 41, 313–331.

Mendonca, C.X. and Jensen, L.S. (1989) Influence of protein concentration on the sulfur-containing amino acid requirement of broiler chickens. *British Poultry Science* 30, 889–898.

Meyers, M.R. (1995) Comparative tolerance and metabolic adaptations to glucose of the barn owl (*Tyto alba*) and chicken (*Gallus domesticus*). PhD Thesis, University of California, Davis, California.

Migliorini, R.H., Linder, C., Moura, J.L. and Veiga, J.A.S. (1973) Gluconeogenesis in a carnivorous bird (black vulture). *American Journal of Physiology* 225, 1389–1392.

Mohamed, K., Larbier, M. and Leclercq, B. (1986) A comparative study of the digestibility of soyabean and cottonseed meal amino acids in domestic chicks and muscovy ducklings. *Annales de Zootechnie* 35, 79–86.

Moran, E.T., Ferket, P.R. and Blackmaan, J.R. (1983) Maintenance nitrogen requirement of the turkey breeder hen with an estimate of associated essential amino acid needs. *Poultry Science* 62, 1823–1829.

Morris, T.R., Al-Azzawi, K., Gous, R.M. and Simpson, G.L. (1987) Effects of protein concentration on responses to dietary lysine by chicks. *British Poultry Science* 28, 185–195.

Muramatsu, T. and Okumura, J. (1985) Whole-body protein turn-over in chicks at early stages of growth. *Journal of Nutrition* 115, 483–490.

Muramatsu, T., Aoyagi, Y., Okumura, J. and Tasaki, I. (1987) Contribution of whole-body protein synthesis to basal metabolism in layer and broiler chickens. *British Journal of Nutrition* 57, 269–277.

Muramatsu, T., Ohshima, H., Goto, M., Mori, S. and Okumura, J. (1991) Growth prediction of young chicks – do equal deficiencies of different essential amino acids produce equal growth responses? *British Poultry Science* 32, 139–149.

Murphy, M.E. (1993) The essential amino acid requirements for maintenance in the white-crowned sparrow, *Zonotrichia leucophrys gambelii*. *Canadian Journal of Zoology – Revue Canadienne de Zoologie* 71, 2121–2130.

Murphy, M.E. (1994) Amino acid compositions of avian eggs and tissues – nutritional implications. *Journal of Avian Biology* 25, 27–38.

Murphy, M.E. (1996) Energetics and nutrition of molt. In: Carey, C. (ed.) *Avian Energetics and Nutrition Ecology.* Chapman and Hall, New York, pp. 158-198.

Murphy, M.E. and King, J.R. (1985) Diurnal variation in liver and muscle glutathionine pools of molting and nonmolting white-crowned sparrows. *Physiological Zoology* 58, 646–654.

Murphy, M.E. and King, J.R. (1987) Dietary discrimination by molting white-crowned sparrows given diets differing only in sulfur amino acid concentration. *Physiological Zoology* 60, 279–289.

Murphy, M.E. and King, J.R. (1990) Diurnal changes in tissue glutathione and protein pools of molting white-crowned sparrows – the influence of photoperiod and feeding schedule. *Physiological Zoology* 63, 1118–1140.

Murphy, M.E. and King, J.R. (1991) Protein intake and the dynamics of the postnuptial molt in white-crowned sparrows, *Zonotrichia leucophrys gambelii*. *Canadian Journal of Zoology – Journal Canadien de Zoologie* 69, 2225–2229.

Murphy, M.E. and King, J.R. (1992) Energy and nutrient use during moult by white-crowned sparrows *Zonotrichia leucophrys gambelii*. *Ornis Scandinavica* 23, 304–313.

Murphy, M.E. and Pearcy, S.D. (1993) Dietary amino acid complementation as a foraging strategy for wild birds. *Physiology and Behavior* 53, 689–698.

Murphy, M.E. and Taruscio, T.G. (1995) Sparrows increase their rates of tissue and whole-body protein synthesis during the annual molt. *Comparative Biochemistry and Physiology A – Physiology* 111, 385–396.

NRC (1994) *Nutrient Requirements of Poultry.* National Academy Press, Washington, DC.

O'Malley, P.J. (1996) An estimate of the nutritional requirements of emus. In: Deeming, D.C. (ed.) *Improving Our Understanding of Ratites in a Farming Environment.* Ratite Conference, Oxford, pp. 92–108.

Otto, M., Snejdarkova, M. and Simon, O. (1989) Utilisation of free and protein dietary lysine in chicks estimated with isotopes. *British Poultry Science* 30, 633–639.

Parish, J.W. and Martin, E.W. (1977) The effect of dietary lysine level on the energy and nitrogen balance of the dark-eyed junco. *Condor* 79, 24–30.

Penz, A.M. and Jensen, L.S. (1991) Influence of protein concentration, amino acid supplementation, and daily time of access to high-protein or low-protein diets on egg weight and components in laying hens. *Poultry Science* 70, 2460–2466.

Robbins, K.R. and Baker, D.H. (1980) Evaluation of the resistance of lysine sulfite to Maillard destruction. *Journal of Agricultural Food Chemistry* 28, 25–29.

Robbins, K.R., Baker, D.H. and Norton, H.W. (1977) Histidine status in the chick as measured by growth rate, plasma free histidine and breast muscle carnosine. *Journal of Nutrition* 107, 2055–2061.

Rosebrough, R.W., McMurtry, J.P., Mitchell, A.D. and Steele, N.C. (1988) Chicken hepatic metabolism *in vitro*: protein and energy relations in the broiler chicken – VI. Effect of dietary protein and energy restrictions on *in vitro* carbohydrate and lipid metabolism and metabolic profiles. *Comparative Biochemistry and Physiology* 90B, 311–316.

Roudybush, T.E. (1986) Weaning of cockatiels. In: *Proceedings 35th Western Poultry Disease Conference*. University of California, Davis, California, pp. 162–170.

Roudybush, T.E. and Grau, C.R. (1991) Cockatiel (*Nymphicus hollandicus*) nutrition. *Journal of Nutrition* 121, S206.

Sasse, C.E. and Baker, D.H. (1972) The phenylalanine and tyrosine requirements and their interrelationship for the young chick. *Poultry Science* 51, 1531–1536.

Sedinger, J.S. (1984) Protein and amino acid composition of tundra vegetation in relation to nutritional requirements of geese. *Journal of Wildlife Management* 48, 1128–1136.

Skadhauge, E. (1983) Formation and composition of urine. In: Freeman, B.M. (ed.) *Physiology and Biochemistry of the Domestic Fowl*, Vol. 4. Academic Press, London, pp. 108–136.

Stevens, L. (1996) *Avian Biochemistry and Molecular Biology*. Cambridge University Press, Cambridge.

Stewart, D.J., Holmes, W.H. and Fletcher, G. (1968) The renal excretion of nitrogenous compounds by the duck maintained on freshwater and on hypertonic saline. *Journal of Experimental Biology* 50, 527–539.

Sugahara, M. and Kandatsu, M. (1976) Glycine serine interconversion in the rooster. *Agricultural Biological Chemistry* 40, 833–839.

Swain, S.D. (1992) Flight muscle catabolism during overnight fasting in a passerine bird, *Eremophila alpestris*. *Journal of Comparative Physiology B – Biochemical Systemic and Environmental Physiology* 162, 383–392.

Taruscio, T.G. and Murphy, M.E. (1995) 3-Methylhistidine excretion by molting and non-molting sparrows. *Comparative Biochemistry and Physiology A – Physiology* 111, 397–403.

Taylor, E.J., Nott, H.M.R. and Earle, K.E. (1994) Dietary glycine: implications in growth and development of the budgerigar (*Melopsittacus undulatus*). *Journal of Nutrition* 124, 2555s–2558s.

Thomas, D.W., Bosque, C. and Arends, A. (1993) Development of thermoregulation and the energetics of nestling oilbirds (*Steatornis caripensis*). *Physiological Zoology* 66, 322–348.

Turcek, F.J. (1966) On plumage quantity in birds. *Ekologia Polska Series A* 14, 617–633.

Ueda, H., Yabuta, S., Yokota, O. and Tasaki, I. (1981) Involvement of feed intake and feed utilization in the growth retardation of chicks given excessive amounts of leucine, lysine, phenylalanine or methionine. *Nutrition Reports International* 24, 135–144.

Underwood, M.S., Polin, D., O'Handley, P. and Wiggers, P. (1991) Short term energy and protein utilization by budgerigars fed isocaloric diets of varying protein concentrations. *Proceedings of the Association of Avian Veterinarians* 91, 227–237.

Wheeler, K.B. and Latshaw, T.D. (1981) Sulfur amino acid requirements and interactions in broilers during two growth periods. *Poultry Science* 60, 228–236.

White, T.C.R. (1993) *The Inadequate Environment: Nitrogen and the Abundance of Animals.* Springer-Verlag, Berlin.

Wiggins, D., Lund, P. and Krebs, H.A. (1982) Adaptation of urate synthesis in chicken liver. *Comparative Biochemistry and Physiology* 72B, 565–568.

Williams, T.D. (1996) Variation in reproductive effort in female zebra finches (*Taeniopygia guttata*) in relation to nutrient-specific dietary supplements during egg laying. *Physiological Zoology* 69, 1255–1275.

Wu, G., Flynn, N.E., Yan, W. and Barstow, D.G. (1995) Glutamine metabolism in chick enterocytes – absence of pyrroline-5-carboxylase synthase and citrulline synthesis. *Biochemical Journal* 310, 1055–1065.

Yamane, T., Ono, K. and Tanaka, T. (1979) Protein requirements of laying Japanese quail. *British Journal of Poultry Science* 20, 379–383.

Yeh, Y.Y. and Leveille, G.A. (1969) Effect of dietary protein on hepatic lipogenesis in the growing chick. *Journal of Nutrition* 98, 356–366.

CHAPTER 7
Lipids

Dietary lipids supply energy, essential fatty acids, and pigments. Lipids are the most concentrated energy source that a bird can consume and are typically digested and metabolized with high efficiency. Lipids are the only dietary component that is deposited intact into tissues with little or no modification. Storage pools of lipids are the primary energy supply that fuels a bird between meals, throughout migration, and during embryonic development.

Essential Fatty Acids

Birds are able to synthesize saturated fatty acids *de novo* and to oxidize them to mono- and diunsaturated fatty acids up to the ninth carbon inward from the carboxyl end (Δ^9). For example, avian hepatic cells can synthesize stearic acid (18 carbons in length) and introduce a double bond between the eighth and ninth carbons to give oleic acid ($C_{18}:1\Delta^9$). Birds lack the enzymatic capacity to introduce double bonds past the Δ^9 (Fig. 7.1). Thus, they cannot use stearic acid to synthesize linoleic acid ($C_{18}:2\Delta^{9,12}$) or α-linolenic acid ($C_{18}:3\Delta^{9,12,15}$). Only plants have the enzymes capable of inserting Δ^{12} or Δ^{15} double bonds into C_{18} fatty acids and consequently linoleic and linolenic acids are essential fatty acids for birds. Once consumed, these two fatty acids can be further metabolized by enzymes within the endoplasmic reticulum of chicken hepatocytes (Figs 7.1 and 7.2). Linoleic acid can be desaturated between the sixth and seventh carbons to γ-linolenic acid ($C_{18}:3\Delta^{6,9,12}$), which may be elongated by two carbons and desaturated again to give arachidonic acid ($C_{20}:4\Delta^{5,8,11,14}$). Arachidonic acid may be further metabolized to C_{22} fatty acids, such as the prostaglandins. Likewise, dietary α-linolenic acid can be elongated and desaturated by hepatocytes to give eicosapentanoic acid ($C_{20}:5\Delta^{5,8,11,14,17}$) and then further metabolized to other C_{22} fatty acids.

Few species of birds other than Galliformes have been examined for their capacity to elongate and desaturate linoleic and α-linolenic acid. Given the variability in this capacity among other vertebrate taxa, it should not be assumed that all avian species are able to synthesize all of the polyunsaturated fatty acids (PUFAs) that are metabolically required. Carnivorous mammals and fish have a dietary requirement for arachidonic acid and it is possible that some avian species that consume diets high in animal fat may have lost their capacity for arachidonic acid synthesis. Thus, all birds have a dietary requirement for linoleic

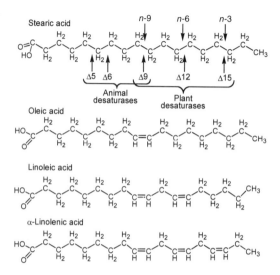

Fig. 7.1. The structure of stearic acid (18 carbons in length) showing the positions of the bonds that are subject to desaturation. The *n* designation indicates the number of carbons from the methyl end. The Δ designation indicates the number of carbons from the carboxyl end. The desaturases of birds can introduce double bonds at the Δ5, Δ6, and Δ9 positions, but not beyond. Birds can synthesize stearic acid and convert it, using their Δ9 desaturase, to oleic acid prior to storage in adipose tissue. Birds lack the Δ12 and Δ15 desaturases needed to synthesize linoleic and α-linolenic acid.

acid and α-linolenic acid, but a prudent diet would also include arachidonic acid and eicosapentanoic acid.

It is common in nutrition to name fatty acids by referring to the position of double bonds relative to the methyl end of the molecule (*n* or ω nomenclature). This is appropriate because the methyl end is not subjected to elongation and desaturation and determines nutritional essentiality. Using this nomenclature, linoleic acid and all of its elongation products are members of the *n*-6 family of fatty acids, and α-linolenic acid and all of its elongation products are members of the *n*-3 family of fatty acids. This precursor–product relationship means that the dietary requirement for linoleic acid is decreased by the consumption of other *n*-6 PUFAs, such as γ-linolenic acid or arachidonic acid. In fact, in Japanese Quail, γ-linolenic can substitute for linoleic acid with a bioactivity of greater than 2 (Murai *et al.*, 1995). Likewise, the consumption of other *n*-3 PUFAs (e.g., eicosapentanoic) decreases the requirement for α-linolenic acid. For example, the fat of marine animals is high in C_{20} and C_{22} PUFAs of the *n*-3 series and they diminish the need for α-linolenic acid. Dietary *n*-3 PUFAs do not decrease the requirement for dietary *n*-6 PUFAs, because these families are not inter-convertible. The *n*-9 family of fatty acids can be synthesized by the bird and these fatty acids are not required in the diet.

The C_{18}, C_{20}, and C_{22} PUFAs are stored in phospholipids of cell membranes, where they contribute to the structural integrity and fluidity. They may be

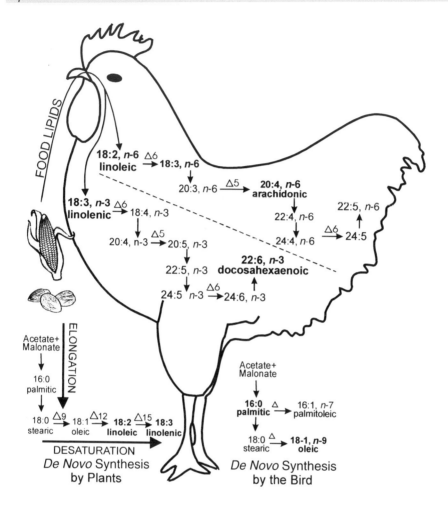

Fig. 7.2. Pathways of fatty acid elongation and desaturation. Birds can elongate and desaturate both dietary and endogenously synthesized fatty acids. Fatty acids shown in bold type are the dominant end products of the individual pathways. (Adapted with permission from Walzem, 1996.)

released by the action of phospholipases as important events in cellular communication. Released PUFAs serve as precursors for the eicosanoids: prostaglandins, leukotrienes, and thromboxanes. In birds, the eicosanoids regulate almost every physiological system, including oviposition, embryonic development, growth, immunity, bone development, thermoregulation, and behavior (Watkins, 1991, 1995).

Requirement and deficiencies

The absolute requirement for linoleic acid is considerably greater than that for α-linolenic acid. Precise requirement levels are difficult to establish, due to the large carry-over from the egg to the chick and long-lasting stores in adipose tissue and cell membranes. Quantitatively, the primary need for essential fatty acids is for the synthesis of phospholipids, which are incorporated into cell membranes and yolk lipids. Thus, nutritional needs are driven by the rate of growth and egg production. Assuming normal enrichment of the yolk and typical reserves at hatching, young growing chickens, Japanese Quail, Northern Bobwhite Quail, and turkeys require linoleic acid at 1% of the diet (NRC, 1994). This level is also adequate for reproduction. Almost all natural foods that meet the energy requirement of birds meet the requirement for these two fatty acids, and frank deficiencies are rare. Seeds, some fruits, and animal prey are particularly rich sources. Artificial diets that utilize partially purified or fat-extracted ingredients (e.g., soybean meal) often require the inclusion of vegetable oil or animal fat to provide essential fatty acids. Penguins consuming smelt, which are very low in linoleic acid, exhibit reproductive problems that can be corrected by supplementation of corn oil (Gailey, 1975).

Deficiencies of linoleic acid are most readily identified by a slow rate of growth and the accumulation of fat in the liver. Diagnosis can be confirmed by biochemical analysis of hepatic phospholipids, which will be low in linoleic and arachidonic acids, but enriched in eicosatrienoic acid (20:3 *n*-9). Chronic deficiency symptoms include increased susceptibility to infectious diseases and greater water consumption. In reproductively active birds, a deficiency is expressed as impaired spermatogenesis in the male, and the production of fewer and smaller eggs in the female. Fertilized eggs often have higher embryonic mortality (Menge *et al.*, 1965; Balnave, 1970). Dietary α-linolenic acid is incorporated into the myelin and synaptosomal lipids and found in highest concentrations within the brain and retina. Dietary α-linolenic acid protects chicks against encephalomalacia induced by vitamin E deficiency; however, the absolute requirement has yet to be determined (Budowski and Crawford, 1986; Anderson *et al.*, 1989).

The ratio of *n*-3 to *n*-6 fatty acids in the diet is an important modulator of many physiological processes and is typically more nutritionally relevant than the absolute concentration of these fatty acids. This is because the ratio of *n*-3 to *n*-6 fatty acids in the diet is reflected in the composition of membranes and consequently the type and rate of the different eicosanoids produced for regulatory purposes. The modulation of the avian immune response by the proportion of dietary essential fatty acids illustrates this important relationship. High ratios of *n*-6 to *n*-3 fatty acids are proinflammatory and low ratios are antiinflammatory. Also, in young growing chickens, feeding dietary *n*-3 fatty acids from fish enhances the antibody response to sheep red blood cell (SRBC) vaccinations but suppresses rates of lymphocyte replication following stimulation. These dietary manipulations occur at levels well above the requirement and markedly impact resistance to infectious diseases (Klasing, 1997).

Birds do not have a requirement for triglycerides *per se*, but triglycerides facilitate the absorption of fat-soluble vitamins and are the densest energy source found in foods. Their presence in the diet can also improve its palatability for some species. Thus, from a practical point of view, triglycerides are an important, though not absolutely essential, component of the diet.

Carotenoid Pigments

Birds are the most conspicuous and colorful of the vertebrates. The unique colors of their feathers, skin, eyes, scales, beaks, and egg yolks are often dependent upon pigments of dietary origin. Dietary pigments utilized by birds for their coloration are polyenes, which as a class, are called carotenoids (Fig. 7.3). Each carotenoid has a specific color, ranging from brilliant reds, oranges, and yellows to violet. The understanding of the specific carotenoids required for appropriate coloration of birds as well as the foods that provide these pigments is an important and developing area of avian nutrition. Although carotenoids are not responsible for all of the pigments that give birds their beautiful appearance, they provide many of the most distinctive and conspicuous colors. Dark colors (e.g., black, brown, gray, and related tints) are provided by melanin and porphyrin pigments, often complexed with trace minerals. Structural elements related to the organization and positioning of keratin fibrils can provide white and blue colors. Carotenoids act synergistically with melanin pigments and structural colors to give green. Even white or dark-colored birds use dietary carotenoids to pigment the retinal oil droplets of their eyes and their egg yolks (Fox, 1979; Durrer, 1986; Brush, 1990).

Carotenoid pigments are synthesized only by plants and are conspicuously present in many blossoms, pollens, seeds, fruits, leaves, and roots (Table 7.1). Many animals (e.g., insects, mollusks, crustacea, fish) concentrate and further metabolize the carotenoids they consume, providing a rich food source for birds. However, some animals (e.g., laboratory rats and mice) selectively excrete carotenoids and are a very poor source. Carotenoids are subdivided into two categories: carotenes and xanthophylls. Carotenes have unmodified cyclohexenyl rings and include α- and β-carotene. β-Carotene is used as a precursor for vitamin A in all species and as a pigment in some. Xanthophylls are oxycarotenoids, because they have alcohol, keto, or ester groups on their terminal cyclohexenyl rings. The positioning and type of these groups determine the color of xanthophylls, and complexing of xanthophylls with proteins may shift the color or give iridescence. Lutein, zeaxanthin, astaxanthin, rhodoxanthin, and capsanthin are prevalent xanthophylls in foods consumed by many captive and free-living birds (Table 7.1). These dietary carotenoids may be used directly to pigment tissues, or they may be metabolized to other carotenoids prior to incorporation into tissues. Consequently, the carotenoids that pigment tissues (Table 7.2) may differ from those consumed. For example, canthaxanthin, a red pigment found in the feathers of flamingos and other pink or red birds, is not produced in high amounts by most plants, with the exception of some algae. Canthaxanthin and astaxanthin are produced commercially and often supplemented to the diets of birds kept in captivity.

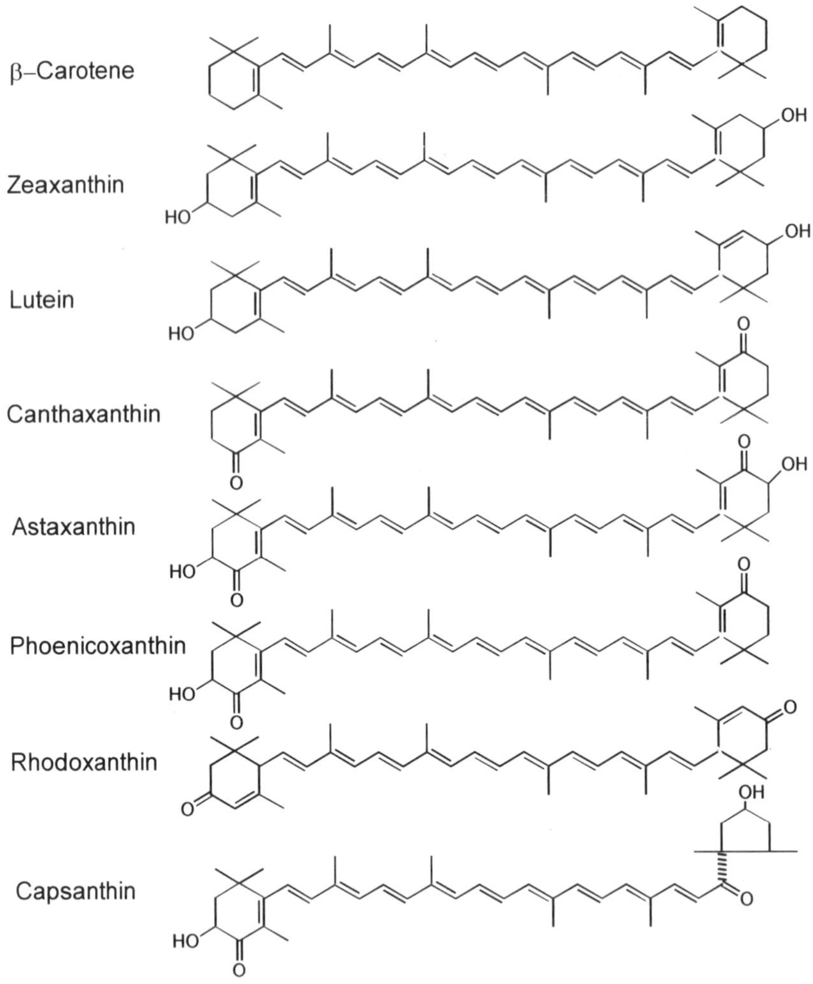

β–Carotene

Zeaxanthin

Lutein

Canthaxanthin

Astaxanthin

Phoenicoxanthin

Rhodoxanthin

Capsanthin

Fig. 7.3. Carotenoids commonly used for pigmentation by birds. In their free form, these pigments have the following color ranges: β-carotene – pale yellow to pale orange; zeaxanthin – golden yellow; lutein – lemon yellow; canthaxanthin – orange to scarlet red; astaxanthin – salmon pink; phoenicoxanthin (adonirubin) – orange to red; rhodoxanthin – deep red; capsanthin – paprika red. Binding of a carotenoid to a protein shifts its color. Additionally, the chiral configuration and isomer mixture influence the precise color.

Digestion and metabolism

The deposition of pigment into specific tissues is dependent upon three primary factors: (i) the quantity of the appropriate carotenoid in the diet; (ii) the bird's capacity to digest and absorb specific carotenoids and metabolize them to the correct chemical form; and (iii) the capacity of specific tissues (e.g., feather follicles, epidermal cells, adipocytes) to take up carotenoids and insert them into the structure of growing tissue. These three process are highly regulated, as is evident by the fact that the feather color changes with age, season, and

Table 7.1. Carotenoid content of some pigment sources (from Borenstein and Bunnell, 1966; Scott *et al.*, 1982; Torrissen *et al.*, 1989).[*]

Pigment source	Carotenoid	Amount (mg kg^{-1})
Krill (*Euphausia* spp.)	Astaxanthin	22–77
Copepod	Astaxanthin	39–84
Shrimp	Astaxanthin	100
Salmon	Astaxanthin	5
Silk caterpillar (*Bombyx mori*)	Total xanthophyll	240
Mussel (*Mytilus californianus*)	Total carotenoids	22–96
Algae (*Spirulina* spp., dry)	β-Carotene	434
	Echinenone	118
	Cryptoxanthin	389
	Zeaxanthin	80
Apples	Total carotenoids	1–5
Apricots	Total carotenoids	35
Black figs	Total carotenoids	9
Cherries	Total carotenoids	5–11
Cranberries	Total carotenoids	5.8
Grapes	Total carotenoids	2
Japanese persimmon	Total carotenoids	51
Strawberry	Total carotenoids	1
Navel orange pulp	Total carotenoids	23
Tomatoes	Total carotenoids	51
Red pepper	Total carotenoids	127–248
Brussel sprouts	Total carotenoids	1–6
Lettuce	Total carotenoids	8–68
Spinach	Total carotenoids	26–76
Alfalfa meal	Total xanthophylls[†]	220–330
Clover (dry)	Total xanthophylls	490
Seaweed	Total xanthophylls	350
Paprika	Total xanthophylls[‡]	275
Chili peppers (dry)	Total xanthophylls	185
Carrots (dry)	Total xanthophylls	65
Corn	Total xanthophylls	20–25
Red sorghum	Total carotenoids	2
Wheat (hard red winter)	Total carotenoids	2
Marigold petals	Total xanthophylls[†]	8000

[*]Values are on a wet basis unless otherwise stated.
[†]Primarily lutein.
[‡]Primarily capsanthin.

physiological state, and it differs between sexes in many species.

Carotenoids are found in the food either in the free form or esterified to fatty acids (Hamilton, 1992; Hencken, 1992). Most carotenoids in fruits and seeds are found in the esterified state and most in stems and leaves are in the free form. Those in the free form are more digestible than those that are esterified. This is

Table 7.2. Distribution of carotenoids in birds (from Fox, 1979; Goodwin, 1984; Brush 1990).

Species	Color	Carotenoids
Caribbean Flamingo	Pink feathers	Astaxanthin, canthaxanthin, adonirubin
Canary	Yellow feathers	Lutein and undefined canary xanthophylls
Canary	Golden-yellow feathers	Zeaxanthin, lutein
Cedar Waxwing	Red feathers	Astaxanthin
Northern Cardinal	Red feathers	Canthaxanthin
Orange Dove	Orange feathers	Rhodoxanthin
Resplendent Quetzel	Red feathers	Canthaxanthin
Roseate Spoonbill	Pink feathers	Canthaxanthin, astaxanthin
Scarlet Ibis	Pink-red feathers	Canthaxanthin, astaxanthin
Toco Toucan	Yellow and orange feathers and beak	Lutein, rhodoxanthin, zeaxanthin
Chicken	Golden-yellow egg	Lutein, zeaxanthin
Chicken	Yellow skin, beak	Lutein

because carotenoid esters must be hydrolyzed by specific intestinal esterases prior to absorption and this cleavage is the rate-limiting step in utilization. In the chicken, esterified carotenoids are less digestible as the chain length of the esterified fatty acid increases. Those free carotenoids in the diet and those released by hydrolysis of dietary carotenoid esters are absorbed along with other lipids in the diet and are transported by lipoproteins in the blood. Dietary and physiological factors that improve fat digestion improve carotenoid digestion. Each carotenoid appears to have its own individual pattern of absorption, plasma transport, and metabolism and there are considerable species differences in the types of carotenoids that are preferentially absorbed and metabolized. For example, canaries and chickens efficiently absorb and utilize xanthophylls to pigment their tissues, but they do not efficiently absorb β-carotene and metabolize it to the appropriate xanthophylls. Conversely, flamingos effectively absorb and utilize β-carotene as a precursor for skin and feather pigments (e.g., astaxanthin), but they do not readily utilize many dietary xanthophylls, including astaxanthin (Fox, 1979).

The liver is especially active at taking up carotenoids and it has a variety of oxidative or reductive enzymes that modify carotenoid structure. This metabolism changes the color of the carotenoids and impacts the capacity of target tissues (e.g., feather follicles) to utilize them as pigments. Very simple chemical changes can make profound color changes. For example, the seasonal change from brilliant red to brilliant yellow in male Scarlet Tanagers is due to a shift from the incorporation of canthaxanthin (4,4'-diketo-β-carotene) into feathers to the incorporation of isozeaxanthin (4,4'-dihydroxy-β-carotene). The female also uses isozeaxanthin to pigment her feathers, but adds a form of melanin, which combines to give a green hue (Brush, 1990).

Within an individual, the deposition of specific carotenoids into newly formed tissue may differ sharply between different parts of the body. This gives rise to color patterns in the plumage or skin. For example, the distinct color patterns of the Gouldian Finch arise from the carotenoid deposition character-istics of feather follicles in specific regions of the body. The feather follicles that produce the yellow ventral feathers integrate lutein, whereas the follicles that produce the red face feathers integrate canthaxanthin. Their lilac and green feathers also contain lutein, but its hue is changed by the way it is complexed with keratin. It is not known if the feather follicles can further metabolize carotenoids in order to change feather color, or if color discrimination only occurs at the level of carotenoid uptake from the blood.

More subtle color differences within an individual or between individuals in a species arise from changes in the amount of carotenoids incorporated, shifting in the location of the carotenoids on the keratin fibrils of the feather, or masking or unmasking of carotenoids with melanin pigments. The sudden appearance of a novel pigment in the diet may result in the expression of unusual coloration responses. For example, female Red Crossbills normally have a dull olive-gray color, but when they are fed high amounts of canthaxanthin they turn the bright red-orange color exhibited by males. The normal color difference between sexes is probably due to their differing capacities to metabolize normal dietary carotenoids to canthaxanthin, and red females occur only when a dietary supply of canthaxanthin bypasses this control point. As the foods normally consumed in the wild do not contain canthaxanthin, the sexes normally have different colors (Brush, 1990; Hill and Benkman, 1995).

Some species are not highly discriminatory in the type or quantity of carotenoids that they absorb and deposit. For example, the color of a chicken's skin, adipose tissue, and egg yolk directly reflects the types and levels of xanthophylls found in its food. The intensity of pigmentation of the yolk and skin of chickens increases and then plateaus as the dietary xanthophyll concentration increases (Fig. 7.4). At dietary levels of between 3 and 10 mg kg^{-1} of xanthophylls (lutein, canthoxanthin, and zeaxanthin), about 40% of the intake is deposited in the yolk, while only 15% is deposited when the daily intake is 50 mg kg^{-1}. Changing the proportion of the specific xanthophylls changes the color of the yolk, such that colors ranging from lemon yellow to golden yellow to orange-red can be obtained. Poultry nutritionists pay careful attention to the amounts of specific carotenoids provided in the feed, so that egg yolks and carcass fat are pigmented with the intensity and hue that is appropriate for various consumer preferences. Some consumers prefer to purchase poultry meat or eggs that have little pigmentation, whereas others associate such products with unwholesomeness. Some consumers want a 'sun'-colored yolk, others prefer a more yellow color. For this reason, poultry nutritionists have developed empirical equations that permit the formulation of specific xanthophyll combinations and levels to give specific coloration of eggs or carcasses destined for specific markets (Marusich and Bauernfeind, 1981; Fletcher, 1989).

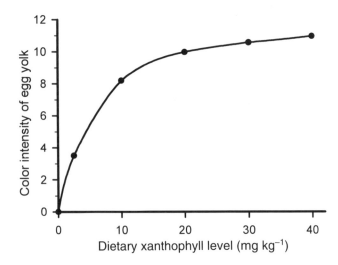

Fig. 7.4. Relationship between the level of xanthophylls in the diet of chickens and the intensity of color in the yolks of their eggs (Adams, 1985).

Environmental conditions impact the ability of the bird to express normal coloration (Marusich and Bauernfeind, 1981; Hudon, 1994; Allen, 1996). For example, Red Crossbills and Common Redpolls lose their distinctive red feathers and replace them with unpigmented ones when stressed due to being kept in captivity, even when the diet is rich in xanthophylls. Presumably, the stresses of captivity modify hormones, such as thyroxine and corticosterone, which impact xanthophyll metabolism. The inflammatory stress of infection causes the release of cytokines that depress the absorption of xanthophylls or interfere with their metabolism. A depression in carotenoids is especially conspicuous during infection with parasites, such as coccidia. Further, xanthophylls in foods or tissues lose their color when they are oxidized, and adequate levels of dietary antioxidants help maintain pigment color. Long-term storage of avian foods results in the oxidation of carotenoids and losses in their pigmentation value (Fig. 7.5).

In addition to their role in providing color, carotenoids can serve an antioxidative function (Lawlor and Obrien, 1995). For example, astaxanthin and, to a lesser extent, β-carotene quench singlet oxygen atoms and scavenge free radicals induced by paraquat in avian cells. The antioxidant activity of some carotenoids is superior to that of vitamin E, though their distribution limits their functionality. Carotenoids also absorb light and have a photoprotective role.

Requirement and deficiencies

Captive birds can live, reproduce, and appear healthy when consuming a diet devoid of carotenoids (but with vitamin A). In captivity, carotenoids are mostly an aesthetically and commercially appealing decoration, which also provides a margin of safety against oxidative stress and photochemical damage, but they

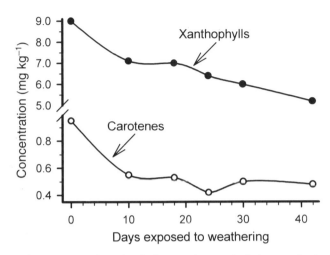

Fig. 7.5. Loss of carotenes and xanthophylls in sorghum grain during weathering in the field (Blessin *et al.*, 1958).

cannot be considered nutritionally essential. In many species of free-living birds, carotenoids are required for breeding success, as poorly colored birds are less likely to breed. Thus, carotenoids might be considered to be required for free-living birds in order to communicate reproductive fitness.

The dietary levels of carotenoids needed for pigmentation of poultry products have been determined for a variety of situations (Marusich and Bauernfeind, 1981; Fletcher, 1989; Hencken, 1992). The requirement for maximal yolk color of chickens laying an egg daily is about 40–50 mg kg^{-1} of total dietary xanthophylls rich in lutein, zeaxanthin, or canthaxanthin. The intensity of yolk pigmentation is directly related to the amount of xanthophyll in the diet (see Fig. 7.4). If laying hens are switched to a xanthophyll-free diet, yolk pigmentation declines over a few days and within a week eggs with white yolks are laid. Clearly, tissue stores of xanthophylls are low relative to the large amounts needed for yolk pigmentation. If dietary carotenoid consumption resumes, yolks reach full pigmentation within a week's time.

Even at high dietary carotenoid levels, poultry divert the majority to yolk production and other tissues lose color because they do not receive a normal (maintenance-level) allocation. The rate of color loss in specific tissues is related to the rate of cell turnover in that area. The skin around the vent loses its carotenoid pigment fastest, followed by skin on the face. The beak begins to lose pigmentation at its base, where it is renewed, and if laying persists for 5 weeks the entire beak becomes unpigmented. The scales of the shanks lose pigment following 4–5 months of continuous egg laying. When laying stops, pigment reappears in the same order in which it disappears. Thus, a bird's current and past laying status can be determined by visual inspection of its pigmentation pattern (Wilson and Wakabayashi, 1956).

Setting a precise dietary level of carotenoids for optimal pigmentation is not currently possible for most species other than poultry. Considerably more research on the specific dietary carotenoids needed and on the quantitative relationship between dietary levels and appropriate pigmentation levels is needed. These relationships may often be more complex than they initially appear. For example, Caribbean Flamingos deposit both canthaxanthin and astaxanthin in their feathers to provide a distinct shade of pink. The canthaxanthin can be synthesized from β-carotene or absorbed from the diet. Dietary astaxanthin is not well absorbed and these birds synthesize it from canthaxanthin. Birds fed astaxanthin as their sole dietary pigment are very poorly colored. However, feeding canthaxanthin at about 25 mg kg^{-1} dry matter can supply most of the pigmentation needs for flamingos, as well as for Scarlet Ibis, House Finches and tanagers, but dose–response relationships have not been adequately characterized. Furthermore, these levels of canthaxanthin do not supply similar coloration to Carmin Bee-eaters. Thus, dietary carotenoid requirements may not always be predictable from the color of feathers or the spectrum of tissue carotenoids (Fox, 1979; Dierenfeld and Sheppard, 1996).

If the diet is deficient in carotenoids, the limited stores are soon depleted and newly formed integument lacks carotenoid pigmentation. Over time, these tissues become the color of background structural components, often a pale yellow to white. For example, the reddish shank color of domestic pigeons maintained on a carotenoid-deficient diet gradually becomes white over a period of about 2 weeks. Similarly, Canaries or flamingos maintained on a carotenoid-deficient diet grow white feathers following a molt (Brockmann and Volker, 1934; Fox, 1979; Hill *et al.*, 1994).

Metabolism and Storage of Triglycerides

Energy is stored in adipose tissue as triglycerides when food is readily available; conversely, triglycerides are mobilized to provide energy when food is limited. Triglycerides are also stored in the yolk of the egg and supply most of the energy to the developing embryo. In addition to serving as an energy source, lipids serve as important structural components of cell membranes. Adipose tissue has insulating properties important in thermoregulation and protection of vital internal organs. Adipose levels are extremely variable among birds, normally ranging between 3 and 50% of body mass, depending on species, diet, and season (Table 7.3). At least 1.5% body fat is required for survival in Eurasian Oystercatchers and Lapwings. At this low level, essentially all of the triglycerides are located in cell membranes and depot fat is not present (Marcstrom and Mascher, 1979).

Transport

Fatty acids, triglycerides, and other lipids are almost completely insoluble in blood plasma and tissue fluids. Free fatty acids are solubilized by binding to fatty

Table 7.3. Lipid content of birds in various physiological states (g lipid g^{-1} lean dry weight) (from Blem, 1976).

Species	In migration		Not in migration
	Spring	Autumn	
Permanent residents			
Black-necked Stilt			0.23–0.53
Mockingbird			0.18
Brown Thrasher			0.28
Carolina Wren			0.35
House Sparrow			0.22–0.57
Yellow Bunting			0.18–0.38
Bullfinch			0.06–0.16
Cardinal			0.43
Short-range migrants			
Mourning Dove			0.31
Savannah Sparrow		0.77	0.33
Tree Swallow	0.34	0.20	0.17–0.43
White-throated Sparrow		0.20–0.68	
Dark-eyed Junco	0.41–0.44		0.27–0.61
Long-range migrants			
Ruby-throated Hummingbird		3.13	
Least Sandpiper			0.25
Arctic Tern			0.30
Red-eyed Vireo	0.54–0.67	0.74–1.39	
Ovenbird	0.45	0.86–1.51	
American Redstart	0.34–0.43	0.73–1.29	
Scarlet Tanager		1.17–2.16	0.22
Bobolink	0.87–0.97	2.13	
Dickcissel		0.37–0.53	0.12–0.66

acid-binding proteins within cells and to albumin in body fluids (Sams *et al.*, 1991). Triglycerides are transported within cells as components of cell membranes and in the blood as components of lipoproteins (Fig. 7.6). In birds, the primary lipoprotein classification consists of portomicrons, very-low-density lipoprotein (VLDL), low-density lipoprotein (LDL), and high-density lipoprotein (HDL). Portomicrons are synthesized during intestinal absorption of dietary fat, circulate in the blood, and are cleared by the liver (Fig. 7.7). The liver repackages dietary lipids and adds lipids synthesized in the hepatocytes to give VLDL, which is secreted into the blood. Apolipoprotein B on the surface of VLDL targets it to peripheral tissues, where the triglycerides can be used as a fuel source or they may be stored. Usually, the peripheral tissues remove only the triglycerides from the VLDL, leaving a smaller remnant, which is remodeled intravascularly to give

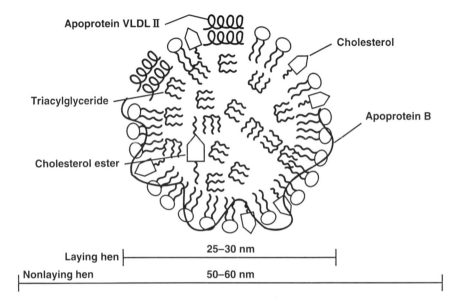

Fig. 7.6. Generalized structure of a triglyceride-rich lipoprotein. Surface components solubilize the nonpolar core components of triglyceride and cholesterol ester within the polar environment of the blood. Scale bars indicate the degree of reduction in the diameter of the very-low-density lipoprotein (VLDL) that is synthesized for yolk deposition. (From Walzem, 1996, with permission.)

a cholesterol ester-rich LDL, which delivers cholesterol to cells. The HDL redistributes lipids among the peripheral tissues and returns lipids to the liver for reuse or excretion (Griffin and Hermier, 1988; Walzem, 1996).

Triglycerides within circulating VLDL are removed within tissues by the action of lipoprotein lipase (Butterwith, 1988; Bensadoun, 1991; Cooper et al., 1992). This enzyme is found in the tissue capillaries and hydrolyzes triglycerides to free fatty acids and glycerol, which diffuse into the cells. Lipoprotein lipase is synthesized by tissues that use fatty acids as fuel (e.g., muscle) or that store them (e.g., adipose tissue). Following a meal, adipocyte lipoprotein lipase levels are very high, promoting the storage of triglycerides. During a fast, lipoprotein lipase activity in adipose tissue falls, but activity is high in muscle, so that triglycerides can be used as a fuel source. During periods of negative energy balance, hormone-sensitive lipase hydrolyzes triglycerides stored in adipose cells and the free fatty acids (nonesterified fatty acids) that are released are transported in the blood bound to albumin. Blood free fatty acid levels increase markedly during flight and during fasting (Savard et al., 1991; Swain, 1992).

Oxidation

Fatty acids that are oxidized can be of direct dietary origin (e.g., following a fatty meal), synthesized *de novo* by the liver (e.g., following a high-carbohydrate meal), or mobilized from adipose tissue (fasting). Most tissues can oxidize fatty acids, with skeletal and heart muscle using the most. When the liver oxidizes large

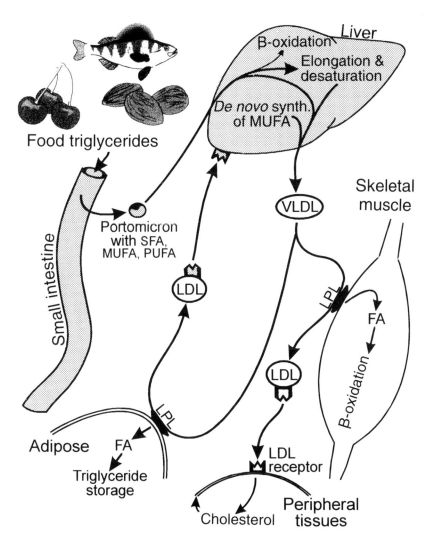

Fig. 7.7. Food triglycerides are absorbed in the small intestine and secreted into the portal vein as portomicrons. Following uptake by the liver, food-derived fatty acids may be selectively oxidized or they may be incorporated into very-low-density lipoproteins (VLDL), along with *de novo* synthesized fatty acids. Elongation and desaturation of dietary fatty acids may occur prior to incorporation in VLDL. Lipoprotein lipase (LPL) in peripheral tissues hydrolyzes triglycerides from VLDL and the resulting free fatty acids (FA) are taken up. Triglyceride-depleted VLDL circulates as low-density lipoprotein (LDL), which can be taken up by tissues through the LDL receptor. SFA, saturated fatty acids; MUFA, monounsaturated fatty acids; PUFA, polyunsaturated fatty acids.

amounts of fatty acids, it diverts much of the acetyl-coenzyme A (CoA) away from the citric acid cycle and synthesizes ketone bodies. In birds, β-hydroxybutyrate is the predominant ketone and acetoacetate is relatively minor. The liver does not utilize ketones and they are released into the blood. Ketones are water-soluble so they can be transported to peripheral tissues without packaging in lipoproteins. The brain and kidney are especially capable of oxidizing ketones, but many tissues can utilize some (Nehlig *et al.*, 1980). The high metabolic rate and relatively low glycogen stores of birds oblige the use of considerable triglycerides and ketones during a fast. During an overnight fast in small birds, such as the Horned Larks, glycogen reserves are used up within the first few hours and free fatty acids, which are mobilized from adipose-tissue triglycerides, become the primary fuel source. The liver produces large quantities of β-hydroxybutyrate within hours after the initiation of the fast. This ketosis occurs much faster in birds than in mammals and probably serves to decrease the use of glucose. Thus, ketosis spares the catabolism of body proteins from use for gluconeogenesis (Bailey and Horne, 1972; Swain, 1992). Glucagon and growth hormone are the two principal hormones that induce lipolysis in avian adipocytes. Growth hormone appears to be of primary importance in regulating energy metabolism during food restriction in American Kestrels (Lacombe *et al.*, 1993). Migration requires even greater mobilization and oxidation of fatty acids and β-hydroxybutyrate than an overnight fast. In Dark-eyed Juncos this mobilization is facilitated by activation of hormone-sensitive lipase in adipocytes (Savard *et al.*, 1991).

The biochemical details of β-oxidation of fatty acids in avian mitochondria and the further oxidation of the resulting acetyl-CoA are similar to those in other vertebrates (Stevens, 1996). Carnitine palmitoyltransferase, the enzyme that activates fatty acids with carnitine so that they can be transported into the mitochondria, is a key regulatory enzyme for fatty acid oxidation. The activity of this enzyme is higher in the flight muscles of hummingbirds than in any other vertebrate studied, reflecting the extreme demands for fatty acids as a fuel for flight (Suarez, 1995).

Synthesis

When energy intake exceeds energy needs, triglycerides may be synthesized from nonlipid substrates. The liver has a very high capacity to synthesize fatty acids from nonlipid substrates, a process called *de novo* synthesis. Adipose tissue, intestine, bone marrow, skin, and skeletal tissue also synthesize fatty acids (Pearce, 1980; Herzberg and Rogerson, 1990). The relative contribution of these tissues is species- and age-dependent. In the chicken, duck, turkey, pigeon, quail, Common Starling and White Wagtail, the liver is about 20 times more active in fatty acid synthesis than adipose tissue. The liver is particularly active in adult females producing eggs. The contribution of other tissues has not been adequately investigated, but in some species nonhepatic tissues are very active. For example, in growing Common Murre chicks, the skin contributes more to fatty acid synthesis than does liver on a whole-body basis.

The biochemical details of fatty acid synthesis are similar to those in mammals, with a few important exceptions (Donaldson, 1979, 1990; Goodridge, 1987). Reducing equivalents (nicotinamide adenine dinucleotide phosphate (NADPH)) needed for the reduction of carbon bonds are not generated at high rates by the pentose shunt and instead are generated by the action of malic enzyme and glucose-6-phosphate dehydrogenase. Fatty acid synthetase, acetyl-CoA carboxylase, and malic enzyme are the primary enzymes of fatty acid synthesis that are regulated by nutrition. These enzymes mediate the cessation of fatty acid synthesis during fasting and the high rate following refeeding. The type of fatty acids synthesized *de novo* in birds differs from that of many mammals. Avian fatty acids are characteristically monounsaturated, due to an active Δ^9-desaturase (stearyl-CoA desaturase), and oleic and palmitoleic acids predominate. In female Japanese Quail, hepatic Δ^9-desaturase activity increases upon sexual maturity, resulting in increased proportions of oleic and palmitoleic acids for incorporation into the egg yolk (Pageaux *et al.*, 1992).

Substrates used for *de novo* fatty acid synthesis depend on the diet. Domestic granivores, such as chickens and pigeons, use dietary carbohydrate, especially glucose, to synthesize fatty acids. High-fructose foods (nectar, fruits) are especially lipogenic. Faunivores may synthesize fatty acids from dietary amino acids. The high level of dietary fat consumed by some faunivores makes high rates of *de novo* fatty acid synthesis unnecessary, and in Common Murres and Barn Owls the capacity for fatty acid synthesis by the liver is low relative to that of grain-consuming chickens (Shaw *et al.*, 1978; Herzberg and Rogerson, 1990; Meyers and Klasing, 1998).

The uropygial gland, or preen gland, is the principal sebaceous gland in most birds and is very active in the synthesis of lipid. It is largest in aquatic species and is absent or very small in ratites, bustards, woodpeckers, and many pigeons and parrots. It synthesizes an oil that is spread over the surface of the feathers during preening and serves to waterproof feathers, enhance their suppleness, and kill contaminating bacteria. The oil consists mostly of waxes and glycerides, which are synthesized *de novo* in the secretory tubules of the gland. The fatty acids present in the waxes are unique, in that many are highly branched and, in the case of many waterfowl, are short in length (Jacob, 1976).

Storage

Most dietary energy that is consumed in excess of immediate needs is stored as triglycerides, particularly in adipose tissue (see Table 7.2). The reason for the conversion of dietary amino acids and carbohydrates to triglycerides prior to storage is to conserve weight and volume and to minimize interference with metabolic processes. Triglycerides are highly reduced relative to protein or glycogen and yield more than twice as much metabolizable energy per gram dry weight. Triglycerides are also hydrophobic and can be stored in a very compact form, with very little associated water. The combination of these two factors allows about tenfold more energy to be realized per gram of tissue when stored fatty acids are oxidized, compared with stored glycogen. The efficiency with which dietary energy is used to synthesize 1 kJ of fat is 74% versus 44% for protein accretion. Once deposited, adipose tissue has a considerably lower energy cost associated with maintenance than protein pools (Pullar and Webster, 1977; Weglarczyk, 1981).

Birds accumulate triglycerides within specific adipose depots in a precise sequence (McGreal and Farner, 1956). For example, White-crowned Sparrows first accrete lipid in subcutaneous layers associated with the feather tracts. Subsequent deposition occurs in the fucular region (claviculocoracoid fat pad) and then in the abdominal and thoracic cavities. Adipose tissue is also deposited within the skeleton, is associated with the mesentery of the intestines, and is deposited in the connective tissue of most organs. Within a fat depot, the adipocytes vary in size, depending on the amount of triglyceride stored. They continue to fill until a maximal amount of triglyceride is accumulated (hypertrophy); additional fat can also be stored in newly recruited adipocytes (hyperplasia), especially in younger birds (Johnston, 1973; Butterwith, 1988; Foglia et al., 1994). A typical adipose cell is about 90% lipid and the triglycerides are relatively inert. The length of time an average triglyceride molecule resides in the cell without being hydrolyzed by hormone-sensitive lipase is known as the turnover rate. In growing chickens, the turnover rate is about 23 days. This very slow turnover rate translates into minimal costs of maintenance for adipose tissue and contributes to the energetic efficiency of storing energy as triglycerides.

NUTRITIONAL INFLUENCES

The amount of body fat is mostly affected by the level of metabolizable energy intake relative to energy expenditure. This relationship depends on appetite and food availability. However, the composition of the diet impacts the amount of fat that is deposited, independent of energy intake. For example, the protein-to-calorie ratio directly affects fat deposition (Fig. 7.8). Wide ratios (high dietary protein) minimize fat storage and enhance muscle deposition, whereas narrow ratios result in more adipose tissue (Rosebrough and McMurtry, 1993). Dietary protein in excess of the requirement for amino acids increases insulin-like growth factor-I levels, and this hormone enhances muscle deposition and inhibits fat deposition.

The balance of amino acids in the dietary protein also influences fat deposition. Poorly balanced proteins often promote fat deposition. A marginal deficiency of methionine or lysine does not reduce energy intake but supports lower levels of skeletal muscle and greater fat deposition. Presumably, energy normally used to deposit and maintain skeletal muscle is deposited as fat, due to the amino acid deficiency.

The amount of dietary fat has little effect on fat deposition as long as the ratio of protein to metabolizable energy of the diet is not changed. In other words, substituting fat for an equivalent number of calories of carbohydrate does not increase fat deposition. This is because dietary fat inhibits de novo fatty acid synthesis from carbohydrate, resulting in similar amounts of fat available for deposition (Laurin et al., 1985). The deposition of dietary fat instead of newly synthesized fatty acids often changes the composition of the triglycerides in the adipose tissue.

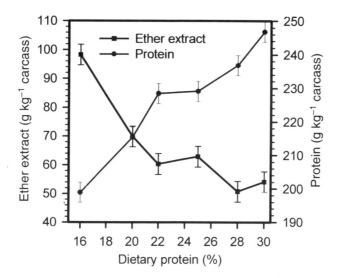

Fig. 7.8. Japanese Quail chicks were fed diets based on corn, soybean meal, sunflower meal, and fish meal to give dietary protein levels ranging from 16 to 30% of the diet, but similar energy levels (12 MJ kg^{-1}) for 5 weeks. Their carcass protein increased and their carcass fat (ether extract) decreased in proportion to the dietary protein content (Kirkkpinar and Oguz, 1995).

DEVELOPMENTAL, SEASONAL, AND REPRODUCTIVE INFLUENCES

Fat deposition follows a particular developmental and seasonal pattern in most birds. This pattern is variable depending on the species, but, in general, fat deposition is often intense during maturation, in preparation for migration or for overwintering in temperate climates, and in females prior to breeding.

Most chicks hatch with very low levels of adipose tissue, but build up fat reserves prior to fledging. Altricial species typically deposit larger lipid reserves than precocial species (Cherel and Le Maho, 1985; Blem, 1990; Ricklefs and Schew, 1994). This is particularly the case for chicks whose parents temporarily neglect them (e.g., Procellariiformes) or who need an extended period of time to develop foraging skills (e.g., Pelecaniformes). For example, chicks of some pelagic species deposit very large amounts of fat, which buffer the variable period between parental feedings. In the case of nestling Leach's Storm-Petrels, twice as much fat can be accumulated as their fat-free dry weight. Once the chicks have reached sufficient size and maturity to fledge, their parents stop feeding them and their fat stores decrease over a period of about 10 days, giving a body weight sufficiently low to permit flight. The extreme case of fat deposition is represented by the chicks of King Penguins, which have sufficient adipose reserves to last more than 5 months.

Winter fat deposition often occurs in small nonmigrating birds of the temperate zone, and tropical birds may deposit fat in preparation for seasonal periods of low food availability (King, 1972; Blem, 1990; Rogers and Smith, 1993; Biebach, 1996). Species that exploit an unpredictable food source during the winter store more fat than species that have a more predictable food supply. In small passerines, the amount of stored fat is a function of the overnight energy expenditure and increases toward midwinter. Up to three-quarters of this reserve may be mobilized during the night to maintain metabolic rate and body temperature. It is then replenished during the next day's feeding. In many small passerines, the amount of fat stored may only sustain energy needs for 1 day or less; however, in others it may last for as much as 4 days. The amount of fat stored tends to increase with body size. Some waterfowl and sea birds can store sufficient fat to last several weeks. When fat reserves are depleted, protein, mostly from skeletal muscle, is depleted to sustain energy needs.

Fat deposition in preparation for migration is common in many birds, especially those that must cross barriers, such as oceans, deserts, or mountains, and those that migrate long distances very quickly. In small birds, fat deposition occurs within the week prior to migration. For example, fat deposition rates of 10% of body weight per day have been recorded in hummingbirds. This feat requires record rates of fatty acid synthesis from glucose and is facilitated by a hepatic fatty acid biosynthesis capacity of a least tenfold that found in mammals (Suarez et al., 1990; Carpenter et al., 1993). Waders, such as Sanderlings and Bar-tailed Godwits, more than double their body weight from fat deposition in the 2 weeks prior to migration. The fat stores are completely utilized by several days of migration. The general relationship between maximum fat deposition for migration (in % lean body mass) and body weight is: $2.92 \times mass^{-0.25}$. Such rapid rates of fat accretion and mobilization are not seen in other vertebrates. Interestingly, migratory birds kept in captivity show the same seasonal fat deposition as their wild conspecifics. Without the energetic demands of migration, captive birds do not show the rapid utilization of these stores (Summers et al., 1987; Lindstrom, 1991; Lindstrom and Piersma, 1993).

Some birds store fat in their adipose tissue in preparation for egg production. Migratory waterfowl increase their food consumption and deposit large fat reserves, which are used partially for migration to breeding grounds and partially for subsequent egg production and brooding (Raveling, 1979; Blem, 1990; Gauthier et al., 1992). Those species that arrive at breeding grounds prior to the growth of spring food supplies rely on these stores almost completely. In these birds, clutch size and incubation success are correlated with fat reserves. Similarly, female Adelie Penguins fast for 2–3 weeks before and during egg laying and all nutrients must be supplied by body reserves (Astheimer and Grau, 1985).

Fat depots provide insurance against variability in food supply and ebbs in food-acquisition capabilities due to weather or sickness. Counterbalancing these infrequent benefits are a variety of associated costs, including pathological costs, impaired reproduction, and increased susceptibility to predators or to injury due to lessened flight capacity and maneuverability (Rogers and Smith, 1993; Witter and Cuthill, 1993; Scott et al., 1994). When wild birds are brought into captivity, they typically deposit greater amounts of fat and become even fatter with subsequent generations. Presumably, the relaxing of food-acquisition, reproductive, and predator pressures permits this expression. Modern poultry, especially broilers, deposit sufficient quantities of adipose tissue to impair egg production and to increase disease susceptibility when permitted to feed ad libitum.

COMPOSITION OF STORED FATS

The types of fatty acids esterified within the triglycerides and phospholipids are extremely variable. Genetic and environmental influences can account for some of this variability but most is due to nutritional factors, particularly the relative importance of hepatic lipogenesis versus the amount and type of dietary fat consumed (see Fig. 7.7). The endogenous fat synthesized by chickens fed a fat-free diet is mostly 16:0 and 18:1, with smaller amounts of 16:1 and 18:0. When fat is consumed, the fatty acids may be deposited in stores, diluting endogenously synthesized fatty acids. Three parameters determine the extent to which the types of fatty acids in a bird resemble that of the diet: (i) preferential oxidation or esterification of some types of fatty acids; (ii) modification of dietary fatty acids by elongation and/or desaturation; and (iii) the amount of dietary fatty acids relative to the amount synthesized *de novo* (Blem, 1976; Bartov, 1979; Walzem, 1996).

Some dietary fatty acids are stored in triglycerides and phospholipids of cells, with little or no modification. However, the extent of modification of dietary fatty acids is greater in birds than in most mammals, due to the route of intestinal absorption. In mammals, dietary fatty acids enter chylomicrons, which leave the intestines via the lymphatics and enter the general circulation, where many tissues take them up without prior modification by the liver. In birds, the absorption of dietary fats as portomicrons into the portal blood allows the liver to buffer the fatty composition of triglycerides prior to the provision of other tissues through exported VLDL. This buffering is due to the preference of the liver for oxidizing some types of fatty acids and esterifying others into triglycerides of VLDL. This discrimination is made by the relative affinities of the fatty acid-binding proteins that deliver fatty acids to the mitochondria and of the acyltransferases that assemble triglycerides from dietary or *de novo* synthesized fatty acids. Dietary linoleic and linolenic acids are good substrates for hepatic diacyltransferase, and they are enriched in VLDL and consequently tissue lipids. This enzyme shows discrimination against PUFAs, such as eicosapentanoic acid, and PUFA content in triglycerides is diluted relative to dietary concentrations. However, this discrimination is incomplete, and birds that consume fish or fish oils have adipose tissue that contains most of the same fatty acids prevalent in the diet, although at lower levels of enrichment. For example, the adipose tissue of Adelie Penguins is enriched in long-chain PUFAs and low in linoleic acid, reflecting the levels found in their diet of cold-water marine fish and krill (Johnson and West, 1973). Short-chain fatty acids, such as lauric and myristic acids, found in the fruits of palms consumed by Oilbirds are deposited in adipose tissue and give its fat a fluid texture. Saturated fats, such as the stearic acid found in many mammalian fats, do not markedly impact the tissue fatty acids of birds, because they are elongated and desaturated in the liver to the same fatty acids typically synthesized *de novo* (e.g., oleic and palmitoleic acids).

The fatty acid composition of adipose tissue obtained by biopsy gives an indication of the dietary history of a bird. The slow turnover rate of triglycerides (23 days in the chicken) means that it takes over 3 weeks for 50% of the fatty acids in adipose tissue to reflect those of the diet. In Indigo Buntings, changes in the fatty acid composition of adipose tissue require about 6 weeks to complete following a dietary change (Johnston, 1973). Thus, the fatty acid composition of adipose tissue buffers short-term dietary aberrations and reflects longer-term patterns. Presumably, cold weather, which is accompanied by nighttime utilization and daily redeposition of body fat, would decrease the time required for equilibration between dietary and adipose fatty acids in small birds.

Fat metabolism and egg formation

The production of a clutch of eggs requires the deposition of large amounts of yolk lipids, mostly during the several days prior to ovulation. Yolk lipids (and proteins) are synthesized in the liver under the influence of estrogen and progesterone and are transferred through the blood to the ovarian follicles. Lipids in the yolk are of two main types: lipoproteins and vitellogenins. In the chicken, lipoproteins contribute about 95% of the yolk lipids. The liver packages and secretes triglycerides and phospholipids in a special yolk-targeted VLDL (VLDLy), which has unique structural and biochemical properties for targeting it to the ovary (Griffin *et al.*, 1982; Barber *et al.*, 1991; Walzem, 1996). This lipoprotein is half the size of normal VLDL and has apoprotein VLDL II on its surface, making it a poor substrate for lipoprotein lipase (see Fig. 7.6). Consequently, the triglycerides in VLDLy are not well used by skeletal muscle or adipose tissue. Its small size permits it to pass through the granulosa basal lamina of the ovarian follicle and bind to the apolipoprotein-B receptor on the oolemma. It is endocytosed intact to form the yolk. The sieving-like action of the ovarian-follicle basal lamina prevents the uptake of portomicrons arriving from the diet. This combination of follicular ultrastructure and VLDLy size allows dietary fat to be modified by the liver prior to inclusion into the yolk of eggs, permitting better control of yolk lipid characteristics by the female. This hepatic modification is not complete, however, and yolk lipid composition still reflects that of the diet, especially in content of PUFAs (Noble *et al.*, 1996).

Vitellogenin is a protein synthesized by the liver of the laying female that complexes with phospholipids and cholesterol. This lipoprotein is taken up by the developing oocyte and cleaved to give phosvitin and lipovitellin (Jackson *et al.*, 1977).

The intensive synthesis of yolk lipoproteins by the liver occurs faster than their mobilization from the hepatocytes, resulting in a transient increase in liver size and lipid content and a change in its color to a yolk-yellow. Further, the rate of clearance of VLDLy by the ovarian follicles is not as rapid as hepatic release, and circulating triglycerides increase from two- to tenfold during egg production.

Embryonic fat metabolism

About 90% of the energy requirement of the developing chick is supplied by yolk lipids (Romanoff, 1960; Vanheel *et al.*, 1981; Noble and Cocchi, 1990). Most of this utilization (80%) occurs during the last third of incubation, when the embryo grows rapidly. The yolk-sac membrane is responsible for the uptake of yolk lipids. This membrane is essentially an extension of the embryo's small intestine and is continuous with it. The inner surface is histologically similar to the intestinal epithelium, with a simple columnar epithelium, villi, microvilli, and a central capillary bed. In the chicken, the yolk-sac membrane completely surrounds the whole yolk by the fifth day of incubation and replaces the vitelline membrane that surrounded the yolk at ovulation. Between days 13 and 21, most of the yolk lipid is absorbed by the membrane. Yolk lipids are accumulated in the yolk-sac membrane and then they are rapidly transferred to the embryo. Unlike the intestines, the yolk-sac membrane absorbs lipid by nonspecific phagocytosis of yolk granules and lipoproteins. Within the membrane, the yolk lipids are metabolized in an effort to accomplish two purposes: (i) to facilitate rapid lipid assimilation by the embryo; and (ii) to supply essential fatty acids at a rate that matches the needs of developing tissues (Fig. 7.9).

The structure of yolk lipids is rearranged by the yolk-sac membrane in order to supply the embryo. The triglycerides and phospholipids that are taken up by the membrane are hydrolyzed and then reesterified, so that docosahexaenoic acid (22:6 *n*-3) is transferred from phospholipids to triglycerides. The free cholesterol found in yolk is esterified to oleic acid by the action of the acyl-CoA-cholesterol acyltransferase found in the yolk-sac membrane. The newly synthesized triglycerides, phospholipids, and cholesterol esters are used to assemble small lipoproteins, which are secreted into the membrane capillaries and delivered to the embryonic liver through the vitelline portal veins. The cholesterol esters accumulate mainly in the liver, where they account for 80% of

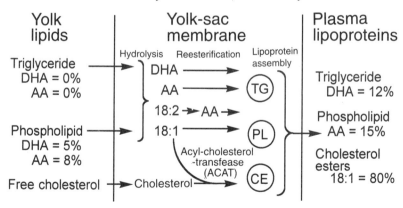

Fig. 7.9. Yolk lipids are taken up by the yolk-sac membrane and metabolized to provide lipoproteins for utilization by the developing embryo. These lipoproteins are enriched in cholesterol esters, docosahexaenoic acid (DHA) and arachidonic acid (AA), which are necessary for functional development (Maldjian *et al.*, 1995). TG, triglyceride; PL, phospholipid; CE, cholesterol esters.

hepatic lipid and 30% of its dry matter. A large proportion of fatty acids taken up by embryonic tissues are oxidized, but some are deposited in membranes and adipose tissue. Docosahexaenoic acid is particularly enriched in cellular membranes (e.g., brain) and in adipose tissue, where it provides a store for the hatchling. The broad tissue distribution of docosahexaenoic acid is made possible by its transfer from yolk phospholipids to the triglycerides of lipoproteins (Maldjian *et al.*, 1995).

The yolk-sac membrane preferentially absorbs yolk lipids over protein, so the yolk that remains at hatching is enriched in protein (Romanoff, 1960; Vanheel *et al.*, 1981; Noble and Ogunyemi, 1989). For example, in the pigeon, lipids account for about 30% of the wet weight of the yolk of unincubated eggs, but only 12% at hatching. About 30% of the original yolk lipids remain at hatching in pigeons and 25% remain in chickens. Following hatching, part of the remaining yolk is absorbed by the yolk-sac membrane in a similar manner as in the embryo. Also, some remnant yolk material is expelled through the yolk stalk into the small intestines during the process of retraction of the yolk into the peritoneal cavity during hatching. This yolk is then digested by adult-type processes. In precocial chicks, the lipid remaining in the yolk, together with cholestoryl esters accumulated in the liver, can provide fuel for several days at thermoneutral temperatures. In less than a week posthatch, the yolk sac and its contents become vestigial.

Excretion and Toxicities

Excretion

Triglycerides and cholesterol circulate as components of lipoprotein complexes, and free fatty acids are bound to albumin. As such, they are too large to be filtered by the glomerulus for excretion from the kidney. Bile secretion into the small intestine is the only appreciable form of lipid excretion. Excess cholesterol, cholesterol esters and other nonpolar lipids from peripheral tissues are transferred to the liver as components of HDL and LDL, where they may be excreted with the bile. The secretions of the uropygial gland may be viewed as an excretory route; however, the lipids in this secretion are synthesized locally and do not represent excretion of excesses from other sources.

Rancidity

Dietary fats and oils can become rancid by two methods: hydrolysis and oxidation. Hydrolytic rancidity results from enzymatic hydrolysis of triglycerides to give free fatty acids, monoglycerides, diglycerides, and glycerol. The enzymes responsible are usually of microbial origin and the rate of hydrolytic rancidity is a function of the population size of contaminating microorganisms. Some foods also have high levels of endogenous lipases, which cause enzymatic hydrolysis. This form of rancidity may affect the palatability of foods but does not markedly influence the nutritional value.

Oxidative rancidity occurs by peroxidation of unsaturated fatty acids and markedly reduces the metabolizable energy value of the lipid, destroys essential fatty acids, and produces potentially toxic end products. In general, the susceptibility of a fat to oxidative rancidity is positively correlated with its level of PUFAs and negatively correlated with the amount of antioxidant. Vitamin E and xanthophylls are the primary natural antioxidants found in foods, and synthetic antioxidants, such as ethoxyquin and butylated hydroxytoluene (BHT), are commonly supplemented to diets to augment natural antioxidants. The essential fatty acids linoleic and α-linolenic acids are susceptible to oxidation due to their unsaturation. The decomposition products resulting from oxidation of PUFAs include a variety of short-chain aldehydes and epoxides, which are cytotoxic and damage the intestinal epithelium. Following absorption they may also damage the liver. Foods that have undergone oxidative rancidity are usually depleted of their vitamin E and they increase oxidation of unsaturated fatty acids in the bird's tissues (Sheehy *et al.*, 1993; Engberg *et al.*, 1996).

Other Nutritional Considerations

Dietary lipids are the primary energy source in the natural diets of some birds, particularly some faunivores, granivores, and frugivores. The caloric value of dietary fats is variable and is primarily dependent on the types of fatty acids that make up the triglycerides (Mabayo *et al.*, 1994). In general, the longer the chain length and the more saturated the fatty acids in a fat source, the greater its gross energy value, but the poorer its digestibility. There is considerable variation in lipid digestibility across species and with age. Lipid digestibility is usually poor in young chicks and improves with age (see Chapter 3). In poultry, dietary fat has been described as having an extracaloric effect on a diet, in that it increases the digestibility of other nutrients by increasing their retention time in the digestive tract (Mateos and Sell, 1980).

For some granivorous and omnivorous species, dietary fat enhances consumption. It has been observed that the addition of fat to low-fat diets enhances intake, even when the caloric content is kept constant. This phenomenon appears to be due to palatability factors and is usually expressed in diets that are artificially low in fat due to the use of fat-extracted ingredients (e.g., soybean meal). The increased intake due to dietary fat is largely independent of the fat type and occurs over a very broad range of fatty acid types.

The amount of dietary fat required for adequate absorption of fat-soluble vitamins and carotenoids appears to be about 2% of dry matter. This level is found in most avian foods and lower values would typically only be found in artificial diets made with fat-extract ingredients.

References

Adams, C.A. (1985) *Pigmenters and Poultry Feeding.* Demin Europa NV, Herentals, Belgium.

Allen, P.C., Danforth, H.D., Morris, V.C. and Levander, O.A. (1996) Association of lowered plasma carotenoids with protection against cecal coccidiosis by diets high in N-3 fatty acids. *Poultry Science* 75, 966–972.

Anderson, G.J., Connor, W.E., Corliss, J.D. and Lin, D.S. (1989) Rapid modulation of the *n*-3 docosahexaenoic acid levels in the brain and retina of the newly hatched chick. *Journal of Lipid Research* 30, 433–440.

Astheimer, L.B. and Grau, C.R. (1985) The timing and energetic consequences of egg formation in the Adelie penguin. *Condor* 87, 256–268.

Bailey, E. and Horne, J.A. (1972) Formation and utilization of acetoacetate and D-3-hydroxybutyrate by various tissues of the adult pigeon. *Comparative Biochemistry and Physiology* 42B, 659–667.

Balnave, D. (1970) Essential fatty acids in poultry nutrition. *World's Poultry Science* 26, 442–449.

Barber, D.L., Sanders, E.J., Aebersold, R. and Schneider, W.J. (1991) The receptor for yolk lipoprotein deposition in the chicken oocyte. *Journal of Biological Chemistry* 266, 18761–18768.

Bartov, I. (1979) Nutritional factors affecting quantity and quality of carcass fat in chickens. *Federation Proceedings* 38, 2627–2630.

Bensadoun, A. (1991) Lipoprotein lipase. *Annual Review of Nutrition* 11, 217–225.

Biebach, J. (1996) Energetics of winter and migratory fattening. In: Carey, C. (ed.) *Avian Energetics and Nutritional Ecology.* Chapman & Hall, New York, pp. 280–323.

Blem, C.R. (1976) Patterns of lipid storage and utilization in birds. *American Zoologist* 16, 671–684.

Blem, C.R. (1990) Avian energy storage. In: Power, D.M. (ed.) *Current Ornithology,* Vol. 7. Plenum Press, New York, pp. 59–113.

Blessin, C.W., Ban Elten, C.H. and Wiebe, R. (1958) Carotenoid content of the grain from yellow endosperm-type sorghums. *Cereal Chemistry* 35, 359–365.

Borenstein, B. and Bunnell, R.H. (1966) Carotenoids: properties, occurrences, and utilization in foods. *Advances in Food Research* 15, 195–276.

Brockmann, H. and Volker, O. (1934) Der gelbe Federfarbstoff des Kanarienvogels und das Vorkommen von Carotinoiden bei Bogeln. *Hoppe-Seyler's Zeitschrift fuer Physiologische Chemie* 224, 193–215.

Brush, A.H. (1990) Metabolism of carotenoid pigments in birds. *FASEB J* 4, 2969–2977.

Budowski, P. and Crawford, M.A. (1986) Effect of dietary linoleic and linolenic acids on the fatty acid composition of brain lipids in the young chick. *Progress in Lipid Research* 25, 615–621.

Butterwith, S.C. (1988) Avian adipose tissue: growth and development. In: Leclercq, B. and Whitehead, C.C. (eds) *Leanness in Domestic Birds: Genetic, Metabolic and Hormonal Aspects.* Butterworths, London, pp. 203–222.

Carpenter, F.L., Hixon, M.A., Beuchat, C.A., Russell, R.W. and Paton, D.C. (1993) Biphasic mass gain in migrant hummingbirds – Body composition changes, torpor, and ecological significance. *Ecology* 74, 1173–1182.

Cherel, Y. and Le Maho, Y. (1985) Five months of fasting in king penguin chicks: body mass loss and fuel metabolism. *American Journal of Physiology* 249, R387–R392.

Cooper, D.A., Lu, S.C., Viswanath, R., Freiman, R.N. and Bensadoun, A. (1992) The structure and complete nucleotide sequence of the avian lipoprotein lipase gene. *Biochimica et Biophysica Acta* 1129, 166–171.

Dierenfeld, E.S. and Sheppard, C.D. (1996) Canthaxanthin pigment does not maintain color in carmine bee-eaters. *Zoo Biology* 15, 183–185.

Donaldson, W.E. (1979) Regulation of fatty acid synthesis. *Federation Proceedings* 38, 2617–2621.

Donaldson, W.R. (1990) Lipid metabolism in liver of chicks: responses to feeding. *Poultry Science* 69, 1183–1187.

Durrer, H. (1986) Coloration. In: Bereiter-Hahn, J., Matoltsky, A.G. and Richards, K.S. (eds) *Biology of the Integument*, Vol. 2. Springer-Verlag, Berlin, pp. 239–247.

Engberg, R.M., Lauridsen, C., Jensen, S.K. and Jakobsen, K. (1996) Inclusion of oxidized vegetable oil in broiler diets – its influence on nutrient balance and on the anti-oxidative status of broilers. *Poultry Science* 75, 1003–1011.

Fletcher, D.L. (1989) Factors influencing pigmentation in poultry. *Poultry Biology* 2, 149–169.

Foglia, T., Cartwright, A.L., Gyurik, R.J. and Philips, J.G. (1994) Fatty acid turnover rates in the adipose tissues of the growing chicken (*Gallus domesticus*). *Lipids* 29, 497–502.

Fox, D.L. (1979) Pigment transactions between animals and plants. *Biological Reviews* 54, 237–268.

Gailey, J.J. (1975) Dietary supplements for penguins at the Baltimore Zoo. *The Keeper* 1, 9–10.

Gauthier, G., Giroux, J.F. and Bedard, J. (1992) Dynamics of fat and protein reserves during winter and spring migration in greater snow geese. *Canadian Journal of Zoology–Revue Canadienne de Zoologie* 70, 2077–2087.

Goodridge, A.G. (1987) Dietary regulation of gene expression: enzymes involved in carbohydrate and lipid metabolism. *Annual Review of Nutrition* 7, 157–185.

Goodwin, T.W. (1984) *The Biochemistry of Carotenoids*, Vol. 2, *Animals*. Chapman & Hall, London.

Griffin, H. and Hermier, D. (1988) Plasma lipoprotein metabolism and fattening in poultry. In: Leclerq, B. and Whitehead, C.C. (eds) *Leanness in Domestic Birds: Genetic, Metabolic and Hormonal Aspects*. Butterworths, London, pp. 175–201.

Griffin, H., Grant, G. and Perry, M. (1982) Hydrolysis of plasma triacylglycerol-rich lipoproteins from immature and laying hens (*Gallus domesticus*) by lipoprotein lipase *in vitro*. *Biochemistry Journal* 206, 647–652.

Hamilton, P.B. (1992) The use of high-performance liquid chromatography for studying pigmentation. *Poultry Science* 71, 718–723.

Hencken, H. (1992) Chemical and physiological behavior of feed carotenoids and their effects on pigmentation. *Poultry Science* 71, 711–715.

Herzberg, G.R. and Rogerson, M. (1990) Tissue distribution of lipogenesis *in vivo* in the common murre (*Uria aalge*) and the domestic chicken (*Gallus domesticus*). *Comparative Biochemistry and Physiology B-Comparative Biochemistry* 96, 767–769.

Hill, G.E. and Benkman, C.W. (1995) Exceptional response by female Red Crossbills to dietary carotenoid supplementation. *Wilson Bulletin* 107, 555–557.

Hill, G.E., Montgomerie, R., Inouye, C.Y. and Dale, J. (1994) Influence of dietary carotenoids on plasma and plumage colour in the house finch – intrasexual and intersexual variation. *Functional Ecology* 8, 343–350.

Hudon, J. (1994) Showiness, carotenoids, and captivity – a comment on Hill (1992). *Auk* 111, 218–221.

Jackson, R.L., Lin, H.Y., Mao, J.T.S., Chan, L. and Means, A.R. (1977) Estrogen induction of plasma vitellogenin in the cockerel: studies with a phosvitin antibody. *Endocrinology* 101, 849–857.

Jacob, J. (1976) Bird waxes. In: Kolattukudy, T. (ed.) *Chemistry and Biochemistry of Natural Waxes*. Elsevier, Amsterdam, pp. 93–146.

Johnson, S.R. and West, G.C. (1973) Fat content, fatty acid composition and estimates of energy metabolism of Adelie penguins (*Pygoscelis adeliae*) during the early breeding season fast. *Comparative Biochemistry and Physiology* 45B, 709–719.

Johnston, D.W. (1973) Cytological and chemical adaptations of fat deposition in migratory birds. *Condor* 75, 108–113.

King, J.R. (1972) Adaptive periodic fat storage by birds. In: *Proceedings XVth International Ornithological Congress*, Leiden, The Netherlands, pp. 200–217.

Kirkpinar, F. and Oguz, I. (1995) Influence of various dietary protein levels on carcase composition in the male Japanese quail. *British Journal of Poultry Science* 36, 605–610.

Klasing, K.C. (1997) Interactions between nutrition and infectious disease. In: Calnek, B.W. (ed.) *Diseases of Poultry*. Iowa State University Press, Ames, Iowa, pp. 73–80.

Lacombe, D., Bird, D.M., Scanes, C.G. and Hibbard, K.A. (1993) The effect of restricted feeding on plasma growth hormone (GH) concentrations in growing American kestrels. *Condor* 95, 559–567.

Laurin, D.E., Touchburn, S.P., Chavez, E.R. and Chan, C.W. (1985) Effect of dietary fat supplementation on the carcass composition of three genetic lines of broilers. *Poultry Science* 64, 2131–2135.

Lawlor, S.M. and Obrien, N.M. (1995) Astaxanthin – Antioxidant effects in chicken embryo fibroblasts. *Nutrition Research* 15, 1695–1704.

Lindstrom, A. (1991) Maximum fat deposition rates in migrating birds. *Ornis Scandinavica* 22, 12–19.

Lindstrom, A. and Piersma, T. (1993) Mass changes in migrating birds – the evidence for fat and protein storage re-examined. *Ibis* 135, 70–78.

Mabayo, R.T., Furuse, M., Murai, A. and Okumura, J. (1994) Interactions between medium-chain and long-chain triacylglycerols in lipid and energy metabolism in growing chicks. *Lipids* 29, 139–144.

McGreal, R.D. and Farner, D.S. (1956) Premigratory fat deposition in the Gambel white-crowned sparrow: some morphological and chemical observations. *Northwest Science* 30, 12–23.

Maldjian, A., Farkas, K., Noble, R.C., Cocchi, M. and Speake, B.K. (1995) The transfer of docosahexaenoic acid from the yolk to the tissues of the chick embryo. *Biochimica et Biophysica Acta – Lipids and Lipid Metabolism* 1258, 81–89.

Marcstrom, V.B. and Mascher, J.W. (1979) Weights and fat in lapwings *Vanellus vanellus* and oystercatchers *Haimatopus ostralegus* starved to death during a cold spell in spring. *Ornis Scandinavica* 10, 235–240.

Marusich, W.L. and Bauernfeind, J.C. (1981) Oxycarotenoids in poultry feeds. In: Bauernfeind, J.C. (ed.) *Carotenoids as Colorants and Vitamin A Precursors*. Academic Press, New York, pp. 320–362.

Mateos, G.G. and Sell, J.L. (1980) Influence of graded levels of fat on utilization of pure carbohydrate by the laying hen. *Journal of Nutrition* 110, 1894–1899.

Menge, H., Calvert, C.C. and Denton, C.A. (1965) Further studies of the effect of linoleic acid on reproduction in the hen. *Journal of Nutrition* 86, 115–122.

Murai, A., Furuse, M. and Okumura, J.I. (1995) Role of dietary gamma-linolenic acid in liver lipid metabolism in Japanese quail. *British Poultry Science* 36, 821–827.

Nehlig, A., Crone, M.C. and Lehr, P.R. (1980) Variations of 3-hydroxybutyrate dehydrogenase activity in brain and liver mitochondria of the developing chick. *Biochimica et Biophysica Acta* 633, 22–32.

Noble, R.C. and Cocchi, M. (1990) Lipid metabolism and the neonatal chicken. *Progress in Lipid Research* 29, 107–140.

Noble, R.C. and Ogunyemi, D. (1989) Lipid changes in the residual yolk and liver of the chick immediately after hatching. *Biology of the Neonate* 56, 228–236.

Noble, R.C., Speake, B.K., McCartney, R., Foggin, C.M. and Deeming, D.C. (1996) Yolk lipids and their fatty acids in the wild and captive ostrich (*Struthio camelus*). *Comparative Biochemistry and Physiology B – Biochemistry and Molecular Biology* 113, 753–756.

NRC (1994) *Nutrient Requirements of Poultry.* National Academy Press, Washington, DC.

Pageaux, J.F., Joulain, C., Fayard, J.M., Lagarde, M. and Laugeir, C. (1992) Changes in fatty-acid composition of plasma and oviduct lipids during sexual-maturation of Japanese quail. *Lipids* 27, 518–525.

Pearce, J. (1980) Comparative aspects of lipid metabolism in avian species. *Biochemical Society Transactions* 8, 295–296.

Pullar, J.D. and Webster, A.J.F. (1977) The energy cost of fat and protein deposition in the rat. *British Journal of Nutrition* 37, 355–363.

Raveling, D.G. (1979) The annual cycle of body composition of Canada geese with special reference to control of reproduction. *Auk* 96, 234–252.

Ricklefs, R.E. and Schew, W.A. (1994) Foraging stochasticity and lipid accumulation by nestling petrels. *Functional Ecology* 8, 159–170.

Rogers, C.M. and Smith, J.N.M. (1993) Life-history theory in the nonbreeding period – trade-offs in avian fat reserves. *Ecology* 74, 419–426.

Romanoff, A.L. (1960) *The Avian Embryo.* Macmillan, New York.

Rosebrough, R.W. and McMurtry, J.P. (1993) Protein and energy relations in the broiler chicken. 11. Effects of protein quantity and quality on metabolism. *British Journal of Nutrition* 70, 667–678.

Sams, G.H., Hargis, B.M. and Hargis, P.S. (1991) Identification of 2 lipid binding proteins from liver of *Gallus domesticus. Comparative Biochemistry and Physiology B – Comparative Biochemistry* 99, 213–219.

Savard, R., Ramenofsky, M. and Greenwood, M.R.C. (1991) A north-temperate migratory bird – a model for the fate of lipids during exercise of long duration. *Canadian Journal of Physiology and Pharmacology* 69, 1443–1447.

Scott, M.L., Nesheim, M.C. and Young, R.J. (1982) *Nutrition of the Chicken*, 3rd edn. M.L. Scott and Associates, Ithaca, New York.

Scott, I., Mitchell, P.I. and Evans, P.R. (1994) Seasonal changes in body mass, body composition and food requirements in wild migratory birds. *Proceedings of the Nutrition Society* 53, 521–531.

Shaw, R.V., Patel, S.T. and Pilo, B. (1978) Glucose-phosphate dehydrogenase and malic enzyme activities during adaptive hyper-lipogenesis in migratory starling and white wagtail. *Canadian Journal of Zoology* 56, 2083–2087.

Sheehy, P.J.A., Morrissey, P.A. and Flynn, A. (1993) Influence of heated vegetable oils and alpha-tocopheryl acetate supplementation on alpha-tocopherol, fatty acids and lipid peroxidation in chicken muscle. *British Poultry Science* 34, 367–381.

Stevens, L. (1996) *Avian Biochemistry and Molecular Biology.* Cambridge University Press, Cambridge.

Suarez, R.K. (1995) Sustaining the highest mass-specific metabolic rates in the animal kingdom. *Physiological Zoology* 68, 23–25.

Suarez, R.K., Lighton, J.R.B., Moyes, C.D., Brown, G.S., Gass, C.L. and Hochachka, P.W. (1990) Fuel selection in rufous hummingbirds – ecological implications of metabolic biochemistry. *Proceedings of the National Academy of Sciences of the USA* 87, 9207–9210.

Summers, R.W., Underhill, L.G., Waltner, M. and Whitelaw, D.A. (1987) Population, biometrics and movements of the sanderling *Calidris alba* in southern Africa. *Ostrich* 58, 24–39.

Swain, S.D. (1992) Energy substrate profiles during fasting in horned larks (*Eremophila alpestris*). *Physiological Zoology* 65, 568–582.

Torrissen, O.J. (1989) Pigmentation of salmonids – carotenoid deposition and metabolism. *Reviews in Aquatic Science* 1, 209–225.

Vanheel, B., Vandeputte-Poma, J. and Desmeth, M. (1981) Resorption of yolk lipids by the pigeon embryo. *Comparative Biochemistry and Physiology* 68A, 641–646.

Walzem, R.L. (1996) Lipoproteins and the laying hen: form follows function. *Poultry and Avian Biology Reviews* 7, 31–64.

Watkins, B.A. (1991) Importance of essential fatty acids and their derivatives in poultry. *Journal of Nutrition* 121, 1475–1485.

Watkins, B.A. (1995) Biochemical and physiological aspects of polyunsaturates. *Poultry and Avian Reviews* 6, 1–18.

Weglarczyk, G. (1981) Nitrogen balance and energy efficiency of protein deposition of the house sparrow *Passer domesticus* (L.). *Ekologia Polska* 29, 519–533.

Wilson, W.O. and Wakabayashi, G. (1956) Identifying non-laying hens. *Poultry Science* 35, 226–227.

Witter, M.S. and Cuthill, I.C. (1993) The ecological costs of avian fat storage. *Philosophical Transactions of the Royal Society of London Series B – Biological Sciences* 340, 73–92.

CHAPTER 8
Carbohydrates

Dietary carbohydrates diverge into two categories, those that are autoenzymatically digestible and those that are not hydrolyzed by the bird's digestive enzymes (dietary fiber). Most avian foods supply starch, sucrose, and glucose as the primary carbohydrates, which supply glucose and other metabolically essential monosaccharides. Dietary fiber consists mostly of carbohydrates, especially glucose polymers, but the monosaccharide linkages cannot be hydrolyzed by the bird's complement of enzymes. Digestion of dietary fiber requires microbial assistance (alloenzymatic digestion) and volatile fatty acids are the major form of usable energy for the bird. Alloenzymatic digestion of dietary fiber results in relatively insignificant amounts of monosaccharide absorption.

Essentiality

Glucose is required by all cells to supply citric acid cycle intermediates and by some cells as an energy source. The essential nature of dietary carbohydrates or their precursors has been demonstrated in chickens (Renner and Elcombe, 1967). Diets that are devoid of carbohydrates and contain fatty acids as the main source of energy result in slow growth rates and depressed plasma glucose levels. If gluconeogenic precursors are added, growth rate and plasma glucose levels are normal. Glucose can be synthesized from dietary fructose, galactose, sorbitol, sucrose, xylose, glycerol, and a variety of amino acids at sufficient rates to prevent a deficiency. Thus, carbohydrates are considered conditionally essential, with the condition that the availability of gluconeogenic precursors is insufficient. The actual requirement level is not known, but there are no natural foods that are deficient in both carbohydrates and gluconeogenic precursors.

Metabolism and Storage

For almost all diets, glucose is the primary carbohydrate that is absorbed from the intestine, and the diets of all species except strict faunivores have a surfeit content of glucose and other carbohydrates. Following a high-carbohydrate meal, the influx of glucose increases blood concentrations. Glucose is rapidly cleared from the circulation by most tissues, particularly the intestine and liver. Glucose concentrations inside of cells is lower than that in the plasma, enabling uptake by facilitated diffusion, using the glucose transporters known as GLUT 1–5 (Wang *et al.*, 1994).

Glucose is the only carbohydrate found at nutritionally relevant concentrations in the blood and is an important form of energy currency, which can be transported between different tissues. Circulating glucose levels are maintained more or less constant in the bird, regardless of dietary level or feeding frequency. Much of this regulation is due to the interplay of a variety of hormones, including insulin, glucagon, pancreatic polypeptide, corticosterone, and thyroxine. These hormones regulate the metabolism of glucose to facilitate its clearance from the blood following a meal and its net synthesis by gluconeogenesis when circulating levels are low. In chickens, more than a third of the glucose absorbed during a meal is converted to lactate in the intestinal wall, buffering the peak influx (Riesenfeld *et al.*, 1982). Circulating levels of glucose are typically higher in birds than in mammals, but there are considerable species differences. The primary uses of glucose include oxidation to supply energy, glycogen synthesis, fatty acid synthesis, and as a supply of carbon skeletons for synthesis of nonessential amino acids, vitamin C, and other metabolites (Fig. 8.1). Some diets, particularly those high in sucrose, have nutritionally relevant levels of fructose, which can be used for the same purposes as glucose. Fructose is efficiently oxidized in the liver and is an even better substrate for fatty acid synthesis than glucose. Relative to glucose, fructose is not a preferred substrate for glycogen synthesis or for oxidation in nonhepatic tissues (Pearce, 1983).

Oxidation

Birds use glucose as a substrate for cellular oxidation, particularly in nerve cells. Following a meal and during acute exertion, skeletal muscle and other tissues may oxidize glucose to supply energy. Oxidation through glycolysis and the citric acid cycle in Aves is essentially identical to that in mammals and supplies about 36 moles of adenosine triphosphate (ATP) per mole. The contribution of the pentose phosphate pathway to glucose oxidation and the generation of reducing equivalents (nicotinamide adenine dinucleotide phosphate (NADPH)) is minimal following hatching. Instead, malate, a citric acid cycle intermediate arising from the metabolism of glucose, is used to generate NADPH (Leveille *et al.*, 1975).

Muscle can oxidize considerable amounts of glucose when it is available. In the case of highly oxidative, or 'dark,' muscles, this occurs following a high-carbohydrate meal and fatty acids are used at other times. In the case of glycolytic, or 'light,' muscles, glucose is the preferred fuel source at all times and can be mobilized from glycogen stores within the muscle if necessary. Dark muscle fibers are rich in mitochondria and metabolize pyruvate, generated by glycolysis through the citric acid cycle. White muscle fibers, being low in mitochondria, convert pyruvate to lactate, which is released into the blood.

Hummingbirds have served as an enlightening model for investigations of glucose oxidation. These studies are facilitated by the very high rates at which hummingbirds utilize energy, especially in flight, and their consumption of an almost exclusively carbohydrate diet. Hummingbirds preferentially utilize glucose as the major fuel for flight when they are in the fed state and have the

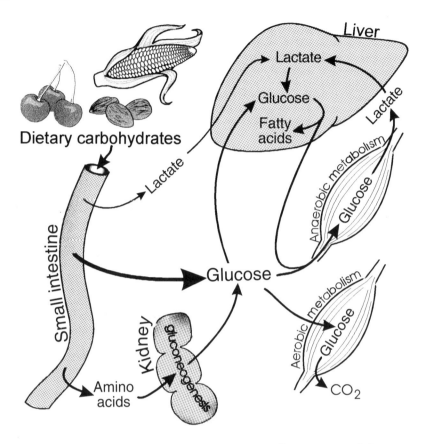

Fig. 8.1. High-carbohydrate meals result in the absorption of large amounts of glucose. This glucose may be used by peripheral tissues to support aerobic metabolism, yielding carbon dioxide (CO_2), or for anaerobic metabolism, yielding lactate. Excess dietary glucose is used as a substrate for fatty acid synthesis. Some dietary glucose is oxidized anaerobically by the intestinal epithelium. If dietary glucose is insufficient to meet tissue needs, lactate and amino acids are converted to glucose by the liver and kidney.

highest activities of the glycolytic enzyme, hexokinase, and the citric acid cycle enzyme, citrate synthase, that have been measured in vertebrate muscle (Suarez *et al.*, 1990; Suarez, 1996). The use of dietary glucose as an energy source is 16% more efficient than using glucose to synthesize fatty acids and then oxidizing the fatty acids. In longer time frames, this advantage is diminished by the greater weight associated with glycogen storage compared with triglyceride storage.

Glycogen synthesis

Dietary glucose that is not oxidized within minutes after a meal can be stored as glycogen. The major glycogen storage areas are the liver and glycolytic muscles. The muscle and liver pools are available for flight or other activities, whereas, during fasting, the liver pool is depleted prior to use of muscle glycogen. Because glycogen is very hydrated, it is a physically bulky (low kJ g^{-1}) energy storage molecule relative to triglycerides. This bulk precludes storage of large amounts and surfeit dietary carbohydrates are converted to fatty acids and stored as triglycerides. The glycogen content of the liver is usually less than 4% and is depleted within a few hours of fasting (Hazelwood and Lorenz, 1959; Blem, 1990; Swain, 1992).

Gluconeogenesis

Gluconeogenesis, the metabolic pathway for the synthesis of glucose from other molecules, is induced when glucose intake is insufficient to meet metabolic glucose demands. This situation occurs several hours postprandial in most birds. The quick transition to gluconeogenesis is due to low levels of storage glycogen coupled with a high metabolic rate. In the fasted state, the primary substrates for gluconeogenesis are glycerol, liberated from lipolysis, and amino acids, liberated from protein degradation. Carnivores, such as Barn Owls and Black Vultures, have continuous gluconeogenesis from amino acids, whether in the fed or fasted state (Migliorini *et al.*, 1973; Meyers, 1995). In the fed state, dietary glucose is insufficient to meet metabolic demands and the gluconeogenic amino acids originate from the diet. In the fasted state, gluconeogenic amino acids are mobilized from body tissues. This continuous use of amino acids for gluconeoge-nesis results in steady blood glucose levels during feeding and fasting and facilitates a meal-eating (feast–famine) consumption schedule. The developing embryo is comparable to faunivores in that yolk and albumen that are 'consumed' have insufficient carbohydrate levels to meet metabolic demands. The embryo is reliant on gluconeogenesis from amino acids to supply glucose. Around the period of hatching, the glycolytic enzymes of the liver are induced to permit adaptation to dietary carbohydrates.

The enzymes for gluconeogenesis are compartmentalized differently among various species (Migliorini *et al.*, 1973; Watford, 1985). Phosphoenolpyruvate carboxykinase (PEPCK), a primary regulatory enzyme, may be found in the mitochondria or in the cytosol. Depending upon its location, different substrates are best utilized for glucose synthesis. In granivorous birds (e.g., chickens, pigeons) an intermitochondrial location prevents effective use of citric acid cycle inter-mediates and amino acids for glucose synthesis, due to a lack of availability of NADPH. Hepatic gluconeogenesis from lactate generates reducing equivalents and is very efficient. In general, the preference of gluconeogenic precursors by the chicken hepatocyte is: lactate $\cong$ glycerol > pyruvate > alanine > serine. The chicken has a distinct cytosolic PEPCK in the kidney, which is responsible for gluconeogenesis from amino acids and citric acid cycle intermediates, such as pyruvate. This kidney form of PEPCK is probably most important in maintaining fasting glucose levels, using amino acids from skeletal muscle, whereas the hepatic form is more important in maintaining glucose levels via the Cori cycle during anaerobic muscle exertion (Langslow, 1978; Meyers, 1995).

In contrast to granivores, carnivorous and piscivorous birds, such as Barn Owls, Black Vultures, and Common Murres, have a hepatic PEPCK that is somewhat evenly distributed between the cytosol and the mitochondria. In these birds, the cytoplasmic form has a very high activity, permitting high rates of gluconeogenesis from amino acids. In these birds, the gluconeogenic amino acids come from the diet when food is available and from muscle when fasting (Migliorini *et al.*, 1973; Herzberg *et al.*, 1988, 1991; Meyers, 1995).

Other uses

Dietary carbohydrates may be used to synthesize important molecules of metabolic and structural significance. Glycolytic and citric acid cycle inter-mediates originating from dietary monosaccharides supply the carbon skeletons for synthesis of many of the nonessential amino acids, ribose, and vitamin C. Dietary carbohydrates can be used to synthesize polysaccharide chains attached to proteins and lipids. Polysaccharides fulfill important structural roles in connective tissue as constituents of chondroitin sulfate, hyaluronic acid, and other mucopolysaccharides. In the chicken, glucose can supply all of the needed monosaccharides. These multiple uses are not known to be substantially different in birds from those in mammals, although data are limited even in the chicken.

Excretion and Toxicities

Normally, monosaccharides are almost completely reabsorbed in the kidney tubules and are not excreted within the intestine. Thus, excretion is not important in the maintenance of body levels and essentially all carbohydrate that is absorbed from the digestive tract is metabolized. Dietary excesses are stored first as glycogen and, when these pools are replete, as triglycerides. Several pathological states can cause disruptions in hormonal regulation of blood glucose levels. Most notably, diabetic conditions have been observed in a variety of birds.

Several types of glucose intolerance can occur in birds (Mayes *et al.*, 1970; Longstaff *et al.*, 1988; Schutte, 1990). Malabsorption can occur if digestive enzymes are lacking. Presumably, all birds lack nutritionally relevant levels of lactase and are intolerant of dietary lactose. Some species lack sufficient sucrase to utilize high levels of dietary sucrose (see Chapter 3). Not all monosaccharides that are absorbed are well tolerated by birds. For example, galactose is efficiently absorbed from the intestine as a monosaccharide but is toxic when fed to chickens at more than 10% of the diet. It competes with glucose for absorption into cells and its metabolites accumulate, blocking glycolysis and causing neurological symptoms and renal damage. Xylose, arabinose, and galacturonic acid are found in nonstarch polysaccharides (e.g., pectin, hemicellulose) of many plant foods and can be released by alloenzymatic digestion. These mono-saccharides are absorbed from the gastrointestinal tract but are excreted by the kidney, making them a poor energy source. In chickens, high dietary levels cause a concomitant increase in urinary output and watery droppings. It is not known if these sugars are better utilized in more herbivorous birds.

A variety of faunivores have been shown to be glucose-intolerant. The infusion of glucose solutions causes greatly elevated blood glucose levels and exaggerated clearance times (Meyers, 1995). This is due to the near absence of glucokinase, the high-Michaelis-constant (K_m) enzyme responsible for hepatic clearance of glucose. The situation is aggravated by continued synthesis of glucose from amino acids despite high levels of glucose (obligatory gluconeogenesis).

Fiber

The term 'dietary fiber' has a variety of meanings to nutritionists and definitions are constantly evolving as analytical techniques and understanding of the physicochemical properties of its constituents improve. In general, dietary fiber is that part of foods of plant origin that cannot be digested autoenzymatically. Dietary fiber is composed predominantly of the nonstarch polysaccharides and lignin, but the exact chemical constituents are extremely variable across plant species and between tissues within a plant (Table 8.1). The dietary fiber in most foods is found in the cell walls, but legumes store considerable nonstarch polysaccharides within the cells of their seeds (Fig. 8.2). The structure of plant cell walls is highly ordered and consists of fibrillar polysaccharides (mainly cellulose), matrix polysaccharides (mainly hemicellulose and pectin), and encrusting

Table 8.1. Primary dietary fiber polymers found in major food groups.

Food group	Polymers present
Legumes	Celluloses, pectins, xyloglucans, galactomannans
Cereals	Celluloses, arabinoxylans, β-glucans, lignin
Fruits	Celluloses, pectins, xyloglucans, cutin, waxes, lignin
Grasses	Celluloses, hemicelluloses, pectins, lignin

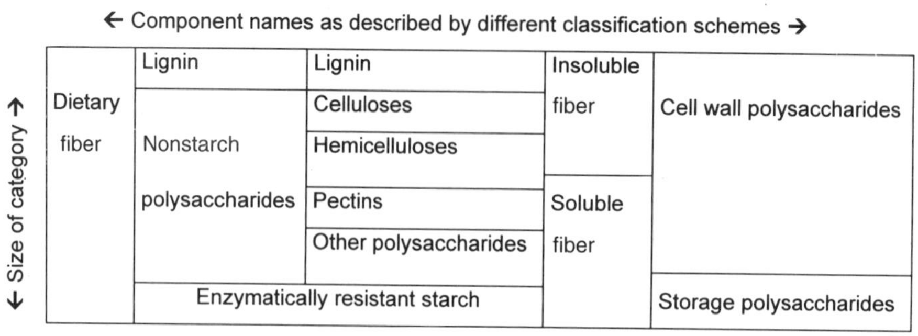

← Component names as described by different classification schemes →

		Lignin	Lignin	Insoluble	Cell wall polysaccharides
↑ Size of category ↓	Dietary fiber	Nonstarch polysaccharides	Celluloses	fiber	
			Hemicelluloses		
			Pectins	Soluble	
			Other polysaccharides	fiber	
		Enzymatically resistant starch			Storage polysaccharides

Fig. 8.2. The relationship between the different categories of dietary fiber.

substances (mainly lignin). The ratios of these components change markedly with maturation of the plant. Techniques based on extraction of food with acid and base detergents are currently the methods of choice for quantifying the general fiber categories in foods. Functionally, dietary fiber may be subdivided into soluble fiber (pectins, gums, β-glucans, and some hemicelluloses) and insoluble fiber (lignin, cellulose, and some hemicelluloses). This further division is nutritionally useful, because the physiological effects of fiber in the intestines diverge according to solubility. Finally, crude fiber is an old term that refers to the remnants of plant material remaining after extraction with acid and alkali during a proximate analysis procedure. Crude fiber underestimates the amount of hemicellulose in food and consequently underestimates the amount of dietary fiber. It also fails to subdivide functionally disparate types of fiber so its use in avian nutrition has decreased.

Physiological effects of fiber

The fate of dietary fiber is highly dependent upon a bird's digestive strategy and the fermentation capacity of its ceca (see Chapters 3 and 4). As described previously, some birds can ferment fiber in their ceca and obtain useful amounts of energy in the form of volatile fatty acids. The soluble components of dietary fiber are most susceptible to fermentation and provide most of the energy, but in some species cellulose can also be fermented (see Chapter 4). Dietary fiber can have a variety of nutritional effects, in addition to the provision of energy. These actions are mostly due to the physicochemical properties of the fiber. The beneficial effects of fiber in human nutrition have been well characterized and are related to the diminution of several chronic diseases of aging. These beneficial effects have yet to be characterized in Aves, where most research, using poultry, ducks, and game birds, has focused on the negative or 'antinutritive' aspects of fiber. When high levels of dietary fiber are consumed, the antinutritive effects usually negate the energy obtained through fermentation in chickens and probably most other nonherbivorous species. The soluble components of fiber are most likely to be implicated in antinutritive activities.

VISCOSITY

Fiber increases the viscosity of the intestinal contents. This increase in viscosity is highly correlated with the solubility of the fiber and not highly dependent upon the exact nonstarch polysaccharides present. Pectins, gums, and β-glucans are the primary viscous components. Generally, high intestinal viscosity impairs the diffusion of substrates and digestive enzymes, hindering their interaction. Gel-forming components of fiber (e.g., gums and pectins) also interact with the glycocalyx of the brush border and increase the thickness of the unstirred water layer. These factors result in decreased digestion of the food and impaired absorption of the end products of digestion. Enzymes of microbial origin are commonly supplemented to commercial diets in an effort to decrease the viscosity of grain-based diets (Bedford and Classen, 1992; Choct *et al.*, 1995; Classen, 1996).

BINDING ACTIVITIES

Some components of fiber, such as pectins, have a high charge density and ionically interact with dietary cations (e.g. manganese (Mn), iron (Fe), zinc (Zn), copper (Cu), which usually decreases their digestibility. Some nonstarch polysaccharides have three-dimensional structures, which allow a chelation of ions and the formation of ionic bridges. Other regions of the molecules have surface activity, causing their association with micelles and bile acids. This physical interference presumably is responsible for the reduction in lipid digestibility caused by dietary fiber (Van der Aar *et al.*, 1983; Smits and Annison, 1996).

INTERACTIONS WITH INTESTINAL MICROFLORA

The amount and type of dietary fiber can markedly shift the types of microflora that inhabit the ceca and the posterior regions of the small intestines. In chickens, the negative nutritional effects of the fiber in rye are at least partly due to the overgrowth of anaerobic bacteria in the ileum. This phenomenon appears to be related to the high viscosity of the soluble components of fiber, causing an increase in the residence time of digesta in the small intestine and leading to lower oxygen tension and a shift in the resident bacterial populations. Some population shifts may lead to the production of bacterial toxins, as well as mucosal immune responses, which negatively impact the bird (Smits and Annison, 1996; Choct *et al.*, 1996).

WATER-HOLDING AND BULKING

Pectins, gums, β-glucans, and some hemicelluloses have very high water-holding capacities. If these components are not fermented, they also provide bulk to the feces. Insoluble fibers can also absorb considerable amounts of water, and they do this without causing a marked increase in the viscosity of the digesta. The insoluble fibers and their associated water and bacteria are the primary contributors to the bulk of feces of florivorous birds. The bulk imparted by insoluble fiber shortens the residence time of the digesta. In some species, this may lead to lower nutrient digestibility but, in others, insoluble fiber has little impact on the utilization of starch, protein, or lipids, beyond nutrient-dilution effects.

References

Bedford, M.R. and Classen, H.L. (1992) Reduction of intestinal viscosity through manipulation of dietary rye and pentosanase concentration is effected through changes in the carbohydrate composition of the intestinal aqueous phase and results in improved growth rate and food conversion efficiency of broiler chicks. *Journal of Nutrition* 122, 560–569.

Blem, C.R. (1990) Avian energy storage. In: Power, D.M. (ed.) *Current Ornithology*, Vol. 7. Plenum Press, New York, pp. 59–113.

Choct, M., Hughes, R.J., Trimble, R.P., Angkanaporn, K. and Annison, G. (1995) Non-starch polysaccharide-degrading enzymes increase the performance of broiler chickens fed wheat of low apparent metabolizable energy. *Journal of Nutrition* 125, 485–492.

Choct, M., Hughes, R.J., Wang, J., Bedford, M.R., Morgan, A.J. and Annison, G. (1996) Increased small intestinal fermentation is partly responsible for the anti-nutritive activity of non-starch polysaccharides in chickens. *British Poultry Science* 37, 609–621.

Classen, H.L. (1996) Cereal grain starch and exogenous enzymes in poultry diets. *Animal Feed Science and Technology* 62, 21–27.

Hazelwood, R.L. and Lorenz, F.W. (1959) Effects of fasting and insulin on carbohydrate metabolism in the domestic fowl. *American Journal of Physiology* 197, 47–56.

Herzberg, G.R., Brosnan, J.T., Hall, B. and Rogerson, M. (1988) Gluconeogenesis in liver and kidney of the common murre (*Uria aalge*). *American Journal of Physiology* 254, R903–907.

Herzberg, G.R., Coady, K., Maddigan, B. and Maccharles, G. (1991) Uric acid synthesis by avian exocrine pancreas. *International Journal of Biochemistry* 23, 545–548.

Langslow, D.R. (1978) Gluconeogenesis in birds. *Biochemistry Society Transactions* 6, 1148–1152.

Leveille, G A., Romsos, D R., Yeh, Y.Y. and O'Hea, E.K. (1975) Lipid biosynthesis in the chick. *Poultry Science* 54, 1075–1093.

Longstaff, M.A., Knose, A. and McNab, J.M. (1988) Digestibility of pentose sugars and uronic acids and their effect on chick weight gain and caecal size. *British Poultry Science* 29, 379–393.

Mayes, J.S., Miller, L.R. and Myers, F.K. (1970) The relationship between galactose-1-phosphate accumulation and uridyl transferase activity to the differential galactose toxicity in male and female chicks. *Biochemistry Biophysical Research Communications* 39, 661–675.

Migliorini, R.H., Linder, C., Moura, J.L. and Veiga, J.A.S. (1973) Gluconeogenesis in a carnivorous bird (black vulture). *American Journal of Physiology* 225, 1389–1392.

Pearce, J. (1983) Carbohydrate metabolism. In: Freeman, B.M. (ed.) *Physiology and Biochemistry of the Domestic Fowl*, Vol. 4. Academic Press, New York, pp. 147–164.

Renner, R. and Elcombe, A.M. (1967) Metabolic effects of feeding "carbohydrate free" diets. *Journal of Nutrition* 93, 31–39.

Riesenfeld, G., Geva, A. and Hurwitz, S. (1982) Glucose homeostasis in the chicken. *Journal of Nutrition* 112, 2261–2266.

Schutte, J.B. (1990) Nutritional implications and metabolizable energy value of D-xylose and L-arabinose in chicks. *Poultry Science* 69, 1724–1730.

Smits, C.H.M. and Annison, G. (1996) Non-starch plant polysaccharides in broiler nutrition – towards a physiologically valid approach to their determination. *World's Poultry Science Journal* 52, 203–221.

Suarez, R.K. (1996). Upper limits to mass-specific metabolic rates. *Annual Review of Physiology* 58, 583–605.

Suarez, R.K., Lighton, J.R.B., Moyes, C.D., Brown, G.S., Gass, C.L. and Hochachka, P.W. (1990) Fuel selection iIn rufous hummingbirds – ecological implications of metabolic biochemistry. *Proceedings of the National Academy of Sciences of the USA* 87, 9207–9210.

Swain, S.D. (1992) Energy substrate profiles during fasting in horned larks (*Eremophila alpestris*). *Physiological Zoology* 65, 568–582.

Van der Aar, P.J., Fahey, G.C., Ricke, S.C., Allen, S.E. and Berger, L.L. (1983) Effects of dietary fibers on mineral status of chicks. *Journal of Nutrition* 113, 653–661.

Wang, M., Tsai, M. and Wang, C. (1994) Identification of chicken liver glucose transporter. *Archives of Biochemistry and Biophysics* 310, 172–179.

Watford, M. (1985) Gluconeogenesis in the chicken: regulation of phosphoenolpyruvate carboxykinase gene expression. *Federation Proceedings* 44, 2469–2474.

CHAPTER 9
Energy

Energy is not a chemically identifiable nutrient but is a property that is realized when nutrients, such as amino acids, carbohydrates, and lipids, are oxidized during metabolism. In birds, the metabolic reactions that oxidize nutrients capture about 40% of the energy in adenosine triphosphate (ATP), which is available for anabolic, catabolic, osmotic, or mechanical work. The remainder of the energy released during nutrient oxidation is in the form of heat. This heat may be useful for maintaining body temperature, but most is lost to the environment, with little benefit to the bird. Ultimately, the energy captured as ATP is released either as heat or as work done on the environment. The joule has been selected by the International System of Units and the US National Bureau of Standards as the official unit for expressing energy (1 joule = 0.239 calories). A wide variety of terms and accounting (budgeting) methods have been applied to energy nutrition of animals (Pesti and Edwards, 1983; Miller and Reinecke, 1984). The nomenclature defined by NRC (1981) has been adopted by most nutritionists and is used here where applicable.

Nutritional energetics is concerned with a careful accounting of the fate of energy consumed in food. The result of this accounting is known as an energy budget. Food energy may be lost from a bird by excretion in urine and feces and through heat produced in metabolism. Some of the consumed nutrients are not oxidized but rather are retained in tissues, feathers, or eggs. The relationships between the various measures of dietary energy and measures of energy requirements are shown in Fig. 9.1. The amount of energy obtained by the complete oxidation of a nutrient to carbon dioxide (CO_2) and oxygen (O_2) is known as its gross energy. Gross energy is sometimes referred to as 'heat of combustion' and is determined in an oxygen bomb calorimeter. The gross energies of starch, grain lipids, and protein are 17.5, 38.9, and 23.7 kJ g^{-1}, respectively. The amount of gross energy consumed by an animal is known as the intake energy. As discussed in Chapter 3, some of this energy is not digestible and is lost in the feces or sometimes in egested pellets. The remaining energy is the digestible energy. It is normally assumed that the energy lost in gases (e.g., methane) produced by birds during the fermentation of food is insignificant, and correction of the digestible energy for these losses is typically omitted. However, the digestion and further metabolism of some nutrients cause significant excretion of energy-containing products that are not of food origin (e.g., uric acid, sloughed enterocytes). The metabolizable energy (ME) is that which remains following correction of digestible energy for these endogenous losses. In other words, the ME is the amount of food energy available to the bird for metabolic

210

Terminology

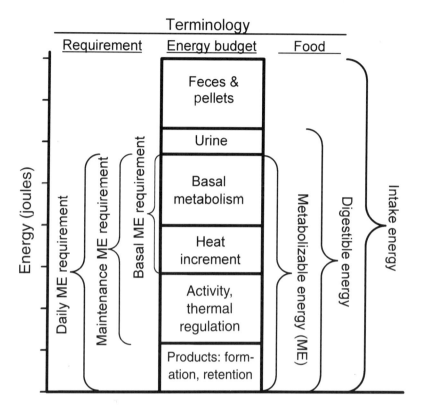

Fig. 9.1. Three sets of terminology are commonly used in avian nutritional energetics. Terminology used to describe a bird's energy budget describes the partitioning of dietary energy into specific digestive losses and metabolic expenditures (see Fig. 9.2). The categories of the energy budget are used to calculate a bird's energy requirement at basal, maintenance, and total daily levels of expenditure. Requirement terminology is expressed as metabolizable energy (ME), usually in kJ ME day^{-1}. The amount of energy in the food consumed by a bird can be expressed as: the gross energy in the food (intake energy); the energy that is actually digested and absorbed (digestible energy); the energy that is available for metabolic purposes (ME).

purposes. As ME is the best measure of food energy that has utility for the bird, it is the currency used to describe the energy content of foods and to express the rate of energy expenditure by the bird for various purposes (Fig. 9.2).

Further losses in dietary ME occur due to inefficiencies in intermediary metabolism, which are necessary to digest, transport, excrete wastes from, and transform nutrients into usable forms. These losses are referred to as the heat increment. The size of the heat increment varies with the composition of the diet. The heat increment is lowest for dietary lipid, because lipids require minimal enzymatic digestion, few energy-requiring transformations prior to use, and no waste-product synthesis. Conversely, the heat increment for protein is high, because of the need for extensive digestion, the high cost of protein synthesis,

Metabolizable
energy (joules)

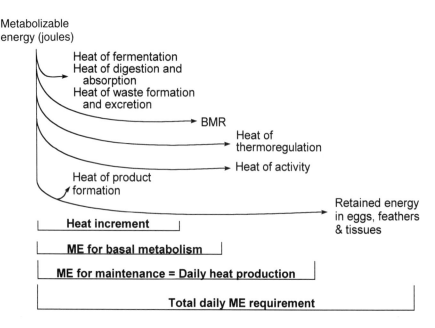

Fig. 9.2. Uses of metabolizable energy in birds. BMR, basal metabolic rate; ME, metabolizable energy.

and the energy needed to synthesize uric acid when amino acids are oxidized to provide energy. In chickens, the heat increment is generally considered to be about 30% of ME for protein, 15% for starch, and 10% for lipid. Black Brants consuming a high-carbohydrate diet have a heat increment that is 20% of dietary ME, and Eurasian Kestrels consuming mice have a heat increment that is 17% of dietary ME. Under conditions of low environmental temperatures, part of the energy attributable to the heat increment may be useful for thermogenesis (Scott *et al.*, 1982; Masman *et al.*, 1989; Sedinger *et al.*, 1992).

The ME content of avian foods is sometimes approximated as the intake energy minus the energy in the excreta (urine, feces, and egested pellets; see Chapter 3). The resulting energy value is referred to as apparent metabolizable energy (AME), because it is biased by energy in the excreta (e.g., sloughed intestinal epithelial cells) that is not attributable to the consumption of the food. Many nutritionists have adopted techniques to correct for this portion of excreta, so that true metabolizable energy (TME) can be determined. The correction for endogenous losses has rarely been applied to nondomestic species, but these losses in Ruffed Grouse are similar in magnitude to those of chickens and domestic ducks. The use of TME is particularly important at low levels of dietary intake and for foods high in fiber or plant secondary products. A further refinement in measurement of food energy content is to correct for the loss or retention of body protein, so that values derived from birds that are growing or losing weight are comparable. Use of nitrogen-corrected ME (ME_n) is especially important for foodstuffs that have poor amino acid balance or levels and foods that have toxins (Sibbald, 1981, 1989; Storey and Allen, 1982; Guglielmo and Karasov, 1993).

Energy Requirements

An extensive body of research has itemized the energy requirements of a wide variety of birds during specific periods of their life. In general, energy requirements are characterized at three levels: basal, maintenance, and daily. The basal requirement is equal to the amount of ME needed by a bird that is quietly resting at thermoneutral temperatures, without depositing tissues, and in a post-absorptive state. The maintenance requirement is equal to the basal requirement plus ME needed for thermoregulation and physical activity. The total daily requirement is the maintenance requirement plus additional ME needed for depositing tissues associated with growth, reproduction, molting, or reserves needed in anticipation of migration or seasonal deprivations. When healthy birds are not depositing tissues, their total daily energy requirement is exactly equal to their maintenance requirement.

Measurement of energy requirements

A bird's energy requirement is typically expressed as ME, so that it can be directly compared with the energy content in foods. The ME requirement of a bird may be determined empirically by measuring the minimal amount of food ME that must be consumed to optimally accomplish a specific function (e.g., egg production, growth). Alternatively, the requirement may be calculated by factorial summation of the energy used for specific metabolic processes inherent in that function (e.g., ME expended for basal metabolism + energy for activity + ME for thermoregulation + ME for egg production). According to the first two laws of thermodynamics, these two methods should give identical results. Requirements estimated by ME intake are commonly used for captive birds, because accurate measures of food consumption and food energy values are possible. However, these empirical estimates suffer from the disadvantage of being applicable only to the species, level of productivity, and environment circumstance in which they were determined. A factorial summation approach is advantageous for comparative avian nutrition because it permits estimation of requirements for species, environments, and processes for which no specific information is available. Budgets that partition a bird's use of ME for identifiable purposes are commonly constructed by ecologists and are used to illuminate trophic and environmental interactions within an ecosystem. Ultimately, budgets based on ME must be converted into actual diets, consisting of variable amounts of individual foods.

A bird's daily ME requirement is equal to the amount of energy expended through the oxidation of nutrients (heat production) plus energy retained in tissues (Fig. 9.2). The amount of energy retained in body tissues, eggs, or feathers is determined by bomb calorimetry. The rate of heat production is commonly determined in captive birds by either direct or indirect calorimetry. Direct calorimetry measures ME expenditure by the rate at which heat is released from a bird. Indirect calorimetry measures the rate of O_2 consumption and CO_2 release, which can be used to calculate the rate of ME expenditure. In free-living birds, the metabolism of a dose of isotope-labeled water is typically used to estimate O_2 consumption and thus ME expenditure.

Basal requirements

The minimal energy expended to maintain ion gradients, protein turnover, respiration, and other vital metabolic processes of a resting, thermally neutral, and postabsorptive bird is known as the basal metabolic rate (BMR). This is not necessarily the absolute minimum amount of dietary energy for survival, because sleeping, torpor, or a prolonged energy deficit can further decrease the metabolic rate. The BMR is best thought of as the minimal energy needed to keep a bird's cells and organs in a state of readiness so that they can function at higher levels when needed (e.g., for flight, reproduction, thermoregulation). From a practical point of view, even captive birds never have daily activity levels that are as low as those used for measurement of BMR, and the dietary ME required for basal metabolism is insufficient for normal daily functions. Nevertheless, knowledge of BMR furnishes a useful reference point when discussing daily energy requirements of captive and free-living birds.

The BMR of a large number of avian species has been measured and found to be highly correlated to body weight (Fig. 9.3). The BMR is usually higher in passerine species than in nonpasserines of similar size (Aschoff and Pohl, 1970). The BMR of adults of nonpasserine species follows the general relationship: kJ energy day^{-1} = 308 × kg body weight$^{0.73}$. The BMR of adult passerines follows the relationship: kJ energy day^{-1} = 480 × kg body weight$^{0.73}$. The body-weight scaling means that BMR increases by only 66% for each 100% increase in body weight. The weight of a bird taken to the exponent 0.73 is often referred to as its metabolic body size and is useful for many types of nutritional comparisons and predictions. Expression of the BMR following division by metabolic body size is referred to as the mass-specific BMR.

The exact equation used to describe the relationship between BMR and body weight depends on the breadth of taxonomic groups included in carrying out the analysis (Laisiewski and Dawson, 1967; Bennet and Harvey, 1987; Nagy, 1987; Daan et al., 1990). On average, the BMR of birds is greater than that of similar-sized mammals, probably because the body temperature of birds is greater than that of mammals. For the smallest birds (3.3 g) this difference is 35% and it is 24% for large birds (17,600 g). Equations developed for individual families usually provide the most accurate allometric relationship. Among nonpasserines, families associated with marine habitats have higher BMR values than those associated with other habitats, and families composed largely of nocturnal species have lower BMR values than their diurnal counterparts.

The mass-specific BMR varies with such factors as time of day, body composition, nutritional status, age, season, and climate. For example, the metabolic rate of birds that are in the active part of their circadian cycle averages about 20% higher than BMR, which is determined during the inactive part of the day (nighttime for most species; Fig. 9.3). Frugivorous species have lower BMR values than predicted by their body weight, though the reason for this has not been established. Body composition affects BMR, because different tissues have different rates of metabolism. High rates occur in the kidney, heart, and intestine; lower rates occur in adipose tissue and skin, and feathers are metabolically inert. Species indigenous to the tropics have lower mass-specific BMR values than closely related species at temperate latitudes, in part due to the smaller liver, kidney, and heart masses afforded by a stable tropical environment (McNab, 1988; Daan *et al.*, 1990; Dawson and O'Conner, 1996).

The mass-specific BMR of chicks is low at hatching, increases during the early growth phase, and then declines to the adult level. The low BMR immediately following hatching is largely due to low body temperature and the provision of supplemental heat through brooding. Altricial hatchlings have a mass-specific BMR that averages only about 43% of adults (Klassen and Drent, 1991; Weathers, 1996).

A bird's calculated or measured BMR does not equal its daily basal requirement for dietary ME, because BMR measurements are determined in fasted birds. Conversion of BMR into a basal ME requirement requires the addition of the amount of energy needed for the heat increment associated with

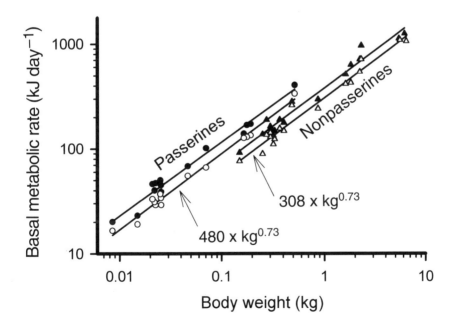

Fig. 9.3. Relationship between the basal metabolic rate and body weight of adult birds. The top regression line is for the active phase and the bottom is for the inactive phase of the daily cycle. Separate regressions are shown for passerine and nonpasserine species. (Data from Aschoff and Pohl, 1970.)

food consumption. Thus, a bird that has a BMR of 100 kJ day^{-1} must consume 125 kJ of dietary ME each day, assuming a heat increment of 20% of ME (100 kJ/0.8). In other words, its basal energy requirement is 125 kJ ME day^{-1}.

The metabolic rate of resting, thermoneutral birds consuming their customary diet is referred to as the resting metabolic rate. The resting metabolic rate does not need to be corrected for the heat increment when used to calculate ME requirements for basal metabolism and is more useful to nutritionists than BMR. Further, measurement of resting metabolic rates is often preferred for practical reasons. For example, some hummingbirds go into torpor when they are fasted long enough to achieve a postabsorptive state. Other birds become agitated when fasted and a resting state is not possible.

Maintenance requirement

The maintenance requirement is often defined as the ME intake at which body weight and composition remain constant in a healthy, non-reproducing bird living in its normal environment. The ME requirement for maintenance includes ME for BMR, heat increment, thermoregulation, and activity (see Fig. 9.2). The maintenance requirement is always greater than the basal requirement, but the exact increase is dependent upon a bird's environment and habits. In the simplest case, birds kept in captivity and indoors need only expend additional energy for minimal activity. In this situation, the maintenance requirement exceeds the basal requirement by only about 50% in chickens. Free-living birds must forage for food, travel to overnight roosts, and thermoregulate. For these birds, the maintenance requirement averages about 2.8 and 2.5 times the basal requirement for passerines and nonpasserines, respectively (Nagy, 1987). But this relationship is highly variable, due to variable environmental temperatures and needs for activity. For example, birds that fly while foraging for food have a maintenance energy requirement that is 16–38% higher than those that are more sedentary. Sea birds that live in cold-water climates have maintenance energy requirements that are 70% higher than those that live in warm water. Birds that provision their nestlings have high maintenance requirements, due to very high levels of activity required to procure and transport sufficient food for fast-growing young. During this crucial period, the parents' maintenance requirement exceeds fourfold basal requirements in many species. The extent of increase in maintenance requirements needed to provision young is diet-dependent. Faunivorous species must expend considerable effort to procure food relative to that expended by many omnivorous, frugivorous, and granivorous species. One of the highest maintenance requirements yet measured is eight times BMR in Blue-throated Hummingbirds. This high ME requirement appears to be driven by aggressive territorial behavior, and not reproductive or climatic events (Williams, 1988; Birt-Friesen et al., 1989; Daan et al., 1990; Weathers and Sullivan, 1993; Powers and Conley, 1994).

EFFECT OF TEMPERATURE

Adult birds are able to maintain their body temperature over a range of environmental temperatures with no impact on their maintenance energy requirements. This is their zone of thermoneutrality (Fig. 9.4). The precise range of thermoneutrality depends on a bird's size, insulation, and behavioral repertoire, as well as environmental factors, such as wind speed, incidence of sunlight, and humidity. Contact with water dramatically increases the temperature at which birds must expend energy for thermoregulation. In general, the smaller the bird, the narrower its thermoneutral zone.

Birds frequently encounter environmental conditions outside of their thermoneutral zone, although migration moderates summer and winter extremes for many species. At cold temperatures, birds augment heat production to maintain their body temperature. Increased heat production occurs predominantly through shivering thermogenesis. Flight muscles, due to their size and central location, provide the largest amount of heat, although leg muscles also contribute. The existence of nonshivering thermogenesis in birds is controversial, but brown adipose tissue and a mitochondrial uncoupling protein are absent. The heat generated by consuming a meal (heat increment) may contribute to thermogenesis, although the extent of this contribution is not clear. In young chicks, the energy generated by the heat increment is dissipated by increased peripheral blood flow (Misson, 1982; Klassen *et al.*, 1989; Brigham and Trayhurn, 1994).

Heat produced through activity can displace some of that needed for thermogenesis in cold climates. The exact amount appears to be species- and situation-dependent. In general, the amount of substitution increases as environmental temperature decreases but is probably always less than 100%. Activity increases airflow over body surfaces, which increases convective and evaporative heat losses. High levels of heat generated during acute bursts of intense activity may be inefficiently used, because they increase a bird's body temperature,

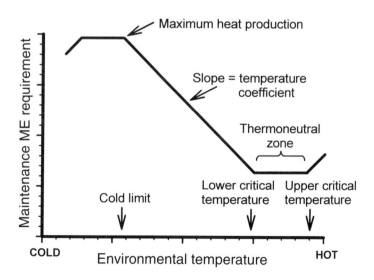

Fig. 9.4. Influence of environmental temperature on a bird's ME requirement for maintenance. Temperatures below the 'cold limit' result in decreased body temperature.

leading to a larger heat gradient and greater losses to the environment (Paladino and King, 1984; Webster and Weathers, 1990; Weathers and Sullivan, 1993).

Under laboratory conditions, the rate at which maintenance energy requirements increase with decreasing temperature is referred to as the temperature coefficient (Kendeigh *et al.*, 1977). The temperature coefficient is dependent upon the species, age, and time of day. However, once it is determined for a particular situation, this coefficient is useful for calculating maintenance ME requirements. Poultry nutritionists routinely monitor the temperature of poultry houses, so that energy requirements can be accurately estimated and used to predict food intake.

Measurements of the energy cost of thermogenesis of free-living birds is difficult and imprecise (Weathers and Sullivan, 1993; Dawson and O'Conner, 1996). During cold weather, small birds and birds that live in the water have maintenance requirements that often exceed three times their basal requirements. For active birds in mildly cold environments, thermoregulation may actually have little net cost, because the heat required is generated by normal activities. For example, the heat production required for thermogenesis by Yellow-eyed Juncos (19.5 g) living at temperatures ranging between −7°C and 14°C is 57% of the total daily expenditure, but, after accounting for the contribution of heat generated by activity and the heat increment, no additional heat needs to be generated by shivering thermogenesis.

Some small birds in cool climates reduce their overnight energy requirement for thermoregulation by decreasing their body temperature through controlled hypothermia (body temperature between 25 and 35°C) or torpor (body temperature below 25°C). For example, thermoregulatory savings due to torpor in hummingbirds can be as much as 75%. Nightly hypothermia in small birds can save 20–40% of thermoregulatory costs. Negating these energetic benefits is an increased susceptibility to predators. Thus, birds generally utilize these strategies only when inadequate dietary energy is available to build the lipid reserves needed to fuel overnight thermogenesis (Reinertsen, 1996).

There is a limit to which birds can increase their heat production in order to defend their body temperature against the cold (Dutenhoffer and Swanson, 1996). The maximal sustainable rate of heat production, often referred to as the summit metabolic rate, is an important determinant of the coldest environmental temperature that a bird can endure. Summit metabolic rates have not been determined for a wide variety of species, but often range between 4 and 8 times the BMR for adults. Young chicks have less capacity for shivering thermogenesis than adults and they are much less capable of maintaining homeothermy. Neonatal gulls, Galliformes, and shore birds can manage only about a twofold increase in heat production. Young altricial chicks have almost no capacity to increase their heat production, and their body temperature closely reflects environmental temperatures (Visser and Ricklefs, 1993; Weathers, 1996).

At hot temperatures, birds employ evaporative cooling by panting, cutaneous evaporation, and, in some species, gular flutter. There is considerable species variability in the amount of energy consumed by these processes, due to behavioral, anatomical, and physiological differences. In some species, energy required for cooling activities in hot environments (> 42°C) can account for one to two times BMR. The actual daily contribution is considerably less than this, because temperatures rarely exceed the upper critical threshold of a bird for more than half of the day. For some birds living in hot arid environments, the increased activity required to procure water lost due to evaporative cooling may contribute more to the maintenance requirement than actual thermoregulatory costs (Dawson and O'Conner, 1996).

EFFECT OF ACTIVITY

Running, flying, swimming, and other activities require large amounts of energy for muscle contraction. It is common for these activities to expend five to ten times BMR. The amount of energy required for flight varies directly with a bird's body weight. For example, a 10 g bird requires 7.7 times its BMR to fly at a rate close to its maximum range speed, whereas a 1 kg bird requires 14.3 times its BMR. The energetic costs of takeoff, gaining altitude, and acceleration are higher yet and short flights require considerable energy. For example, the requirement of European Robins for short flights of only a few meters is 23 times BMR. Although flight entails a very high rate of energy consumption, it is more efficient than running the same distance (Schmidt-Nielsen, 1984; Tatner and Bryant, 1986; Norberg, 1996).

For many birds, activity is the major component of their maintenance ME requirement. The vigorous activity necessary to find sufficient food to provision nestlings is the period of peak daily ME requirement in the life of many altricial and semiprecocial birds. In some species, especially faunivores, this period represents a nutritional bottleneck that may drive important reproductive characteristics, including brood size and chick growth rates (Weathers, 1992).

Daily energy requirements

The daily ME requirement of a bird is equal to its maintenance ME requirement plus additional ME needed for the deposition of products, such as body tissue, feathers, or eggs. The requirement for depositing products has two primary components: the gross energy actually deposited (retained energy); and the energy which is expended to synthesize the products (heat of product formation). Both of these components vary with the type of tissue deposited (Table 9.1).

ENERGY REQUIRED FOR GROWTH

Each species of bird grows with its own characteristic rate and proportion of lean to adipose tissue. The ME required for growth can be calculated from knowledge of the amount of protein and lipid deposited daily. Once a tissue has been synthesized, the continued input of energy needed to maintain it is considered to be part of the basal requirement.

Energy content of tissues.   Body tissue is almost entirely water, protein, lipid, and minerals (ash). The gross energy content of avian lipid is about 38 kJ g^{-1} dry matter, that of protein in birds is about 23 kJ g^{-1} dry matter, and that of ash is essentially zero.

Table 9.1. Energetics of growth, egg production, and molt in chickens (see text for references).

Function	Retained energy (kJ g^{-1} dry matter)		Heat of formation (kJ g^{-1} dry matter)		Total cost of deposition (kJ g^{-1} dry matter)		Efficiency of deposition (kJ kJ^{-1})	
	Protein	Lipid	Protein	Lipid	Protein	Lipid	Protein	Lipid
Growth	23	38	29	14	52	52	44%	74%
Egg production	24	40	24	17	48	57	50%	79%
Molting	26	–	104	–	130	–	20%	–

As a chick grows, the ratio of protein to fat that is deposited decreases, resulting in a linear increase in the energy concentration (kJ g^{-1}) of its tissues. The exact balance of lean:lipid deposition varies with species and diet (see Chapter 7), but a general equation useful for predicting the energy content of a growing chick at any given size is: kJ g^{-1} wet mass = 3.51 + 4.82 × chick body weight/adult body weight (Weathers, 1996).

Efficiency of energy use for tissue deposition. The deposition of lipid is relatively efficient compared with that of protein. Protein accretion requires a considerable energetic investment for the acylation of amino acids to transfer ribonucleic acid (tRNA) and other events required for translational fidelity. The energy expended (heat of product formation) to synthesize tissue protein is 1.25 kJ kJ^{-1} deposited or 28.75 kJ g^{-1} protein. The energy expended to synthesize tissue lipids from a mixture of dietary carbohydrate and lipid is 0.36 kJ kJ^{-1} deposited or 14 kJ g^{-1} lipid. Consequently, the efficiency with which dietary ME is used to synthesize 1 kJ of lipid is 74% and the efficiency for 1 kJ of protein is 44% (Pullar and Webster, 1977; Chwalibog, 1991).

Total metabolizable energy required for growth. The greater energy concentration of lipid relative to protein (38 versus 23 kJ g^{-1}) offsets the greater efficiency of lipid deposition, giving almost identical total costs for deposition (energy retained + heat of product formation) of about 53 kJ ME g^{-1} dry matter. Most body lipid is deposited in adipose tissue, which is low in water content (≈10%), whereas lean tissue (e.g., skeletal muscle) is high in water content (≈75%). Thus, in the live bird, the total energy required to deposit 1 g of lean tissue and its associated water is only 13.3 kJ, compared with 47.7 kJ to deposit 1 g of adipose tissue.

The ME required for growth usually does not contribute more than 25% to the total daily energy requirement, when averaged over the entire growth period (Winjnandts, 1984; Drent *et al.*, 1992). The ME required for basal metabolism represents about half of the daily requirement and the remainder is for

thermoregulation and activity. However, the proportion of the energy requirement partitioned to growth is not constant throughout the growth period. It is
usually greatest in the hatchling and decreases with age. In very young altricial
nestlings, the energy costs for growth are often more than 50% of the daily ME
requirement, because thermoregulation and activity are minimal and the fractional growth rate is at its highest. For example, in Long-eared Owl chicks, the
ME required for growth is 65% of the daily ME requirement at 3 days of age and
then declines to a negligible contribution as adult body weight is reached
(Fig. 9.5).

ENERGY REQUIRED FOR REPRODUCTION

The energy needed for testicular growth and subsequent sperm production is
negligible and probably does not increase daily requirements by more than 1% in
most species (Walsberg, 1983). As this energetic effort is relatively insignificant,
it will not be further discussed. The energy needed for defending breeding
territories, finding mates, and courtship behavior can be quite high in some
species of birds in the wild but is usually considered as part of the activity
(maintenance) requirement.

Females invest considerable energy into gamete formation. The energy
required for reproduction can be broken down into two components: the energy
retained in tissues (ovary, oviduct, eggs) plus the energy which is expended to
synthesize these components (heat of product formation). The time required for
the development of the reproductive tract and for the synthesis of eggs is variable

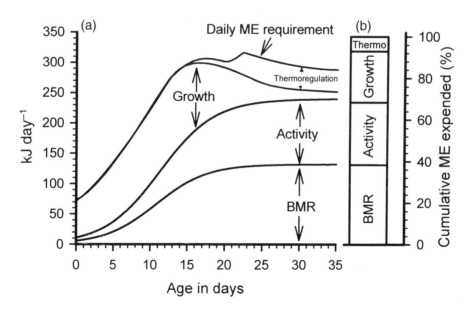

Fig. 9.5. Energy budget of the growing Long-eared Owl. (a) The daily ME intake is
partitioned among its component uses. (b) Cumulative use of ME for specific functions.
Thermo, thermogenesis. (Data from Wijnandts, 1984.)

among species. The size, number, and composition of eggs laid in a clutch are also variable among species. All of these factors markedly impact the contribution of reproduction to the daily ME requirement.

The energy content of eggs.   Though variable, physical and energetic characteristics of eggs follow some predictable patterns (Rahn *et al.*, 1975; Sotherland and Rahn, 1987; Carey, 1996). Egg size varies directly with body size. Data from 809 species in 17 orders give the relationship: egg mass (g) = 0.277 × body mass$^{0.77}$. Females of precocial species lay eggs that have larger yolks and that are more energy-dense than similar-sized altricial females. Yolk size of altricial eggs can be described by the relationship: yolk mass (g) = 0.239 × body mass$^{0.99}$; and for precocial eggs: yolk mass (g) = 0.346 × body mass$^{1.02}$. Thus, a precocial egg contains an average of about 45% more yolk than an altricial egg of the same size. Consequently, precocial birds generally invest more energy into egg production (but not chick rearing). Although these allometric equations have little utility for estimating egg size and composition of individual species, they illustrate general patterns and interesting exceptions to these patterns. For example, some species invest more energy into a smaller number of eggs and others lay many, smaller eggs. Kiwis (Brown) are predicted to lay eggs that are about 3% of their body weight, but their single egg is actually 15% of adult body weight.
    The energy content of an egg can be easily calculated, using established values for lipid and protein or for yolk and albumen (Sotherland and Rahn, 1987). The gross energy content of lipid and protein in eggs is about 40 and 24.2 kJ g^{-1}, respectively. Albumen is about 92% protein on a dry-matter basis, with the balance being mostly carbohydrate and ash, with less than 2% lipid. Thus the energy content of albumen is similar to that of protein, at 22.5 kJ g^{-1} dry matter. The lipid:protein ratio of yolk is variable, but an average yolk contains 33.1 kJ g^{-1} dry matter. Averaged across species, the gross energy content of an egg can be roughly estimated by multiplying the dry egg content (minus shell) by 29 kJ g^{-1}, but estimates calculated from the percent albumen and yolk are more accurate. The energy contents of eggs from a variety of species are given in Table 9.2. The number of days over which the egg is produced is also variable, with most of the yolk synthesized during the period of rapid follicular growth (3–14 days; Table 9.2) and the albumen synthesized during a period of about a day (see Chapter 6).

Efficiency of nutrient use for egg formation.   For the chicken, the relative efficiency of deposition of egg protein is 50% and for lipid it is 79% (Chwalibog and Thorbek, 1989). Consequently, the total cost of deposition (energy retained in deposited lipid and protein + heat of product formation) of 1 g of protein is 48.1 kJ and for 1 g of lipid it is 50.6 kJ. The deposition of energy into the egg is 15% more efficient than the deposition of energy into body tissue. The metabolic basis for this greater efficiency is not known, but it may be due to the fact that egg proteins and lipids are not metabolically active.

Energy for the reproductive tract.   The energy content of the oviduct plus a functional, but nonovulating ovary is directly proportional to body weight,

Table 9.2. Egg energy content and approximate length of the rapid follicular growth phase of different avian species (data summarized from Walsberg, 1983).

	Body mass (g)	Egg mass (g)	Rapid follicular growth phase (days)	Egg energy content kJ g^{-1}	Egg energy content kJ egg^{-1}	Egg:BMR*
Kiwi (Brown)	2098	350	$7\frac{1}{2}$	11.45	4014	11.50
Pelican, Brown	3510	92.1	–	5.74	528	0.68
Leach's Storm-Petrel	43.5	10.2	–	7.61	77.6	2.52
Grey-lag Goose	3250	200	–	7.70	1537	2.10
Canada Goose	1890	197	7	7.55	1487	3.62
Mallard	1870	79.9	6–7	7.3	586	1.20
Wood Duck	665	42.6	7	8.67	369	1.62
Chicken	2500	58.0	7–8	6.95	402	0.67
Common Quail	97	9.9	–	6.87	68.0	1.22
Turkey	4200	65.2	–	6.91	451	0.51
Ruddy Turnstone	97.1	17.1	5–6	–	–	–
Western Gull	1137	86.8	10–11	6.12	531	1.57
Herring Gull	894	94.0	11–13	7.02	576	2.03
Rock Dove	330	17.4	5–8	4.51	78.5	0.55
Turtle Dove	150	9.2	5–7	5.21	47.9	0.63
Great Tit	18.5	1.6	3–4	4.62	7.7	0.29
American Robin	74.9	6.7	–	4.4860	30.1	0.41
House Sparrow	26.0	2.8	–	4.60	12.88	0.38
White-crowned Sparrow	27.0	2.4	4	–	–	–

*Represents energy content of egg (kJ) as fraction of 1 day's basal metabolism (kJ).

according to the relationship: kJ in tissues = ln 0.66 + 0.94 body weight (Walsberg, 1983). From an accounting perspective, the energy requirement for the initial accretion of the inactive ovary and oviduct of the developing female is charged to growth. The energy required for the rapid growth of the reproductive tract immediately prior to ovulation is usually allocated to egg production. After a clutch is laid, the ovary and oviduct partially regress, followed by subsequent regrowth during a period of 3–30 days prior to the next egg-production cycle. The deposition of these tissues represents about 2–9% of the basal requirement, depending on the species.

Calculation of the daily energy expenditure for egg production. As described for amino acid requirements (Chapter 6), the daily requirement for egg production varies daily, depending on the clutch size and the interval of lay. From a nutritional perspective, it is useful to calculate the requirement as the average over the laying period and also as the peak daily requirement. For species that lay large clutches (more than seven eggs), the peak requirement may last for many days. The timing of the peak energy demand is slightly different for altricial and precocial species. This is because the energy required for yolk deposition is spread over 4 or more days prior to ovulation, whereas the energy required for albumen deposition is needed during the day prior to ovulation. Because altricial species have proportionally more albumen, their energy requirement is shifted slightly towards the time of ovulation. A factorial-summation approach can be used to calculate the exact requirement for egg production on any given day by summing the energy needed for accretion of yolk, albumen, and other egg components (Fig. 9.6). Using such an approach, Ricklefs (1974) estimated that the peak daily energy requirements for egg production were about 39% of basal requirements in hawks and owls, 45% in passerines, 126% in Galliformes, 140% in shore birds, 170% in gulls, and 180% in ducks. Among waterfowl, Ruddy Ducks have one of the highest energy requirements for reproduction, at 280% of BMR (Alisauskas and Ankney, 1994).

In many species, not all of the ME used to synthesize eggs arrives from the diet on the same day that it is utilized. Often, energy reserves (adipose tissue) are mobilized to buffer peak energy requirements (see Chapter 8). In other species, food intake increases, so that ME requirements are met from the diet. During egg production, the requirement for essential amino acids increases more than that for energy and a laying female's food choice often changes to items that have a higher protein:energy ratio.

ENERGY REQUIRED FOR MOLT

The composition of feathers is relatively constant across species and consequently the energy content of feathers is fairly uniform. Both feathers and

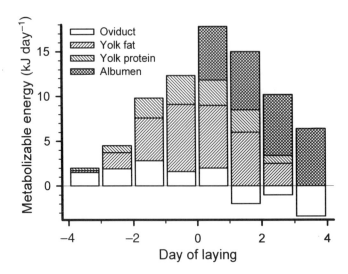

Fig. 9.6. Daily requirement for metabolizable energy to support egg production in House Sparrows laying a four-egg clutch. The negative energy values for the oviduct for days 1–4 are due to catabolism of oviduct protein. (Data from Krementz and Ankney, 1986.)

feather sheaths have a gross energy content of about 22 kJ g^{-1} dry matter (Murphy and King, 1992; Murphy, 1996). The total weight of feather sheaths is typically about 20% of the total feather weight. As feathers are the component of molt that are conveniently measured, a value of 26.4 kJ g^{-1} feathers may be used to account for both the feather and its sheath. The rate, frequency, and extent of molt and feather replacement are very diverse across species. The peak and average energy requirements for molt are dependent upon the molting pattern. Even during a rapid molt, the energy deposited in the integument is less than 6% of BMR, which is trivial compared with that needed for rapid growth or egg production. However, most of the energy required for molt cannot be accounted for by that actually deposited in feathers, sheaths, and skin (retained energy) and is due to the heat of formation (Fig. 9.7).

The total ME requirement for feather deposition of 20 species ranges from 90 kJ g^{-1} for Long-eared Owls to 1225 kJ g^{-1} for European Goldfinches (Lindstrom *et al.*, 1993). The daily ME required for molt is highly dependent upon metabolic body size and follows the relationship: kJ g^{-1} feather + sheath = 270 × mass-specific BMR (in kJ g^{-1} day^{-1}). Clearly, the efficiency of use of ME for deposition of feathers and other structures in the integument is extremely low: 29.3% for Long-eared Owls and only 2.1% in European Goldfinches. Much of this inefficiency can be explained by the energetic costs associated with the high rates of protein turnover necessary to match the dietary influx of cysteine with the demands of feather accretion. Additionally, the cost of maturation and maintenance of feather follicles and their associated blood supply may account for a significant fraction of the increased energy requirement (Murphy and Taruscio, 1995; Taruscio and Murphy, 1995). For many birds, the energy costs of

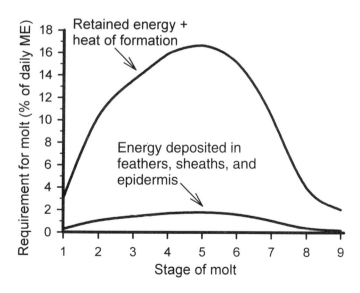

Fig. 9.7. Fraction of the daily energy requirement used for molt by White-crowned Sparrows (data from Murphy and King, 1992).

a full molt are partially compensated by decreased activity, because flying and swimming are curtailed. In birds in cold or wet environments, changes in insulation due to molting may increase energy requirements for thermoregulation.

The energetic costs of molt change with the intensity of the molting process. For example, in White-crowned Sparrows, the energy required for molt is low at early and late stages and peaks midmolt at 58% of BMR, which is 20% of maintenance ME requirements (Fig. 9.7). It is interesting that feather synthesis requires high rates of protein accretion,with essentially no lipid accretion; yet the poor energetic efficiency of this process causes the change in the daily ME requirement to be similar in magnitude to the change in the amino acid requirement. For this reason, the nutritional cost of molting can often be accommodated by increased food intake without a marked change in the type of food consumed.

ENERGY REQUIRED FOR MIGRATION
Migration is a period of exceptionally high energy demands. The migratory distances, rates, and strategies of various populations of birds are exceptionally variable, even within a species. Therefore, the energy costs of migration are not easily generalized. From a nutritional perspective, the energy costs of migration may be largely met during premigratory fattening (Chapter 7). Birds also undergo muscle hypertrophy during the premigratory period. Although the energetic efficiency of this lipid and protein deposition is not known, it probably does not differ much from that of growth. Thus, the increased energy requirement can be calculated from the daily amount of protein and lipid deposited in preparation for migration in the same manner as discussed above for growth. The increased energy requirement for premigratory fattening is predominantly met by an increase in food intake. However, the daily ME requirement increases much more than the daily amino acid requirement and a change in food selection is common. Many species switch to foods that have a high energy : protein ratio, such as fruits and seeds, during premigratory fattening (Bairlein and Gwinner, 1994; Beibach, 1996; Berthold, 1996).

Energy deficiency
Because birds adjust their food intake as ME requirements change, deficiencies usually only occur when the ME density of the diet is too low (see below) or when the food supply is inadequate in quantity. Often, inadequate supplies of food result in deficiencies in other nutrients, such as amino acids or calcium, in addition to energy. The impact of an uncomplicated energy deficiency on poultry is very well defined, but studies with wild birds have emphasized food deprivation without dissecting the effect of an energy deficiency from that of other nutrients. Lack of dietary control and knowledge of first limiting nutrients causes conflicting results and varied interpretations (Ankney *et al.*, 1991; Arnold and Rohwer, 1991; Drobney, 1991). Just because ME intake is below the requirement does not

necessarily mean that it is the first limiting nutrient for a bird, especially during growth or egg production.

A marginal deficiency in ME intake is often manifest initially as lower voluntary activity. Continued shortage of dietary ME impacts productive processes. For chicks, low ME results in slowed growth rates, with a greater depression in the deposition of adipose than lean tissue. Low ME intake may delay or completely prevent the initiation of egg production. In species, such as waterfowl, that rely heavily on endogenous energy stores for egg production, low dietary ME intake has an impact through decreasing energy reserves. For these species, it is primarily egg number that is impacted by low energy status. Birds, such as chickens, that obtain most of their energy for egg formation directly from the diet often produce smaller eggs with smaller yolks when consuming marginally low levels of ME. In chickens, a further decrease in ME intake causes a decrease in the rate of egg laying.

Birds need quality feathers for insulation, sexual display, and flight, so it is not surprising that feather replacement is a very high-priority process. The onset of molt in many species is not delayed by even severe restrictions in ME intake. In fact, energy deprivation accelerates the onset of molt in chickens and quail that are laying eggs. There appears to be a very effective mechanism to mobilize tissue stores or to sacrifice functional tissues to finish a molt. However, restricted ME intake increases the time required for a complete molt and decreases the size of feathers in White-crowned Sparrows (Murphy, 1996).

When ME intake is reduced below ME expenditure, body energy stores are mobilized (Boismenu *et al.*, 1992; Groscolas and Cherel, 1992; Cherel *et al.*, 1994; Sartori *et al.*, 1995). In experiments with quail, geese, and penguins the type and amount of stores mobilized during fasting can be divided into three phases. The first period (phase I) lasts only a few days and is characterized by an initial high rate, at which energy stores are mobilized, followed by a rapid decrease in their use. During phase I, glycogen stores are depleted and both triglycerides and proteins are mobilized and used for energy. The second period (phase II) is longer and characterized by a slow and steady decline in the rate of mobilization of energy stores. During phase II, use of body protein is particularly decreased and triglycerides supply most of the energy. Basal energy requirements are decreased during phase II. The third period (phase III) is characterized by the depletion of triglyceride stores and the use of body protein as the primary energy source. The loss of body proteins impairs important functions and soon results in death. Biochemical and regulatory details of the mobilization and oxidation of these nutrients have been discussed in the previous three chapters.

The length of phase II is dependent on the bird's size and the amount of its lipid stores. King Penguin chicks store large amounts of triglycerides, and fasts of 5 months or more are tolerated prior to entry into phase III. Very small birds usually have such low triglyceride stores that phase II may not even occur and the bird goes directly from phase I to phase III. These birds are fasting-intolerant and death occurs in a few days.

Other Nutritional Considerations

Factors that affect the metabolizable energy value of foods

There is considerable variability in the ME of foods, depending on factors inherent in the food itself and the species consuming the food. The primary factor that explains species-to-species variation is the digestive anatomy and strategy (see Chapters 3 and 4). The primary factor inherent in the food that explains the ME level is the amount and type of fiber and the protein content. Poultry nutritionists have developed a variety of equations to predict the ME value of any food from its gross energy content. Application of these formulas has met with variable results due mostly to laboratory errors in measurement of food composition and because of the presence of secondary plant metabolites. According to Sibbald (1989):

> while the search for indirect assays continues, it seems unlikely that one will be found which is an adequate replacement for biological assay. This is not to say that existing indirect assays are without value but rather to stress that the quality of the data which they yield is inadequate for many purposes and that occasionally they will produce very erroneous values.

To date, the most useful equations that have been developed are those that predict the ME content of one specific food (e.g., corn, wheat, soybeans) based on compositional analysis. These food-specific equations are considerably more accurate than general equations and have been applied to many species of domestic birds (Japanese Quail, turkeys, ducks) without significant error.

Dietary energy concentration and density

Because birds adjust their voluntary food intake to satisfy a demand for energy, it is not possible to express an energy requirement on a concentration basis (e.g., kJ ME kg^{-1} diet), but rather the energy requirement is expressed on a daily basis (kJ day^{-1}). Daily requirements change rapidly during the growth period and due to changes in reproductive status, environmental temperatures, and activity level. Thus, it is often desirable to know the range of dietary energy concentrations (kJ ME kg^{-1} diet) that a bird can accommodate and still meet its energy needs via adjusting its intake. Birds have a finite intestinal volume and a limited range of digesta passage rates. Diets may contain so much indigestible bulk or water as to surpass the capacity of the bird's digestive system to process sufficient quantities of food to meet its daily energy requirement. In other words, there is a minimal dietary ME concentration that is necessary for a bird to be able to consume enough food to meet its daily energy requirement. In actual practice, digestion is volume-, not weight-, limited and nutrient density is a better indicator of intake limitations than nutrient concentration. The energy density of a diet is the kJ of diet vol^{-1} of diet and is usually expressed on a kJ ME cm^{-3} basis. The lower critical ME density is the lowest dietary ME density which will permit a bird to meet its energy requirement. This density is dependent upon a bird's digestive strategy (Chapter 3) and its physiological state. In general, the lower critical ME density is lowest for frugivores and herbivores, intermediate for omnivores, and highest for faunivores. High daily energy requirements increase the lower critical ME density. Chickens cannot meet their daily energy requirement during their early growing period at ME densities below 6.27 kJ cm^{-3}, but adult chickens at maintenance can meet their requirement at less than 4 kJ cm^{-3} (Scott *et al.*, 1982).

In wild birds, it is often of interest to determine the factor that limits the maximum amount of activity, coldest environmental temperatures, or fastest growth rates that can be sustained (Suarez, 1996). It has been debated if these energetically taxing events are limited by intake and digestive parameters or by physiological limits to cellular processes and metabolic rates. Certainly, intake and digestive constraints can be limiting for all of these processes if the dietary energy density is low. The highest dietary ME density that can be utilized is dependent on a bird's capacity to digest and assimilate dietary lipids.

Optimal foraging theory

Energy is unique among the nutrients in that its acquisition requires high amounts of its expenditure. In captive birds, food must be digested and assimilated, resulting in expenditure of energy as the heat increments. Wild birds must search to locate food. Once located, faunivores must pursue and kill their prey, often expending considerable energy in the process. Many florivores and faunivores prepare food by removing poorly digestible components prior to consuming it, again expending energy. Often, birds defend food supplies from other competing birds and animals, at additional energetic cost. Much of the energy in easily obtainable abundant foods, such as leaves, cannot be digested and the rewards of consumption are low. Conversely, animal prey and nectar are easily digested and are a rich source of energy, but they are often widely dispersed and require relatively high levels of activity to obtain. It seems reasonable that a bird should adopt a strategy that best balances the energetic demands of foraging relative to the energetic rewards of consuming various foods. Optimal solutions to this cost–benefit relationship form the basis of optimal foraging theory. According to this concept, a bird's morphological, physiological, and behavioral traits have been shaped by evolutionary pressures in a way that maximizes net energy acquisition. Critics point out that only a small part of a bird's foraging characteristics can currently be explained by such theories. Also, other nutrients, such as amino acids or calcium, are frequently the limiting nutrients during the most nutritionally demanding periods of reproduction and growth of many species. Nevertheless, a robust literature has developed that attempts to relate the energetic efficiency of foraging to the fitness of an individual. It is hoped that mathematical models based on optimal foraging theory will one day be useful for predicting dietary choices and foraging patterns of birds (Lovvorn and Gillingham, 1996; Maurer, 1996).

The rate at which an average bird acquires energy during foraging is related to its body weight and follows the relationship: kJ h^{-1} = 2.02 × body weight$^{0.68}$ (Bryant and Westerterp, 1980). Dividing this rate of energy acquisition by the maintenance requirement reveals that an average bird in an average environment must forage about 5 h day^{-1} to meet its maintenance energy requirement.

It is interesting that the rate of energy acquisition through foraging scales to a similar exponent as BMR, making the number of hours required daily for foraging independent of body size. Consequently, small birds typically forage a similar number of hours each day as large birds (Maurer, 1996).

Relationship between basal metabolic rate and metabolizable energy requirements

It is common to express the ME needed for functions such as activity, egg production, or thermoregulation as multiples of the BMR. While this approach often serves as a useful approximation, it is based on similarities in allometric scaling of these functions and has little physiological basis. The underlying metabolism needed for maintaining cells and organs is often very different from the metabolic processes that are recruited when these same cells are engaged in a high level of functional activity. For example, basal metabolism of pectoral-muscle cells is primarily related to the turnover of proteins and the maintenance of ion gradients. During flight, the energy needed to support these functions is minor compared with that used to drive the forceful contraction of actin-myosin filaments. Similarly, the primary metabolic processes of a quiescent cell in the magnum of the oviduct are unrelated to the extraordinary energy investment in albumen synthesis by the same cell during egg laying. This dramatically different metabolic emphasis causes calculations of daily energy requirements based on BMR to be tenuous and useful only for broad interspecies comparisons. For practical nutritional applications, it is desirable to use actual experimental measurements of daily requirements. Requirement calculations based on factorial summation of energy needs for specific functions (energy budgets) is a less accurate alternative, but typically provides better results than estimates based on multiples of BMR. Ultimately, dynamic models based on metabolic and physiological principles should be developed and applied to predict requirements (Baldwin, 1995).

References

Alisauskas, R.T. and Ankney, C.D. (1994) Costs and rates of egg formation in ruddy ducks. *Condor* 96, 11–18.

Ankney, C.D., Afton, A.D. and Alisauskas, R.T. (1991) The role of nutrient reserves in limiting waterfowl reproduction. *Condor* 93, 1029–1032.

Arnold, T.W. and Rohwer, F.C. (1991) Do egg formation costs limit clutch size in waterfowl? A skeptical view. *Condor* 93, 1032–1038.

Aschoff, J. and Pohl, H. (1970) Rhythmic variations in energy metabolism. *Federation Proceedings* 29, 1541–1552.

Bairlein, F. and Gwinner, E. (1994) Nutritional mechanisms and temporal control of migratory energy accumulation in birds. *Annual Review of Nutrition* 14, 187–215.

Baldwin, R.L. (1995) *Modeling Ruminant Digestion and Metabolism.* Chapman & Hall, London.

Bennet, P.M. and Harvey, P.H. (1987) Active and resting metabolism in birds: allometry, phylogeny and ecology. *Journal of Zoology London* 213, 327–363.

Berthold, P. (1996) *Control of Bird Migration.* Chapman & Hall, London.

Biebach, J. (1996) Energetics of winter and migratory fattening. In: Carey, C. (ed.) *Avian Energetics and Nutritional Ecology.* Chapman & Hall, New York, pp. 280–323.

Birt-Friesen, B.L., Montevecchi, W.A., Cairns, D.K. and Macko, S.A. (1989) Activity-specific metabolic rates of free-living northern gannets and other seabirds. *Ecology* 70, 357–367.

Boismenu, C., Gauthier, G. and Larochelle, J. (1992) Physiology of prolonged fasting in greater snow geese (*Chen caerulescens atlantica*). *Auk* 109, 511–521.

Brigham, R.M. and Trayhurn, P. (1994) Brown fat in birds – a test for the mammalian bat-specific mitochondrial uncoupling protein in common poorwills. *Condor* 96, 208–211.

Bryant, D.M. and Westerterp, K.R. (1980) Energetics of foraging and free existence in birds. In: van Nohring, R. (ed.) *Acta XVII Congressus Internationalis Ornithologici.* Verlag der Deutschen Ornithologen-Gesellschaft, Berlin.

Carey, C. (1996) Female reproductive energetics. In: Carey, C. (ed.) *Avian Energetics and Nutritional Ecology.* Chapman & Hall, New York, pp. 324–374.

Cherel, Y., Gilles, J., Handrich, Y. and Lemaho, Y. (1994) Nutrient reserve dynamics and energetics during long-term fasting in the king penguin (*Aptenodytes patagonicus*). *Journal of Zoology* 234, 1–12.

Chwalibog, A. (1991) Energetics of animal production – research in Copenhagen, review and suggestions. *Acta Agriculturae Scandinavica* 41, 147–160.

Chwalibog, A. and Thorbek, G. (1989) Fasting heat production in chickens. *Archives Geflugelk* 53, 54–57.

Daan, S., Masman, D. and Groenewold, A. (1990) Avian basal metabolic rates – their association with body composition and energy expenditure in nature. *American Journal of Physiology* 259, R333–R340.

Dawson, W.R. and O'Conner, T.P. (1996) Energetic features of avian thermoregulatory responses. In: Carey, C. (ed.) *Avian Energetics and Nutritional Ecology.* Chapman & Hall, New York, pp. 85–124.

Drent, R.H., Klassen, M. and Zwaan, B. (1992) Predictive growth budgets in terns and gulls. *Ardea* 80, 5–17.

Drobney, R.D. (1991) Nutrient limitations of clutch size in waterfowl: is there a universal hypothesis? *Condor* 93, 1026–1028.

Dutenhoffer, M.S. and Swanson, D.L. (1996) Relationship of basal to summit metabolic rate in passerine birds and the aerobic capacity model for the evolution of endothermy. *Physiological Zoology* 69, 1232–1254.

Groscolas, R. and Cherel, Y. (1992) How to molt while fasting in the cold – the metabolic and hormonal adaptations of emperor and king penguins. *Ornis Scandinavica* 23, 328–334.

Guglielmo, C.G. and Karasov, W.H. (1993) Endogenous mass and energy losses in ruffed grouse. *Auk* 110, 386–390.

Kendeigh, S.C., Dol'Nik, V.R. and Gavrilov, V.M. (1977) Avian energetics. In: Pinowski, S.C. and Kendeigh, S.C.D. (eds) *Granivorous Birds in Ecosystems.* Cambridge University Press, Cambridge, pp. 127–204.

Klassen, M. and Drent, R. (1991) An analysis of hatchling resting metabolism: in search of ecological correlates that explain deviations from allometric relations. *Condor* 93, 619–629.

Klassen, M., Bech, C. and Slagsvold, G. (1989) Basal metabolic rate and thermal conductance in Arctic Tern chicks and the effect of heat increment of feeding on thermo-regulatory expenses. *Ardea* 77, 193–200.

Krementz, D.G. and Ankney, C.D. (1986) Bioenergetics of egg production by female house sparrows. *Auk* 103, 299–305.

Lasiewski, R.C. and Dawson, W.R. (1967) A re-examination of the relation between standard metabolic rate and body weight in birds. *Condor* 69, 13–23.

Lindstrom, A., Visser, G.H. and Daan, S. (1993) The energetic cost of feather synthesis is proportional to basal metabolic rate. *Physiological Zoology* 66, 490–510.

Lovvorn, J.R. and Gillingham, M.P. (1996) Food dispersion and foraging energetics – a mechanistic synthesis for field studies of avian benthivores. *Ecology* 77, 435–451.

McNab, B.K. (1988) Food habits and the basal rate of metabolism of birds. *Oecologia* 77, 343–379.

Masman, D., Daan, S. and Dietz, M. (1989) Heat increment of feeding in the kestrel, *Falco tinnunculus*, and its natural seasonal variation. In: Bech, C. and Reinersten, R.E. (eds) *Physiology of Cold Adaptation in Birds*. NATO ASI Series, Vol. 173, Plenum Press, New York, pp. 123–136.

Maurer, B.A. (1996) Energetics of avian foraging. In: Carey, C. (ed.) *Avian Energetics and Nutritional Ecology*. Chapman & Hall, New York, pp. 250–279.

Miller, M.R. and Reinecke, K.J. (1984) Proper expression of metabolizable energy in avian energetics. *Condor* 86, 396–400.

Misson, B.H. (1982) The thermoregulatory responses of fed and starved 1-week-old chickens (*Gallus domesticus*). *Journal of Thermal Biology* 7, 189–192.

Murphy, M.E. (1996) Energetics and nutrition of molt. In: Carey, C. (ed.) *Avian Energetics and Nutritional Ecology*. Chapman & Hall, New York, pp. 158–198.

Murphy, M.E. and King, J.R. (1992) Energy and nutrient use during molt by white-crowned sparrows *Zonotrichia leucophrys gambelii*. *Ornis Scandinavica* 23, 304–313.

Murphy, M.E. and Taruscio, T.G. (1995) Sparrows increase their rates of tissue and whole-body protein synthesis during the annual molt. *Comparative Biochemistry and Physiology A – Physiology* 111, 385–396.

Nagy, K.A. (1987) Field metabolic rate and food requirement scaling in mammals and birds. *Ecological Monographs* 57, 111–128.

Norberg, U.M. (1996) Energetics of flight. In: Carey, C. (ed.) *Avian Energetics and Nutritional Ecology*. Chapman & Hall, New York, pp. 199–249.

NRC (1981) *Nutritional Energetics of Domestic Animals*. National Academy Press, Washington, DC.

Paladino, F.V. and King, J.R. (1984) Thermoregulation and oxygen consumption during terrestrial locomotion by white-crowned sparrows. *Physiological Zoology* 57, 226–236.

Pesti, G.M. and Edwards, H.M. (1983) Metabolizable energy nomenclature for poultry feedstuffs. *Poultry Science* 62, 1275–1280.

Powers, D.R. and Conley, T.M. (1994) Field metabolic rate and food consumption of two sympatric hummingbird species in southeastern Arizona. *Condor* 96, 141–150.

Pullar, J.D. and Webster, A.J.F. (1977) The energy cost of fat and protein deposition in the rat. *British Journal of Nutrition* 37, 355–363.

Rahn, H., Pagnaelli, C.V. and Ar, A. (1975) Relation of avian egg weight to body weight. *Auk* 92, 750–765.

Reinertsen, R.E. (1996) Physiological and ecological aspects of hypothermia. In: Carey, C. (ed.) *Avian Energetics and Nutritional Ecology*. Chapman & Hall, New York, pp. 125–157.

Ricklefs, R.E. (1974) Energetics of reproduction in birds. In: Paynter, R.A.J. (ed.) *Avian Energetics*, Vol. 15. Nuttall Ornithological Club, Cambridge, Massachusetts, pp. 152–292.

Sartori, D.R.S., Migliorini, R.H., Veiga, J.A.S., Moura, J.L., Kettelhut, I.C. and Linder, C. (1995) Metabolic adaptations induced by long-term fasting in quails. *Comparative Biochemistry and Physiology A – Physiology* 111, 487–493.

Schmidt-Nielsen, K. (1984) *Scaling: Why is Animal Size So Important?* Cambridge University Press, Cambridge.

Scott, M.L., Nesheim, M.C. and Young, R.J. (1982) *Nutrition of the Chicken*, 3rd edn. M.L. Scott & Associates, Ithaca, New York.

Sedinger, J.S., White, R.G. and Hauer, W.E. (1992) Heat increment of feeding and partitioning of dietary energy in yearling black brant. *Canadian Journal of Zoology – Revue Canadienne de Zoologie* 70, 1047–1051.

Sibbald, I.R. (1981) Metabolic plus endogenous energy and nitrogen losses of adult cockerels: the correction used in the bioassay for true metabolizable energy. *Poultry Science* 60, 805–811.

Sibbald, I.R. (1989) Metabolizable energy evaluation of poultry diets. In: Cole, D.J.A. and Haresign, W. (eds) *Recent Developments in Poultry Nutrition*. Butterworths, London, pp. 12–26.

Sotherland, P.R. and Rahn, H. (1987) The composition of birds eggs. *Condor* 89, 48–65.

Storey, M.L. and Allen, N.K. (1982) Apparent and true metabolizable energy of feedstuffs for mature, non-laying female Embden Geese. *Poultry Science* 61, 739–745.

Suarez, R.K. (1996) Upper limits to mass-specific metabolic rates. *Annual Review of Physiology* 58, 583–605.

Taruscio, T.G. and Murphy, M.E. (1995) 3-Methylhistidine excretion by molting and non-molting sparrows. *Comparative Biochemistry and Physiology A – Physiology* 111, 397–403.

Tatner, P. and Bryant, D.M. (1986) Flight costs of a small passerine measured using doubly labeled water: implications for energetic studies. *Auk* 103, 169–180.

Visser, G.H. and Ricklefs, R.E. (1993) Development of temperature regulation in shorebirds. *Physiological Zoology* 66, 771–792.

Walsberg, G.E. (1983) Avian ecological energetics. In: Farner, D.S., King, J.R., and Parkes, K.C. (eds) *Avian Biology*, Vol. 7. Academic Press, New York, pp. 161–220.

Weathers, W.W. (1992) Scaling nestling energy requirements. *Ibis* 134, 142–153.

Weathers, W.W. (1996) Energetics of postnatal growth. In: Carey, C. (ed.) *Avian Energetics and Nutritional Ecology*. Chapman & Hall, New York, pp. 461–496.

Weathers, W.W. and Sullivan, K.A. (1993) Seasonal patterns of time and energy allocation by birds. *Physiological Zoology* 66, 511–536.

Webster, M.D. and Weathers, W.W. (1990) Heat produced as a by-product of foraging activity contributes to thermoregulation by verdins, *Auriparus flaviceps. Physiological Zoology* 63, 777–794.

Wijnandts, H. (1984) Ecological energetics of the long-eared owl (*Aiso otus*). *Ardea* 72, 1–92.

Williams, T.M. (1988) Field metabolism of tree swallows during the breeding season. *Auk* 105, 706–714.

CHAPTER 10
Minerals

Minerals serve a wide variety of structural and functional purposes. Bone and eggshells owe their rigidity to calcium salts. Electrolytes, such as sodium, potassium, and chloride, regulate osmotic balance and participate in pH homeostasis. Some minerals are primary regulators of cellular processes, some serve as activators or catalysts of enzymes, and some are necessary components of deoxyribonucleic acid (DNA) and ribonucleic acid (RNA).

At least 13 minerals are required for the optimal health and productivity of birds. The dietary requirements follow a pattern similar to their concentration in the bird's body (Fig. 10.1). Minerals that serve structural or osmotic functions are required in relatively large amounts in the diet and are referred to as the macrominerals. The macrominerals include calcium, phosphorus, sodium, potassium, chloride, and magnesium. Minerals that are required at relatively low dietary concentrations are referred to as the trace minerals. They include copper, iodine, iron, manganese, selenium, and zinc. Cobalt and sulfate are semiessential,

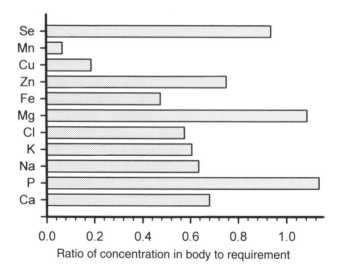

Fig. 10.1. The ratio between the concentrations of trace minerals in the body of 4-week-old Japanese Quail (g kg^{-1} body weight) and the dietary requirement (g kg^{-1} dry matter). Se, selenium; Mn, manganese; Cu, copper; Zn, zinc; Fe, iron; Mg, magnesium; Cl, chloride; K, potassium; Na, sodium; P, phosphorus; Ca, calcium.

in that they are required in the diet but are normally supplied through the essential nutrients vitamin B_{12} and methionine, respectively. Silicon, molybdenum, boron, chromium, nickel, fluoride, and vanadium have some characteristics of essential minerals and are referred to as ultratrace minerals. However, requirements for ultratrace minerals are only observed in birds that are fed experimental diets formulated with highly purified ingredients, provided ultra-pure water, and housed in dust-free environments (Nielson, 1986).

The Requirement and Bioavailability of Minerals

The concentrations of minerals in foods are extremely variable. For example, the selenium content of a single food item can vary by as much as 100-fold, depending on the selenium level in the soil. Further, the digestibility of a mineral, its utilization for specific metabolic processes, and its rate of endogenous excretion following absorption are dependent upon the chemical form in which it is found in the diet. Using the selenium example, four valence states may occur, each with very different bioavailabilities. Other factors inherent in a food, such as level of fiber, chelators, other minerals, and pH, markedly impact the digestion and metabolism of a mineral. For this reason, the value of a food as a mineral source is dependent upon the concentration and chemical form of the mineral, as well as a wide variety of food-specific factors. Because of these uncertainties, the mineral requirements of birds have been determined using highly digestible mineral sources and dietary conditions that give near-maximal absorption and utilization. Usually, the requirement is set at the mineral dietary level that optimizes a physiological process that is dependent upon that mineral. For example, the calcium requirement is usually determined by adding incremental levels of calcium carbonate ($CaCO_3$) to a deficient diet and determining the level that maximizes bone ash content. However, birds eat a wide variety of calcium sources that differ from $CaCO_3$ in the proportion of calcium that is absorbed and the fate of this calcium following absorption. The term mineral bioavailability is used to express the nutritional value of dietary minerals in a manner that considers both digestibility and the metabolic fate of the mineral. The bioavailability of the highly digestible mineral used to determine requirements is usually given a value of 100% and the efficacy of other minerals are set relative to this standard. For example, the bioavailability of calcium in foods or supplements is often compared with that of $CaCO_3$ and the response criterion is usually bone ash content. Typically, the increase in bone ash that results from incremental increases in calcium is used to calculate a relative bioavailability (Fig. 10.2a). Mineral bioavailability is usually determined at dietary levels that are below the requirement, because the percent of dietary mineral that is absorbed usually decreases with the amount that is in the diet (Fig. 10.2b).

It should be realized that the term 'bioavailability' is a relative and not an absolute term. A mineral source such as $CaCO_3$ may by definition have a bioavailability of 100% when only 50% is actually absorbed from the digestive tract. For minerals, the digestibility, or true availability, is always less than the bioavailability. But requirements are set on a bioavailable, not on a digestible or metabolizable, basis.

All of the macrominerals and trace minerals are important in practical avian nutrition. Many of the ultratrace minerals are fascinating from a biochemical viewpoint but rarely become nutritionally important. Of all of the minerals, the requirement for calcium is most variable, both between species and within a species, depending on physiological state. Variations in the calcium requirement of 20-fold between maintenance and egg production occur in some species. Unlike calcium requirements, the trace-mineral requirements are not known to vary more than a few-fold between species or within a species due to physiological state. Thus, the very accurate data on the trace-mineral requirements of poultry and game birds are commonly applied to all bird species, albeit with a margin of safety. However, it has been argued that wild animals should have lower mineral requirements than domestic animals. According to this hypothesis, domestic animals have been selected for rapid and efficient growth while being fed diets that have a surfeit of minerals, and this may have resulted in animals that inefficiently absorb, utilize, and store these minerals (Robbins, 1993). Fortunately, experimental evidence on trace-mineral requirements of domestic chickens and turkeys versus captive nondomestic counterparts does not indicate major differences. Almost no quantitative information is known about the mineral

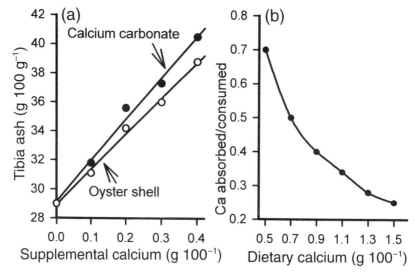

Fig. 10.2. (a) Relationship between the amount of calcium supplemented to a corn–soybean meal-based diet and the ash content of the tibia of growing Japanese Quail (K.C. Klasing, unpublished). The slope of this relationship for calcium carbonate is 28.5, which by definition is equal to a bioavailability of 100%. The slope of this relationship for oyster-shell particles is 24.5, giving a bioavailability of 86% (24.5 ÷28.5). (b) The relationship between the level of dietary calcium and the amount absorbed by growing Japanese Quail. At dietary levels that are well below the requirement of 1 g 100 g^{-1} diet, absorption is high and, at levels well above the requirement, absorption is very low.

requirements of faunivorous, frugivorous, nectarivorous, or herbivorous species. Comparative avian nutritionists need to better define the mineral requirements of species that have digestive strategies that diverge from that of granivorous domestic species.

Absorption, Storage, and Metabolism of Minerals

The absorption and excretion of minerals is regulated to avoid deficiencies and toxicities. Macrominerals are often present in the diet at levels that are near the requirement and are absorbed via specific, energy-requiring transport systems at rates that are proportional to need. Most trace minerals are present in the diet at levels well above the requirement, and the intestine functions to actively exclude these minerals from being absorbed at the same rate and efficiency as other nutrients. The biochemistry of trace-mineral absorption points to toxicity avoidance as the predominant role of the intestines, and specific absorptive pathways for many of the trace minerals are secondary or absent.

Dietary minerals that are complexed to simple carbohydrates (e.g., citrate), amino acids (e.g., methionine, histidine, cysteine), and some chelators (e.g., ethylenediaminetetraacetic acid (EDTA)) are highly soluble in the small intestine and are usually more bioavailable than noncomplexed minerals. Mineral complexes are generally less susceptible to interactions with other minerals in the intestines (Kratzer and Vohra, 1986).

Nutritionally important interactions frequently occur between minerals. These interactions will be discussed in more detail below, but a few generalizations are warranted:

1. Interactions may occur in the intestines and impact absorption.
2. They can occur through competition for binding to various metalloproteins and impact metabolism.
3. They may act at the kidney or liver and impact excretion.

With some interactions, high levels of one mineral increase the requirement for one or more other minerals (e.g., high calcium increases the phosphorus, iron, and manganese requirement). In other cases, high levels of one mineral can decrease the requirement for another mineral or reduce the toxicity of another mineral (e.g., high copper protects against a zinc toxicity).

There are specific storage pools for calcium, phosphorus, iron, and zinc, which buffer dietary inadequacies. Bone serves as a nonspecific reservoir for many of the other minerals (e.g., magnesium, sodium, chloride, manganese, iodine, fluoride), which can be tapped during generalized nutritional deprivation. For most minerals, these storage pools can buffer long periods of nutritional inadequacy in adults. However, during the period of rapid growth of chicks and during the production of a long sequence of eggs, storage pools of macrominerals and zinc may be exhausted in a matter of a few days.

Table 10.1. Distribution of minerals in the chicken egg (from Richards and Packard, 1996).[*]

	Macromineral (mg)			Trace mineral (μg)	
Element	Yolk	Albumen	Element	Yolk	Albumen
Phosphorus	111.6	5.2	Iron	1017	53
Calcium	26.8	2.4	Zinc	571	5
Potassium	20.1	49.3	Copper	37	10
Sodium	12.1	57.1	Iodine	32	2
Magnesium	2.6	3.7	Manganese	10	0.05
			Selenium	9	2

[*]Values are for a 58 g egg with 19 g of yolk, 34 g of albumen, and 4.8 g eggshell, containing 1526 mg calcium.

Mineral Nutrition of the Embryo

All of the minerals required by the developing embryo are supplied by specific storage pools in the egg, which were deposited at the time of its formation (Table 10.1). The embryo obtains its mineral requirement through the coordinated actions of extraembryonic membranes (Grau *et al.*, 1979; Richards and Packard, 1996). Most of the minerals are concentrated in yolk, but a few, such as sodium and potassium, are concentrated in albumen, and the largest deposit of calcium is in the eggshell. Many of the minerals found in yolk were transferred from the liver of the laying female to the developing follicle, in association with vitellogenin. Vitellogenin is cleaved in the follicular membrane to give phosvitin and lipovitellin. Phosvitin avidly binds iron, calcium, and magnesium and it contains covalently bound phosphorus. Lipovitellin binds phosphorus, zinc, copper, and iron. The trace-mineral status of a laying female determines the amount of each trace mineral transferred to the egg and the trace-mineral stores of the hatchling. When adequate, the stores of some trace minerals can support rapid growth in the young hatchling for a period of up to several weeks. The macromineral content of eggs is influenced little by the nutritional status of the mother and typically does not provide long-lasting stores for the hatchling.

Calcium

Calcium is the most prevalent mineral in the body and is required in the diet in a greater amount than any other mineral. The high calcium requirement of growing chicks is driven by the need for skeletal mineralization. In egg-laying females, most dietary calcium is used for shell formation. From a nutritional viewpoint, calcium is the most challenging mineral, because the requirement is extremely variable, depending upon a bird's physiological state, and because many foods are likely to be deficient in calcium. Calcium is also one of the most metabolically active minerals and its metabolism is tightly regulated.

Absorption

Dietary calcium is absorbed from the duodenum and jejunum (Norman and Hurwitz, 1993; Hurwitz *et al.*, 1995). When dietary calcium levels are low, most calcium is absorbed by active transport. The efficiency of absorption is controlled by the levels of parathyroid hormone and 1,25-dihydroxy vitamin D_3. High levels of these hormones occur when blood ionized calcium (Ca^{2+}) levels are low. They induce the synthesis of calbindin, which binds calcium and facilitates its transport across the intestinal epithelial cells. Active calcium transport has four primary steps: (i) energy-dependent uptake of Ca^{2+} across the enterocyte membrane; (ii) binding of Ca^{2+} to calbindin within endocytic vesicles; (iii) fusion of vesicles with lysosomes; and (iv) movement of lysosomes along microtubules and exocytosis of the contents at the basal lateral membrane. In chickens fed a low-calcium diet, about 70% of calcium absorption is vitamin D-dependent. At surfeit levels of dietary calcium, diffusion-based pathways that are vitamin D-independent become predominant.

Calcium is absorbed in the ionic form and factors that impact the solubility and ionization of calcium impact its uptake (Guinotte *et al.*, 1995; Soares, 1995). Inorganic forms of calcium, such as $CaCO_3$, limestone, oyster shell, and calcium phosphates, are readily solubilized by the acid environment of the proventriculus and gizzard. Hydrated calcium salts are often more digestible than anhydrous forms, because they are more soluble. Free fatty acids may form insoluble soaps with calcium, inhibiting its absorption. Calcium in plants often exists in complexes with phytic acid or oxalate, which reduces its digestibility (Fig. 10.3). Phytic acid refers to myoinositol hexaphosphoric acid and its less-phosphorylated relatives. High levels of phytic acid found in many seeds can complex calcium from other foods consumed in the same meal and render it poorly digestible. One mole of phytic acid can complex up to 6 mol of calcium to form insoluble phytates at the pH of the intestine (Kratzer and Vohra, 1986).

In egg-laying Japanese Quail, the availability of calcium from limestone and oyster shell is 72 and 68% of the total consumed, respectively (Kim *et al.*, 1985). Calcium availability in quail is dependent on its particle size and is optimal at 40–80 mesh. The calcium availability of bone is similarly high when the particle size is optimum (e.g., bone meal), but not when bone particles are large. Carnivores that consume intact prey do not digest the intact bones efficiently and much of it is egested in pellets. The apparent metabolizability of rodent bones consumed by raptors averages about 40%. Because of their greater proventricular acidification, Falconiformes (falcons, eagles, hawks) are more efficient at digesting bone than Strigiformes (owls; Duke *et al.*, 1975; Campbell and Koplin, 1986).

The availability of calcium from large particles (grit) is dependent upon the digestive physiology of the bird. Most granivorous and herbivorous birds have the capacity to maintain large particles in their gizzard, where they can be slowly eroded over a period of days. For example, Ring-necked Pheasants quickly pass

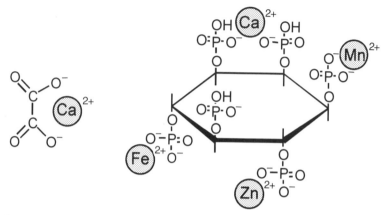

Calcium oxalate   Phytic acid–mineral complex

Fig. 10.3. Phytic acid (myoinositol 1,2,3,4,5,6-hexakis(dihydrogen phosphate)) contains phosphorus as a covalently bound part of its structure and forms ionic complexes with copper, zinc (Zn), iron (Fe), manganese (Mn), and calcium (Ca). When phytic acid is complexed with divalent cations, its solubility and digestibility are markedly decreased. Oxalic acid forms an insoluble complex with calcium, preventing absorption.

limestone particles to their gizzard, where they remain until they are completely solubilized. The rate of solubilization is about 80% of the particle per day (Korschgen et al., 1965).

Metabolism

Calcium constitutes more than a third of the total mineral content of an adult bird. The skeleton contains about 98% of a bird's calcium, most of which is in the form of hydroxyapatite, $Ca_{10}(PO_4)_6(OH)_2$, with small amounts of noncrystalline calcium phosphate and calcium carbonate. Bone serves important structural and metabolic functions. Metabolically, bone provides a labile pool of calcium and phosphate. The hydoxyapatite form of calcium phosphate provides the structural rigidity of bone, and yet it is readily solubilized to provide its minerals to other tissues when needed. Contrary to popular belief, the proportion of the body weight that is skeleton is similar in birds and mammals. Across avian species, the proportion of body weight that is comprised of bone increases with body size (Prange et al., 1979).

In the hen, about 25% of the calcium in the blood circulates as free Ca^{2+} and the remainder is found bound to proteins, such as albumin, or complexed with citrate, phosphate, or sulfate. The low levels of calcium in plasma (≈ 5 meq l^{-1} in nonlaying birds) and cell cytosol (≈ 1 meq l^{-1}) are precisely regulated, because of their important role in intracellular communication, macromolecule interactions, and blood clotting. This regulation is accomplished by the vitamin D endocrine system, consisting of parathyroid hormone, calcitonin, and vitamin D, which control the rate at which calcium is absorbed from the intestine, deposited or mobilized from bone, and excreted by the kidney (Parsons and Combs, 1981; Wideman, 1987; Norman and Hurwitz, 1993).

SKELETAL GROWTH

Longitudinal bone growth is initiated by the accretion of cartilage in the epiphyseal growth plate found on each end of long bones (Leach and Gay, 1987; Hurwitz and Pines, 1991; Bain and Watkins, 1993). This cartilage is degraded by infiltrating osteoblasts, which in turn deposit collagen and hydroxyapatite within the template previously created by cartilage. The width of the bone is increased by the deposition of calcium phosphate on to a collagen matrix located on the outer (periosteal) surface of the bone by osteoblasts. Osteoclasts are found on the inner (endosteal) surface of the bone and function to absorb it. This process, known as bone modeling, permits simultaneous increase in the diameter and length of the bone. Local differences in the rate of osteoblastic and osteoclastic activity sculpt the bone's morphology in response to functional demands. The active generative–degenerative process that occurs in the epiphyseal growth plate is a predominant site of pathology when calcium is deficient (Fig. 10.4).

After the cessation of growth, bone continuously undergoes a remodeling

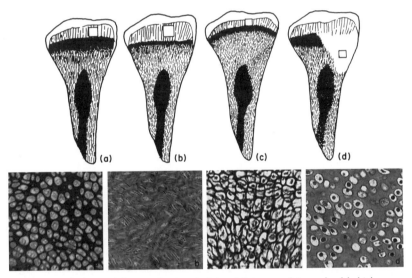

Fig. 10.4. Morphology of the proximal end of the tibiotarsus of 4-week-old chickens and histological sections of the growth plate (265 ×). (a) Normal bone and chondrocytes. (b) Rickets typical of either a calcium or a vitamin D deficiency; note increase in width of the growth plate and disorganized proliferating chondrocytes. (c) Chondrodystrophy typical of a manganese deficiency; note the narrow growth plate and the lack of extracellular matrix surrounding the chondrocytes. (d) Tibial dyschondroplasia, in which the growth plate is very wide and the contents of chondrocytes are condensed. Deficiency of copper or an inappropriate balance of electrolytes or calcium and phosphorus may induce this condition or exacerbate genetic causes. (Used with permission from Leach and Gay, 1987.)

process, in which osteoclasts cluster on bone surfaces and absorb bone. Once a cavity is excavated, osteoblasts are recruited and fill in the space. This closely coupled process of bone resorption and accretion results in continual turnover of the calcium, phosphorus, and other minerals, making them available to buffer dietary shortages.

MEDULLARY BONE

At the onset of sexual maturity, females of many avian species deposit medullary bone in the marrow cavity of long bones. Medullary bone is non-structural, has a spongy appearance, due to an interlaced pattern of thin spicules, and functions as a labile and dynamic reserve of calcium for eggshell formation. Medullary bone is formed on the endosteal surface by osteoblasts, stimulated by androgens and estrogens released by maturing ovarian follicles. The femur and tibiotarsus are rich in medullary bone, and it may also be present in many other bones, such as the ilia, ischia, pubis, sternum, and ribs. In some species, such as pigeons and Canaries, spicules of medullary bone fill the marrow cavity. The large surface area of medullary bone is richly populated by foci of osteoclasts and osteoblasts, which mediate high rates of bone turnover. Medullary bone supplies calcium for eggshell formation at periods when dietary supply is insufficient. A daily cycle has been described in chickens, where medullary-bone calcium is accreted in the morning and used in the nighttime. During the morning, the egg has not yet arrived in the shell gland and dietary calcium is available for medullary-bone accretion. Nighttime mobilization occurs due to the lack of dietary input and the high demand by the shell gland for eggshell deposition. Calcium is also drawn from cortical bone to maintain medullary bone mass. If dietary calcium is insufficient, the mass and density of medullary bone may decline and eventually egg laying is curtailed. In calcium-replete chickens, the total amount of medullary bone is sufficient to supply calcium for about one eggshell; thus, it is most appropriately thought of as a calcium buffer and not a long-term storage pool (Clunies *et al.*, 1992; Dacke, 1993; Etches, 1996).

In some species, such as Japanese Quail, eggs are normally laid during the afternoon and the diet delivers calcium during the most active period of shell formation (Miller, 1977). Nevertheless, medullary bone is still utilized to supply calcium during the early periods of shell deposition and to meet any dietary shortfalls during the day. Following cessation of egg production, medullary bone disappears over the next week.

DEPOSITION OF CALCIUM IN EGGS

Calcium mobilized from bone and supplied by diet is sequestered by the shell gland and secreted into the fluid in the lumen of the shell gland for formation of the shell (Johnson, 1986; Etches, 1996). Ionized calcium is transported by calbindin from the blood and across the secretory cells of the tubular glands. The ability of the shell gland to accumulate Ca^{2+} is remarkable and throughout the period of calcification a supersaturated concentration of Ca^{2+} is maintained in the fluid in the lumen of the shell gland. Carbonic anhydrase, found in the shell gland, catalyzes the formation of carbonate ion (CO_3^{2-}), which precipitates with calcium to give calcium carbonate crystals, which form the shell. In the process of forming CO_3^{2-}, hydrogen ion (H^+) is generated and must be buffered by local and systemic buffering systems.

The amount of Ca^{2+} in the blood is sufficient to sustain shell accretion for less than 20 min, whereas the calcification process requires about 15 h. The drain on plasma Ca^{2+} causes the release of parathyroid hormone and the activation of 1,25-$(OH)_2$ vitamin D_3. These hormones continue to maximize intestinal calcium absorption and cause the mobilization of stores in medullary and cortical bone. Mobilization of calcium from bone also results in the mobilization of phosphorus, some of which is excreted. Thus, eggshell accretion increases the phosphorus requirement, albeit much less than the increase in the calcium requirement.

A small amount of calcium (< 1%) is found in the egg yolk. Calcium is complexed with very-low-density lipoprotein (VLDL) and vitellogenin as they are synthesized by the liver and, through this association, is taken up by the follicles for incorporation into yolk. When vitellogenin is cleaved by the follicle, most of the calcium is associated with phosvitin.

CALCIUM METABOLISM OF THE EMBRYO

Calcium homeostasis of the embryo is tightly regulated by the vitamin D endocrine system. The early embryo obtains its calcium from phosvitin in the yolk granules via uptake by the endodermal cells of the yolk-sac membrane. As incubation progresses, the developing embryo requires large amounts of calcium for skeletal development. At this time, the chorioallantoic membrane attaches to the shell membrane and the $CaCO_3$ in the shell membrane is mobilized. Shell calcium is dissolved by carbonic acid, produced by the enzyme carbonic anhydrase located in the villus cavity cells of the chorioallantoic membrane. The calcium ionized from the shell diffuses through the shell membranes and is captured by calcium-binding proteins on the surface of capillaries of the chorioallantoic membrane for transport into the blood. Toward the end of embryogenesis, calcium in excess of immediate needs is mobilized from the shell and transferred back into the yolk for use following hatching. Altricial species, such as pigeons and Yellow-headed Blackbirds, begin mobilizing calcium from the shell later during development and at a slower rate than precocial species. At hatching, these altricial species have less mineralized bone than precocial species (Packard and Packard, 1991; Hart *et al.*, 1992; Packard, 1994; Richards and Packard, 1996).

Requirement

The amount of dietary calcium needed to maximize bone or eggshell mineralization and strength is greater than that needed for other functions and is typically used as the response criterion for setting the requirement. Requirement levels are based on the assumption that all of the calcium consumed has a bioavailability similar to that in $CaCO_3$. Food sources with lower bioavailability should be discounted proportionally.

MAINTENANCE

The maintenance calcium requirement is that which is needed to replace the small amounts of calcium lost from endogenous sources each day. Most of the endogenous loss occurs in the feces and urine. Loss through these routes is dependent upon other dietary factors, especially phosphorus levels and dietary acidity. Maintenance calcium requirements for birds are not generally known but are less than 0.2% of the diet in adult chickens, and may be less than 0.02% if phosphorus levels are low. Dietary calcium deficiencies are not usually observed in granivores at maintenance that are consuming seeds with about 0.1% calcium, because the phosphorus levels of these seeds are also low (Wilson *et al.*, 1969; Norris *et al.*, 1972; Rowland *et al.*, 1973).

GROWTH

The amount of calcium needed for growth has been determined by empirical methods, which establish the minimal dietary level that maximizes bone ash and bone breaking strength. The requirement is highest early in life, when the fractional growth rate is highest, and decreases as adult body weight is reached (NRC, 1994). For example, turkey poults require 1.2% dietary calcium during their first month and the requirement drops to 0.55% during their sixth month of life (Table 10.2). The rate of skeletal growth of altricial hatchlings is considerably faster than that of precocial birds, but the requirement has not been investigated. Presumably, the combination of a faster growth rate and lower calcification of the skeleton at hatching causes altricial species to have greater requirements than precocial species. The natural foods of most altricial granivores and insectivores contain insufficient calcium for skeletal growth. Parents usually supplement the diet of their nestlings with high-calcium sources, such as mollusk shells, eggshells, and bone fragments (Graveland and Van Gijzen, 1994).

EGG LAYING

The increase in calcium requirement for egg production can be estimated from the number of eggs laid, the amount of calcium in those eggs, and the pattern of egg laying (e.g., daily, alternate days). Bone acts as a buffer for daily calcium needs, so the requirement of birds that lay on alternate days is roughly half of that of birds laying similar-sized eggs daily. Small birds lay proportionally larger eggs than large birds, and small eggs have proportionally more shell. Thus, the calcium requirement for egg production of small birds is greater than that of large birds. Similarly, precocial species generally lay larger eggs than similar-sized altricial species and they have a correspondingly higher calcium requirement. The daily increase in the calcium requirement for various species at known egg production rates and sequences can be calculated if the following assumptions are made: (i) 99% of the calcium in an egg is found in the shell and shell is 95% ash, of which about 32% is calcium; (ii) dietary calcium is 70% digestible; and (iii) metabolic fecal and urinary losses are trivial. This last assumption is known to be valid for poultry and Japanese Quail, where endogenous calcium losses account for less than 0.2% of the requirement.

Most birds that lay only one egg can utilize medullary and some cortical bone

Table 10.2. Dietary calcium (Ca) and phosphorus (P) requirements of birds (from NRC, 1994, except as noted).

Species	Calcium (%)	Phosphorus* (%)	Ratio (Ca/P)
Ring-necked Pheasant			
0–4 weeks	1.0	0.55	1.8
4–8 weeks	0.85	0.5	1.7
9–17 weeks	0.53	0.45	1.2
Laying	2.5	0.40	6.3
Japanese Quail			
Growing	0.8	0.3	2.7
Laying	2.5	0.35	7.1
Muscovy Ducks[†]			
3–8 weeks	0.84	0.4	2.1
8–12 weeks	0.43	0.26	1.7
White Pekin Duck			
0–2 weeks	0.65	0.4	1.6
2–7 weeks	0.6	0.3	2.0
Laying	2.75	–	
Turkey			
0–4 weeks	1.2	0.6	2.0
4–8 weeks	1.0	0.5	2.0
8–12 weeks	0.85	0.42	2.0
12–16 weeks	0.75	0.38	2.0
16–20 weeks	0.65	0.32	2.0
20–24 weeks	0.55	0.28	2.0
Laying	2.25	0.4	5.6
Chicken			
0–3 weeks	1.0	0.45	2.2
3–6 weeks	0.9	0.35	2.6
6–8 weeks	0.8	0.3	2.7
Laying, every day	3.25	0.25	13.0
Laying, alternate days[‡]	1.88	0.25	7.5

*Nonphytate phosphorus.
[†]From Leclercq *et al.* (1990) and based on levels needed to optimize bone characteristics.
[‡]Calculated based on a maintenance calcium requirement of 0.5%.

to supply the acute demand to calcify its shell. Most large birds can lay a second egg, using bone reserves. The increase in calcium requirement in these birds is spread over the week prior to laying and is relatively small. Birds that lay clutches larger than two to three eggs must slightly increase their calcium intake while forming medullary bone and they must markedly increase calcium intake during the days that eggs are laid (Krementz and Ankney, 1995). The number of eggs that could be synthesized by utilizing all of the calcium in the entire skeleton is about five for Zebra Finch and ten for chickens. When fed adequate levels of dietary calcium, these birds are capable of laying continuously for many weeks with little or no loss in skeletal mass. The amount of calcium in the seeds and insects supplied by the natural diets of many birds is grossly inadequate for the production of large clutches and they must select calcium-rich foods as a supplement. When given the opportunity, birds often consume concentrated calcium sources, especially in the evening. Mollusk shells, for example, are slowly dissolved by acid conditions in the gizzard during the evening hours, supplying calcium for shell accretion and sparing the mobilization of bone (MacLean, 1973; Graveland and Van Gijzen, 1994; Houston *et al.*, 1995).

The calcium requirement for many species has been determined empirically by feeding diets with incremental levels of calcium and examining shell quality and egg numbers (Table 10.2). Continuous daily egg production by poultry, ducks, quail, and pheasants requires between 2.25 and 3.25% calcium in grain-based diets. Birds that are smaller, more active, and altricial may require considerably lower levels (% of the diet) of calcium for continuous egg production. For example, Cockatiels and Budgerigars can lay large clutches of eggs with normal shells while consuming diets with as little as 0.35% and 0.8% calcium, respectively (Earl and Clarke, 1991; Roudybush, 1996).

Deficiency

A calcium deficiency may occur due to a low dietary level of calcium or to excess dietary phosphorus. Insufficient vitamin D may cause a secondary calcium deficiency by impairing calcium absorption and bone formation. Increasing dietary calcium above normal required levels ameliorates but does not completely correct this secondary deficiency.

Adult birds that consume a calcium-deficient diet mobilize bone at a rate faster than it is deposited. Bone eventually becomes weak and porous. Bone ash and calcium contents may decrease to less than half of normal and appetite is suppressed. Long bones may become deformed or broken by the weight of the bird, the pull of muscles, or stress from movement. Depletion of cortical bone due to the consumption of a calcium-deficient diet by laying females leads to an osteoporosis-like condition. In chickens, this condition is referred to as cage-layer fatigue, because it is especially prevalent in individuals that are not very active (e.g., caged birds) and is most severe in the long bones of the legs. In females, a marginal calcium deficiency results in decreased eggshell calcification and a more severe deficiency reduces the number of eggs laid. The calcium depleted from cortical bone during egg production can be repleted over a period of a week or two following cessation of laying, if dietary calcium becomes adequate.

In growing chicks, a calcium deficiency causes skeletal abnormalities, including rickets, dyschondroplasia, lameness, enlarged painful joints, and misshapen bones. The epiphyseal plate of long bones may be large and uncalcified (Fig. 10.4).

Carnivores and piscivores consuming a diet of muscle or organs without bone often develop a calcium deficiency. A deficiency may even occur in adults at maintenance, due to a combination of the low levels of calcium (< 0.1%) and the high concentrations of phosphorus found in the soft issues of animal prey.

The excessively low ratio of calcium to phosphorus can quickly deplete bone stores. Insectivores or granivores kept on strict diets of insects or seeds, respectively, without a supplemental form of calcium, also develop calcium deficiency for the same reason. For example, Budgerigars consuming commercial bird seed with a calcium-to-phosphorus ratio of 1:37 quickly develop signs of a calcium deficiency unless given access to supplemental calcium (e.g., cuttlebone). In all cases the deficiency syndrome is called nutritional secondary hyperparathyroidism, because of the hypertrophy of the parathyroid and high levels of circulating parathyroid hormone that occur in an attempt to correct the nutritionally induced hypocalcemia. The more advanced stages of calcium deficiency result in cramps of the large muscle groups. The local muscle spasms progress to tetanic grand mal seizures, which are sometimes called 'fits.' At this stage of deficiency, intravenous calcium is required for recovery (Wallach and Fleig, 1970; Arnold *et al.*, 1973; Randall, 1981; Goodman, 1996).

Growing chicks, and especially laying females, have a specific appetite for calcium. If given a choice between two foods that are identical in every respect except calcium level, they will select a calcium-adequate food more frequently than a deficient one. Poultry and wild birds have been observed selecting high-calcium sources, such as mollusk shells or limestone, when laying eggs. In fact, calcium-rich items are especially chosen on days in which an eggshell is being formed. For example, caged Zebra Finches increase their intake of cuttlefish bone by a factor of three times on days when they are laying eggs. Chickens laying eggs will overconsume dietary energy in order to meet their requirement, although increased food consumption cannot compensate for low calcium levels very long (Gilbert, 1983; Houston *et al.*, 1995).

Calcium deficiencies in wild-bird populations are sometimes the major limitation to reproductive success. For example, acid rain decreases calcium levels and availability in soils, causing a decline in the calcium content of plants and arthropods and a decrease in the abundance of snails in some forests in Europe. Species, such as Great Tits, that rely on these calcium sources lay eggs with shells of such poor quality that embryos dry up, due to excessive evaporation (Drent and Woldendorp, 1989; Gravelan, 1996).

Another example of a calcium deficiency in wild birds was found in South Africa. Cape Griffon (Vulture) populations living in predominantly ranching areas devoid of large predators had access to the soft tissues of livestock carcasses, but the bones were too large to be transported to their nests and their hatchlings quickly developed rickets and other bone problems. Vulture populations in wildlife reserves had access to small bone fragments from hyena kills. They stored small bone fragments into their crops, transported them to their nests, and regurgitated them to their young. This form of supplementation resulted in normal bone growth. Providing Cape Vultures in ranching areas with 'bone restaurants,' consisting of manually crushed bones, eventually eliminated the deficiency problem and increased population numbers (Richardson *et al.*, 1986).

Toxicity

When dietary levels of calcium are in excess, absorption is minimal and most dietary calcium is excreted in the feces. Prolonged dietary excesses can lead to hypercalcemia, rickets, and precipitation of calcium urates in viscera, causing gout, and precipitation in the kidney, causing nephrosis. An excess of calcium interferes with the digestibility of other minerals, such as phosphorus, magnesium, manganese, and zinc, causing secondary deficiencies. High dietary levels of these minerals decrease the toxicity of calcium.

In growing chickens, calcium levels of greater than 1.5% are usually considered excessive. Higher levels may be tolerated if the ratio of calcium to phosphorus is maintained near optimal (Shafey, 1993). Calcium toxicities occur in captive birds, such as poultry, waterfowl, and game birds, when diets formulated for breeding females (2–4% calcium, 0.3% phosphorus) are fed to young chicks. Separate breeder and chick starter diets are an absolute necessity for these birds. However, the high-calcium breeder diets are generally tolerated by adult males, especially if phosphorus levels are adequate. For many aviary birds (e.g., psittacines, finches), a diet can easily be formulated that is adequate, but not excessive, in calcium for both the breeding female and her hatchlings.

Some calcium sources used to supplement diets have toxic levels of other minerals. For example, dolomitic limestone has toxic levels of magnesium, and some calcium phosphate sources may have toxic levels of vanadium or fluoride (Stillmak and Sunde, 1971; Scott *et al.*, 1982).

Calcium-to-phosphorus ratio

Dietary calcium and phosphorus interact during absorption, metabolism, and excretion. The calcium-to-phosphorus ratio of bone is slightly greater than $2:1$ and changes little over time. This ratio is considered to be optimal for the diet of poultry, but ratios between $1.4:1$ and $4:1$ are well tolerated if vitamin D is adequate (Shafey, 1993). Faunivorous mammals can usually tolerate lower ratios and herbivorous mammals can tolerate higher ratios, but similar comparative information for birds is not known.

Dietary calcium sources

Calcium is usually low in grains (0.02–0.10% of dry matter) and insects (0.01–0.4%) and relatively high in vegetative portions of plants (Table 10.3). Alfalfa and clover may contain calcium at 1.2–1.8% of dry matter. Whole-vertebrate prey contain high calcium levels (1.5–5.0%), but a very low level without bones (0.02–0.1%). The bioavailability of calcium in common supplements is: calcium carbonate, 100%; alfalfa, 88%; bone meal, 100%; calcium sulfate 90%; dolomitic limestone, 66%; limestone, 89%; eggshell, 100%; defluorinated phosphate, 94%; oyster shell, 100% (Soares, 1995).

Phosphorus

About 85% of a bird's phosphorus is found in bone. The rest of the phosphorus is found in phospholipids, in nucleic acids, and as part of a huge number of molecules important in intermediary metabolism. High-energy phosphate bonds are involved in the short-term storage of energy as adenosine triphosphate (ATP) and in the transfer of energy during metabolism. Inorganic phosphates (PO_4^{2-}) are present in cells and the blood and serve as an important buffer system.

Absorption

Phosphate absorption from the small intestine can occur by passive diffusion or by an energy-dependent process that is coupled to the cotransport of sodium. The rate of sodium-dependent transport is increased by 1,25-dihydroxy vitamin D_3. Once absorbed, some of the phosphorus is lost, due to endogenous excretions into the intestines and urine. In growing chickens consuming a diet with a ratio of calcium to phosphorus of 2:1, about 20% of the absorbed phosphorus is lost in endogenous excretions. This amount increases as the calcium to phosphorus ratio narrows, probably because more phosphorus is absorbed (Almasri, 1995).

Dietary phosphorus may be present in inorganic or organic forms (Ravindran *et al.*, 1995; Soares, 1995). Most common forms of inorganic phosphorus found in foods are readily absorbed from the diet. Organic phosphorus in the form of phospholipids or phosphoproteins is readily digested and absorbed, but not all organic phosphorus is utilized by birds. More than half of the phosphorus in the seeds of plants is poorly utilized, because it is a component of phytic acid (see Fig. 10.3). To be nutritionally utilized, phytic acid must be enzymatically hydrolyzed by phytases to produce phosphoric acid and orthophosphate salts. Phytases are present in the small intestine of chickens and Japanese Quail, but their activity is insufficient to permit complete utilization of the phosphorus in phytic acid. Phytic acid complexed with minerals, such as calcium, is particularly resistant to hydrolysis. For this reason, phosphorus in phytic acid is not usually considered to be available to birds and plant phosphorus levels are often expressed on a nonphytate basis. When the proportion of total phosphorus that is in phytic acid is not known, a value of 60–70% is typically used for seeds. Vegetative parts of plants and tubers typically have a much lower proportion of their phosphorus in phytic acid (< 25%). Phytic acid forms complexes with cations, such as Ca^{2+}, magnesium (Mg^{2+}), potassium (K^+), manganese, iron, and zinc, and reduces their availability as well. Some seeds may contain high levels of phytases, which have some beneficial effect in releasing phytate phosphorus if food is stored in a moistened state in the crop. Intestinal microbes can produce high amounts of phytase and presumably those species that have extensive cecal fermentation can better utilize phytic acid.

High levels of dietary calcium decrease the absorption of phosphorus by forming precipitates in the intestines. Inorganic phosphorus found in rocks and soils is relatively unavailable. For example, calcium pyrophosphate and calcium metaphosphates are completely unavailable. However, heat treatment and

Table 10.3. Calcium (Ca) and phosphorus (P) content of some avian foods.[*]

Food	Calcium (%)	Phosphorus (%)	Ratio (Ca/P)
Alfalfa leaves	1.96	0.30	6.53
Almond	0.25	0.53	0.46
Anchovy	0.41	0.51	0.80
Apple	0.13	0.12	1.08
Barley	0.05	0.40	0.13
Bean, red kidney	0.12	0.45	0.27
Beef, flank	0.05	0.71	0.06
Bone meal	29.80	12.50	2.38
Calcium carbonate	38.00	0.00	> 100.00
Calcium phosphate, dibasic	22.00	18.70	1.18
Calcium phosphate, monobasic	16.00	21.00	0.76
Carrot	0.28	0.27	1.03
Chicken, 7 days old, whole	2.21	2.06	1.07
Chicken, meat	0.03	0.69	0.04
Clam, meat	0.03	0.75	0.05
Clover leaves	1.71	0.35	4.89
Corn	0.02	0.31	0.06
Cricket	0.34	0.86	0.40
Earthworm	0.20	1.00	0.20
Fairy shrimp	0.30	1.40	0.21
Figs	0.16	0.10	1.64
Grapes	0.06	0.04	1.33
Halibut, meat	0.05	0.74	0.06
Herring, whole	2.40	0.95	2.52
Meat and bone meal	8.99	4.46	2.02
Mealworm	0.27	0.44	0.60
Midge larvae	0.50	1.30	0.38
Millet, pearl	0.06	0.35	0.17
Oats	0.07	0.30	0.23
Orange	0.29	0.14	2.05
Oyster, meat	0.61	0.93	0.66
Peanut, shelled	0.05	0.28	0.18
Rat, weanling, whole	2.50	1.68	1.49
Rice	0.09	0.09	1.00
Rye	0.07	0.36	0.19
Salmon, coho, meat	0.79	0.94	0.85
Sardines, with bones	1.14	1.31	0.88
Silk moth, larvae	0.21	0.54	0.39
Snail (excluding shell)	4.20	0.90	4.67
Sorghum	0.05	0.34	0.15
Soybean meal	0.23	0.73	0.32
Spinach	1.00	0.55	1.82
Squid	0.11	1.21	0.09
Sunflower seeds, dehulled	0.12	0.84	0.14
Wheat, hard red winter	0.06	0.42	0.14

[*]Expressed as % of dry matter.

hydration of these rock phosphates to mono-, di-, and tricalcium phosphates renders the phosphorus almost completely available. Hydrated forms of calcium phosphates are more soluble in the acid environment of the proventriculus and gizzard and are more available than anhydrous forms. Calcium phosphates that are not solubilized are excreted in the same chemical form as consumed (Gillis *et al.*, 1962; Rucker *et al.*, 1968; Rao *et al.*, 1995).

Requirement

Bone ash or bone breaking strength is frequently used as a response criterion in studies to determine the phosphorus requirement. Stated requirements assume that all of the phosphorus consumed has a bioavailability that is similar to that of dicalcium phosphate. Sources with lower bioavailability should be discounted proportionally. The requirement for phosphorus is highest in the growing chick and gradually declines as skeletal growth subsides (see Table 10.2). A maintenance level of about 0.1% of the diet is adequate for chickens when calcium levels are 0.2%.

The phosphorus requirement increases during egg laying to an extent which is much greater than can be accounted for by the phosphorus content of the egg. This is because the high demand for calcium to synthesize eggshells causes bone to be mobilized. The ratio of calcium to phosphate in bone (2.5 : 1) is much lower than that in eggshell (20 : 1) and much of the excess phosphate that is liberated is excreted by the kidneys (Wideman, 1987; Clunies *et al.*, 1992). This endogenous loss results in an increase in the phosphorus requirement. However, the increase in the phosphorus requirement for egg production is considerably less than the increase in the calcium requirement and a wide calcium-to-phosphorus ratio is warranted. For example, in actively laying Japanese Quail hens, the optimum ratio is 7 : 1. The phosphorus requirement of birds of any age is increased by high dietary calcium levels or a vitamin D deficiency.

Deficiency

In young growing chicks, a moderate deficiency of phosphorus, or a wide calcium-to-phosphorus ratio causes loss of appetite, slower growth, and poor bone mineralization. A more severe deficiency causes rickets, weakness, and eventually death. In laying females, a phosphorus deficiency reduces food intake, which may impair egg production in some species. In Northern Bobwhite Quail, a phosphorus-deficient diet decreases egg fertility as well as the number of eggs in a clutch. A phosphorus deficiency can be diagnosed by low serum inorganic-phosphorus concentrations. Unlike in a calcium or vitamin D deficiency, the parathyroid gland does not increase in size and may even atrophy. Phosphorus deficiencies are relatively rare in most wild birds, because of the high level of phosphorus found in many plant and animal food sources. However, strict granivores consuming seeds that are high in phytate (e.g., cereal grains) together with high-calcium grit may display signs of a phosphorus deficiency. Most commercial diets for captive birds are based on high-phytate grains and must be supplemented with phosphorus to prevent a deficiency (Cain *et al.*, 1982; Wilson and Duff, 1991; NRC, 1994).

Toxicity

Excessively high levels of phosphorus can cause a calcium deficiency. In birds laying eggs, excess phosphorus causes a thinning of the eggshell, independent of calcium status. Some inorganic phosphate supplements contain potentially toxic levels of other minerals, such as vanadium, fluoride, iron, and cadmium (McDowell, 1992; Sullivan and Douglas, 1994).

Dietary sources

All foods of animal origin are excellent sources of highly available phosphorus. Plant vegetative parts are good sources of phosphorus, but seeds may be marginal or deficient in phosphorus. High-phosphorus supplements and their bioavailability include: bone meal, 94%; dibasic calcium phosphate ($CaHPO_4$), 85–100%; monobasic calcium phosphate ($Ca(H_2PO_4)_2$), 90–100%; defluorinated rock phosphate, 70–96%; phosphoric acid, 128%; fish meal, 100%; meat and bone meal, 100%; calcium metaphosphate, 0%; alfalfa meal, 77%. Monobasic calcium phosphate contributes more acidity to a diet than equimolar amounts of dibasic calcium phosphate. In commercial avian diets, phosphorus is the most expensive mineral that is supplemented. Additionally, oversupplementation results in high levels of phosphorus excretion, which pollutes streams, rivers, and estuaries. For these reasons, phosphorus levels are often minimized and phytase enzymes of microbial origin are sometimes supplemented to the diet to increase the digestibility of phytate (Keshavarz, 1994; Soares, 1995; Mitchell and Edwards, 1996).

Electrolytes: Sodium, Potassium, and Chloride

The concentration and balance of electrolytes in a bird's intracellular and extracellular fluids are critical for life and are therefore tightly regulated. Among dietary minerals, sodium (Na^+) and K^+ are the major cationic electrolytes and chloride (Cl^-) is the major anionic electrolyte. A bird's dietary requirement for these minerals must be considered in two ways: the minimum amount of each needed for specific cellular functions; and the balance between different electrolytes necessary to optimize important osmotic and acid–base relationships.

Specific functions of sodium, potassium, and chloride

Dietary sodium, potassium, and chloride are readily absorbed in the intestines. As described in Chapter 3, these electrolytes are actively secreted in the upper regions of the gastrointestinal tract and net absorption occurs in the rectum and cecum. The final body level is determined by coordination of excretion and absorption, which are regulated by aldosterone, angiotensin, and antidiuretic hormone. All three of these minerals are important in maintaining ionic balance in body fluids. Sodium is the predominant extracellular cation and functions to maintain plasma volume and osmolarity. Potassium is the predominant intracellular cation and functions to maintain intracellular fluid volume and osmolarity. The potential energy resulting from the separation of these two cations across the cell membrane drives nerve impulses and other forms of cellular communication. Chloride is the primary extracellular anion and also participates in maintenance of plasma osmolarity, fluid volume, and nerve impulses. Concentration of Cl^- by active transport in gastric glands results in acid production in the proventriculus. This hydrochloric acid (HCl) is an important component of gastric digestion. All three of these minerals have regulatory and cofactor roles with a wide variety of enzymes. They also function in active cotransport of many nutrients and waste products across cell membranes and provide the osmotic drive for water movement.

Dietary electrolyte balance

Maintenance of constant internal pH is critical to the function of all physiological processes and the structural integrity of macromolecules. When mineral salts of organic acids (e.g., potassium citrate, sodium bicarbonate) are consumed, the metabolism of the organic acids to carbon dioxide (CO_2) and water (H_2O) consumes H^+ and contributes to metabolic alkalosis. Likewise, metabolism of mineral salts of organic bases and sulfur-containing amino acids consumes hydroxyl ions, causing acidosis. Most foods have an excess of organic acids over organic bases, so the difference in charge is balanced by an excess of mineral cations (Na^+, K^+, Ca^{2+}, Mg^{2+}) over mineral anions (Cl^-, $H_2PO_4^-$, HPO_4^{2-}, SO_4^{2-}). The organic acids (e.g., citric, aconitic, carbonic, oxalic, and malic acids) are very diverse and difficult to quantify. Thus, they are often referred to as undetermined anions, because they are not actually measured analytically, but their presence is inferred by the mismatch in the quantity of mineral anions versus mineral cations. As most of the dietary organic acids and bases are associated with Na^+, K^+, and Cl^-, these three electrolytes are an analytically convenient indicator of the acid–base balance of a diet. Dietary electrolyte balance in milliequivalents (meq) is usually expressed as: $Na^+ + K^+ - Cl^-$. However, the inclusion of Ca^{2+}, Mg^{2+}, $H_2PO_4^-$, HPO_4^{2-}, and SO_4^{2-} in the equation may often be appropriate. The higher the cation excess, the more alkaline the diet and, conversely, the lower the cation excess, the more acidic the diet (Teeter *et al.*, 1985; Austic and Patience, 1988; Mongin, 1989).

The appropriate electrolyte balance is dependent on the physiological state of the bird and the species. The physiological state is important, because processes such as bone formation, eggshell deposition, ketosis, lactic acidosis, and respiratory rate affect the bird's acid–base balance (Table 10.4). Chickens have been well studied and a balance in the range of 200–300 meq kg^{-1} is optimal for growth. A lower balance results in poor bone mineralization and a variety of structural anomalies of the leg bones and joints (e.g., tibial dyschondroplasia). Formation of eggshells consumes CO_3^{2-} and causes a metabolic acidosis, resulting in an optimal dietary electrolyte balance that is higher than that needed for growth. A low dietary balance (< 200 meq kg^{-1}) results in thinning of eggshells.

In hot weather, birds may cool themselves by panting. In the process, they expire more CO_2 than they produce, leading to a respiratory alkalosis. A lower balance of dietary electrolytes ameliorates some of the negative impact of

Table 10.4. Examples of acid- or base-generating processes occurring in avian metabolism.

Metabolic process	Equation
Complete oxidation of sulfur amino acids	Methionine $+ O_2 \rightarrow$ uric acid $+ CO_2 + H_2O$ $+ 2 H^+ + SO_4{}^{2-}$
Metabolism of salts of organic acids	Na citrate $+ O_2 \rightarrow CO_2 + H_2O + Na^+ + OH^-$
Metabolism of salts of organic bases	$NH_4Cl + CO_2 \rightarrow$ uric acid $+ H^+ + Cl^-$
Ketosis	Palmitate $+ 5 O_2 \rightarrow 4$ acetoacetate $+ 4 H_2O +$ $4 H^+$
Lactic acidosis	Glucose $\rightarrow$ lactate $+ 2 H^+$
Bone formation	$10 Ca^{2+} + 4.8 HPO_4{}^{2-} + 1.2 H_2PO_4{}^- + 2 H_2O$ $\rightarrow$ hydroxyapatite $+ 9.2 H^+$
Shell formation	$Ca^{2-} + CO_2 + H_2O \rightarrow CaCO_3 + 2 H^+$
Panting	$HCO_3{}^- \rightarrow OH^- +$ respired CO_2

respiratory alkalosis. The consumption of acidified diets or water increases a bird's resistance to heat stress.

To date, all of the studies on dietary electrolyte balance have been carried out on the granivorous and omnivorous birds commonly kept in captivity (poultry, game birds, pigeons) and other dietary groups have not been considered. Studies on acid–base balance in a limited number of avian species indicate that herbivores typically consume a very alkaline diet (high electrolyte balance) and excrete an alkaline urine (Long, 1982). Faunivores typically consume a very acidic diet (low electrolyte balance) and excrete an acid urine. Presumably the optimal electrolyte balance is lower for faunivores than for herbivores.

Requirement

The requirement of poultry, quail, pheasants, and ducks for sodium and for chloride is between 0.1 and 0.2% of the dry matter during growth and egg production. The maintenance requirement is about 0.05%. Ruffed Grouse apparently require lower dietary sodium levels than domestic poultry. Marine birds have high obligatory salt excretion via their salt glands and presumably have sodium and chloride requirements that are higher than those of poultry (NRC, 1994; Jakubas *et al.*, 1995).

A molar ratio of sodium to chloride of 1.5:1 (1:1 on a weight:weight basis) is typically recommended, although the dietary potassium level modifies the ideal ratio by affecting the overall electrolyte balance. High levels of dietary sodium or potassium increase the chloride requirement. Conversely, high levels of dietary chloride increase the sodium and potassium requirements.

Deficiencies

A sizable pool of sodium, potassium, and chloride is found in bone. This pool becomes available when bone is mobilized during chronic food deprivation, but is not readily mobilized when normal quantities of an electrolyte-deficient diet are consumed. Chronic consumption of a diet low in chloride results in hypochloremia, hemoconcentration, dehydration, and reduced growth or egg production. Chloride-deficient chickens or turkeys that are stressed by noise or handling may go into tetany, with their legs extended to the rear. The primary signs of insufficient dietary potassium are reduced appetite, hypokalemia, muscle weakness, respiratory distress, and cardiac weakness. A sodium deficiency causes decreased plasma fluid volume, low blood pressure, decreased bone strength, poor growth, and a reduction in reproductive capacity (Leach and Nesheim, 1963; Scott *et al.*, 1982).

Birds have a specific appetite for sodium chloride (Cade, 1964). When deficient, they develop a craving and actively seek out salt sources. All species that have been examined can easily choose between diets identical in all aspects except sodium chloride concentration in order to meet their requirements and avoid excesses. When a choice is not given, birds typically do not increase their consumption of low-salt foods in order to obtain their electrolyte requirement. In other words, they will not overconsume metabolizable energy in order to meet their requirement for sodium or chloride.

Florivorous birds are much more at risk for a sodium deficiency compared with faunivorous birds. This is because the vegetative components of young plants are typically low in sodium and relatively high in potassium. Foods of animal origin are usually adequate in all three electrolytes. However, captive marine birds (albatrosses, penguins) kept in fresh water and fed freshwater fish or marine invertebrates may develop a sodium deficiency. This phenomenon is partly due to the fact that marine birds have high obligatory rates of sodium chloride excretion via their salt glands. Fish and invertebrates commonly fed to captive animals have often lost some of their electrolytes, due to freeze-thawing and washing. Further, parents that regurgitate food for their young lose electrolytes in the process and this exacerbates marginal dietary levels (Frings and Frings, 1959; Gailey-Phipps, 1982).

Diarrhea and other gastrointestinal disturbances increase the excretion and decrease the absorption of electrolytes, markedly increasing the requirement. Electrolyte replacement therapy is a common component of the veterinary care given to birds suffering from intestinal infections.

Toxicity

Birds that have salt glands are very resistant to toxicities of sodium or chloride (Chapter 3). Birds without salt glands can tolerate high levels of dietary sodium, potassium, or chloride if fresh water is available. High rates of water consumption permit high levels of electrolyte excretion by the kidney. For example, Ring-necked Pheasants and Northern Bobwhite Quail can tolerate a diet with 5% salt by increasing their water consumption and urine production. Salt toxicosis usually causes dehydration, diarrhea, ataxia, nervousness, and edema. A bird's tolerance for any of the three electrolytes is intimately related to the dietary concentration of the other two. If the diet is marginally adequate in potassium and chloride, sodium may become toxic at much lower levels. Similarly, potassium is more toxic when sodium is present at low levels than when it is in surfeit.

High levels of any of these electrolytes result in the excretion of dilute urine and wet droppings (Scott *et al.*, 1960, 1982; Austic and Patience, 1988; Wages *et al.*, 1995).

Young chickens and ducklings are more sensitive to salt toxicosis than older birds. This age-related susceptibility even occurs in Mallards and American Black Ducks, which have salt glands, because this organ is relatively inactive in young chicks. Altricial chicks are often more sensitive to salt toxicosis than precocial chicks. The parents of altricial chicks often practice behaviors to minimize salt intake and maximize water intake of their chicks, especially in arid or high-saline environments. These behaviors include selecting foods low in salt and high in moisture, washing foods in fresh water, feeding high-moisture crop or esophageal secretions, and regurgitating water to their chicks. For example, White Ibis feeding in highly saline lakes and estuaries can consume high-salt foods, such as crabs and mollusks, if they can occasionally obtain fresh water. However, they must provision their young with low-salt foods taken from freshwater lakes. Nestlings usually do not have a freshwater source and suffer from a salt toxicity if fed the same foods consumed by their parents (Johnston and Bildstein, 1990).

Electrolytes in the water are much more toxic than comparable levels in the food. When water levels are high, attempts to increase water consumption result in more electrolyte consumption. In poultry, tolerable water levels are about ten-fold lower than tolerable feed levels (NRC, 1994).

Dietary sources

Most plant vegetation, seeds, and fruits contain relatively low amounts of sodium, but usually are adequate or rich in potassium. Foods of animal origin are typically good sources of electrolytes. Supplementation of sodium and chloride is usually accomplished by providing common table salt (NaCl). Potassium can be supplemented as potassium chloride, but it is rarely needed. Efforts to manipulate the acid–base balance of a diet require the use of sodium bicarbonate, potassium bicarbonate, or calcium chloride as dietary supplements. Wild birds often consume ashes, salt water, mineral soils, or rock salt placed on roads to melt snow to supplement a sodium-poor diet.

Magnesium

Magnesium acts as a cofactor or an activator of many critical enzymes, including the reactions involving ATP that energize all major metabolic pathways. Magnesium is absorbed throughout the intestines by active transport. More than 50% of dietary magnesium is taken up in the duodenum and the first portion of the jejunum of the chicken. Dietary factors, such as high phosphate or fat, that decrease the ionization of magnesium from salts or organic complexes impair its bioavailability. Magnesium ions in the blood are taken up by all cells of the body and, like potassium, have a high intracellular-to-extracellular concentration gradient. Magnesium in excess of the minimum requirement is stored in the bone. Further excesses of magnesium are excreted through the kidney. The renal tubules have considerable abilities to reabsorb magnesium and endogenous losses are kept very low when dietary levels are marginal.

Most of the magnesium in bird eggs is found in the shell and becomes available to the developing embryo as it withdraws calcium (Tao *et al.*, 1983; McDowell, 1992).

Requirement, deficiency, and toxicity

Very little attention has been paid to the magnesium requirement of birds. Experiments in poultry and quail indicate that the requirement does not exceed 0.06% of the diet dry matter at any stage of the life cycle (Tao *et al.*, 1983; NRC, 1994). Magnesium deficiencies are rare, because most foods have concentrations that are well above the requirement. The most probable reason for a deficiency is high dietary calcium or phosphorus levels. However, most calcium and phosphorus supplements are high in magnesium and such an occurrence is unlikely. When a deficiency is induced by prolonged feeding of a diet consisting of purified ingredients devoid of magnesium, symptoms that develop are indicative of a disruption in ATP metabolism: lethargy, low basal metabolic rate, gasping for breath, ataxia, slow growth, and periods of coma. In Japanese Quail chicks, dietary magnesium levels at twice the requirement increase the magnesium stores in bone sufficiently to permit a magnesium-free diet to be tolerated for about a month before deficiency symptoms develop. Ducklings fed a magnesium-deficient diet from hatching begin to die within 2 weeks and show signs of neuromuscular hyperirritability. Low serum magnesium levels are diagnostic of a deficiency. A deficiency in laying females can be diagnosed by low yolk magnesium levels. (Bird, 1949; Gardner *et al.*, 1960; Edwards and Nugara, 1968).

In growing chickens, a magnesium toxicity can be detected at about 1% dietary magnesium (McWard, 1967). Excess magnesium interferes with calcium absorption and metabolism and symptoms are similar to a calcium deficiency: poor bone mineralization, thin eggshells. Excess magnesium also acts as a laxative and causes wet droppings. High dietary calcium and phosphorus levels mitigate a magnesium toxicity.

Dietary sources

Magnesium is chelated in the porphyrin moiety of chlorophyll in plant cells and is associated with a very large number of enzymes in animal cells. Thus, most unprocessed foods have adequate magnesium to meet known requirements in all parts of the life cycle. Particularly rich dietary sources include limestone (especially dolomitic limestone), bone, and vegetative parts of legumes. Seeds are relatively low in magnesium, but are still adequate.

Copper

Copper is a component of a variety of intracellular and extracellular enzymes, including cytochrome oxidase, lysyl oxidase, superoxide dismutase, and tyrosinase. The apparent digestibility of copper in chicks is low, typically less than 30%. Most of the absorption of dietary copper occurs in the proventriculus and duodenum. High levels of dietary calcium, phosphorus, phytic acid, oxalate, and tannins decrease copper absorption by forming insoluble complexes in the intestines. Copper homeostasis is maintained predominantly by controlling the rate of absorption. When dietary copper levels are high, intestinal epithelial cells synthesize metallothionein, a cysteine-rich protein that tightly binds copper and retards its absorption. High levels of dietary zinc or cadmium impair copper absorption by inducing metallothionein, interfering with transfer across the enterocyte.

Copper in excess of immediate requirements is stored complexed to metallothionein in the liver and other tissues. For example, in Greater Flamingos and Little Egrets, liver and kidney metallothionein levels increase as copper levels increase. Further excesses of copper are excreted through the bile. Most plasma copper is found as a component of ceruloplasmin. Ceruloplasmin is necessary for iron transport and also plays a role in the acute phase of the immune response. The need to synthesize ceruloplasmin during an infection increases the copper requirement (Cosson, 1989; Noy *et al.*, 1994; Baker and Ammerman, 1995a; Koh *et al.*, 1996).

Requirement, deficiency, and toxicity

The copper requirement does not exceed 8 mg kg^{-1} dry matter in any avian species examined to date. Young growing birds have a higher copper requirement than adults and are more likely to show a deficiency (NRC, 1994).

Copper deficiency symptoms include anemia, hemorrhaging, lameness, infertile eggs, and poor feather pigmentation (Scott *et al.*, 1982). Anemia is due to low levels of ceruloplasmin, which cause poor iron utilization. Hemorrhaging is due to defective elastin synthesis, resulting in poor elasticity and integrity of blood vessels. Death from a copper deficiency is often due to an aortic aneurysm, which develops in a weakened and inelastic aorta. The lameness associated with a copper deficiency is due to a failure in the cross-linking of collagen in bone, particularly in the growth plate. The epiphyseal cartilage becomes thickened and poorly vascularized (Fig. 10.4d). The resulting bone and joint abnormalities cause an enlargement of the hocks and bowing of the legs, giving a condition known as perosis. The copper-containing enzyme, tyrosinase, is responsible for melanin synthesis and a copper deficiency induces pigmentation defects. The eggs from copper-deficient females are often infertile and may have abnormal shells, due to defective collagen in the shell membrane. Plasma copper levels or, better, ceruloplasmin levels are diagnostic of a copper deficiency. However, plasma copper and ceruloplasmin levels of deficient birds increase toward normal levels during the acute phase of an infection (Koh *et al.*, 1996).

Relatively high levels of dietary copper are generally well tolerated by poultry, quail, and ducks. Chronic intake of high dietary copper (> 250 mg kg^{-1}) results in accumulation in the liver, may slow growth, and causes gastroenteritis and anemia. The gastroenteritis is usually most severe in the gizzard, where the koilin layer may be thickened and may hemorrhage (Wight *et al.*, 1986). High dietary zinc protects against copper toxicity, probably by inducing metal-lothionein production, impairing absorption, and facilitating detoxification.

Dietary sources

Copper levels in plants vary from 1 to 50 mg kg^{-1} dry matter, depending upon the species and soil factors. The copper in plant vegetative parts may be poorly available, and high levels of molybdenum and sulfur found in some plants interfere with copper utilization. Cereal grains contain marginal levels of copper (4–8 mg kg^{-1}), but seeds of cultivated legumes usually contain higher levels. The amount of copper in animal and insect tissues is variable, depending on their diet. However, the copper present in foods of animal origin is highly available and levels are usually more than sufficient to meet avian requirements. The bio-availability of copper from plant sources is usually considerably lower than that from animal sources (Aoyagi *et al.*, 1993).

Copper is commonly supplemented to diets fed to captive birds as copper sulfate, copper chloride, or copper oxide. In general, cuprous (Cu$^+$) compounds are more available than cupric (Cu^{2+}) compounds (Baker and Ammerman, 1995a). Chelates or complexes of copper with organic molecules are often used when there is a concern about poor bioavailability or interactions with other minerals. Copper sulfate is sometimes added to diets at levels far in excess of the requirement (100–250 mg kg^{-1}) because it has an antimicrobial effect. In particular high dietary copper is used to treat crop mycosis.

Iron

Dietary iron is found in both organic and inorganic forms and in either the ferrous (Fe^{2+}) or ferric (Fe^{3+}) valence state. The digestibility of dietary iron is extremely variable and influenced by a wide variety of bird and dietary factors. Iron absorption occurs throughout the digestive tract, but the duodenum and jejunum are the most important sites. Ferric iron in the food must be converted to the ferrous form in the proventriculus and gizzard by the action of hydrochloric acid prior to absorption. Ascorbic acid or cysteine facilitates the reduction of iron to the ferrous state and improves absorption. Heme iron is absorbed intact, without release from the heme moiety, and utilizes a transport mechanism that is distinct from that used for ionic iron. Iron absorption is tightly controlled to avoid excess uptake. When body stores of iron are adequate, most of the iron that enters the enterocyte is bound to ferritin and is not transferred to the blood. The ferritin iron remains in the enterocyte until shed into the lumen of the gastrointestinal tract. This mucosal block is important, because iron homeostasis is not effectively achieved by urinary or biliary excretion (McDowell, 1992; Saiz *et al.*, 1993; Henry and Miller, 1995).

Ferrous iron is oxidized to the ferric state following transfer to the blood and is tightly complexed by transferrin. Cells that need iron express a transferrin receptor and internalize the Fe^{3+}–transferrin complex. Erythropoietic cells of the bone marrow are particularly active at iron uptake and use it for the synthesis of hemoglobin. Cells require iron as a cofactor for various enzymes (oxidases and oxygenases), for electron transport (cytochromes), and for oxygen binding (hemoglobin, myoglobin). Iron is also an important chromophore in red- and black-pigmented feathers.

Iron stores are found in the liver in association with ferritin and hemosiderin. Ferritin usually predominates at normal iron levels and hemosiderin is formed to buffer excess iron. Iron in the body is always very tightly complexed by proteins and very little free iron is present. This is necessary because free ionic iron catalyzes the generation of reactive oxygen molecules, via the Fenton reaction, causing oxidative damage to cells and connective tissue. The very tight chelation of iron in body fluids makes it the first limiting nutrient for the replication of bacteria and parasites in a bird's tissues. Withholding iron from pathogens is an important component of disease resistance. During the acute phase of an immune response, the liver produces large amounts of transferrin, which transfers iron out of tissue fluids and sequesters it into compartments that are nutritionally unavailable to bacteria and parasites. Hepatic secretion of hemopexin permits heme from damaged cells to be scavenged so that it does not contribute to the nutrition of pathogens. Attempts to reverse this 'anemia of infection' by iron injections or very high dietary supplementation increase the mortality and morbidity due to a variety of disease organisms (Hallquist and Klasing, 1994).

Iron is found in egg yolk in association with phosvitin. Very little iron is found in the albumen. However, albumen contains large amounts of ovotransferrin, which differs from transferrin only in its glycosylation. Ovotransferrin complexes essentially all free iron and protects the egg from bacterial infections (Romanoff and Romanoff, 1949; Morgan, 1975; Spick *et al.*, 1988).

Requirement, deficiency, and toxicity

The iron requirements of poultry and Japanese Quail have been estimated at between 50 and 120 mg kg^{-1} dry matter in a corn–soybean meal-based diet for all stages of the life cycle (NRC, 1994). Iron requirements are influenced by its chemical form in the diet and by the amounts and proportions of other dietary components. Iron in foods of animal origin is considerably more available than in those of plant origin. Roughly, plant iron is less than 10% available and iron in animal foods is about 30% available. High levels of dietary phytic acid, tannins, fat, manganese, zinc, or copper decrease the absorption of inorganic iron and increase the apparent requirement. Heme iron is much less affected by these dietary factors. The requirement is increased by biting insects or parasites that cause substantial blood loss (Bafundo *et al.*, 1984; Baker and Halpin, 1991; Pimentel *et al.*, 1992).

Iron is stored in the liver and an adequate diet can buffer a dietary deficiency for many weeks. Birds raised on concrete or in wire cages are most susceptible, because they do not have access to the soil, which provides supplemental iron. Fast-growing, granivorous chicks in an iron-poor environment are most susceptible to a deficiency. An iron deficiency is first indicated by hypochromic, microcytic anemia. In other words, the circulating red blood cells are low in number, pale red, due to low hemoglobin, and small in size. A chronic iron deficiency results in a reduction in the color intensity of red and black feathers in many, but not all, species. Iron-deficient females lay eggs that undergo embryonic mortality, due to anemia during the later stages of development (Davis *et al.*, 1962; Morck and Austic, 1981).

Iron toxicity can occur from very high levels of dietary iron (2000 mg kg^{-1}). The primary diagnostic indication is hemochromatosis, which is the deposition of microscopically visible brownish-yellow granules of hemosiderin in lysosomes of hepatic cells. As storage increases, the lysosomes become damaged and ionic iron is released, causing oxidative damage to cell membranes and proteins. In advanced stages, the liver becomes cirrhotic.

A syndrome known as iron storage disease occurs in some species of birds at relatively low levels of dietary iron. The primary symptom is hemochromatosis, but other symptoms of this syndrome are splenomegaly, cardiomegaly, ascites, anorexia, depression, and skin pigmentation. This disease is common in mynas, toucans, hornbills, tanagers, birds of paradise, and starlings kept in captivity, but is rare in their wild counterparts. The contribution of dietary iron to this disease is controversial. It is interesting that iron storage disease occurs predominantly in frugivores and insectivores and that many fruits and insects are a poor source of iron. Apparently, susceptible species do not sufficiently decrease iron absorption when storage pools are replete, leading to iron accumulation in the liver. The stresses of captivity and infection also appear to be a major contributor to this disease. In mammals, stress causes the induction of lactoferrin, which mediates a flux of iron to the liver, while transferrin levels decrease, so that iron absorption from the gastrointestinal tract is reduced. Birds do not make a functional equivalent to lactoferrin, and transferrin levels increase during stress, which apparently increases the flux of iron to the liver and simultaneously increases absorption (Kincaid and Stoskopf, 1987; Ward *et al.*, 1991; Dierenfeld *et al.*, 1994; Hallquist and Klasing, 1994; Cork *et al.*, 1995).

Dietary sources

Most grains contain 30–60 mg kg^{-1} iron and seeds of legumes are typically higher (McDowell, 1992). The vegetative parts of most legumes and grasses contain iron, at 50–700 mg kg^{-1} dry matter. Fruits are especially low in iron, with about 15–40 mg kg^{-1} dry matter. The high variability in the iron content of plants is partly species-dependent but is also due to soil attributes, such as content and chemical form of iron, pH, and levels of other minerals. Vertebrate prey usually contain about 500 mg kg^{-1} iron. Calcium phosphate, limestone, and many other commercial mineral supplements contain very high levels of iron (2000–5000 mg kg^{-1}). Drinking water may also contain nutritionally significant levels of iron.

Iron is commonly supplemented as ferrous sulfate, ferric chloride, ferric citrate, ferric ammonium citrate, and iron amino acid complexes. Iron oxide is poorly available to all birds that have been examined (Henry and Miller, 1995).

Zinc

Zinc is absorbed by passive diffusion throughout the small intestine, with the duodenum being the most important site. Zinc traverses the epithelial cells bound to cysteine-rich intestinal protein (CRIP). When body stores and dietary zinc are both high, intestinal cells synthesize large amounts of metallothionein, which tightly binds zinc and prevents its binding to CRIP for subsequent transport to the blood. Zinc absorption is decreased by high dietary levels of phytic acid, calcium, phosphorus, copper, cadmium, or chromium. Calcium and phosphate form insoluble precipitates with zinc in the intestinal lumen. Copper, cadmium, or chromium induce metallothionein in the intestinal epithelium, blocking zinc absorption. Zinc complexed to some small ligands, such as EDTA, citrate, or amino acids (histidine, cysteine, methionine), is less influenced by the negative effects of other minerals or phytic acid (Kratzer and Vohra, 1986; Hempe and Cousins, 1992; Baker and Ammerman, 1995b).

Once absorbed, zinc is bound to albumin and the albumin-bound zinc is readily transferred to the tissues. The highest uptake of zinc occurs in bone, but the liver, intestines, pancreas, and spleen also accumulate zinc. In these latter tissues, part of this zinc is bound to metallothionein and serves as a storage pool. However, this pool does not buffer a dietary zinc deficiency for very long. For example, in growing Japanese Quail and chickens stores last for only about 4 days (Fig. 10.5). Bone stores last a few days longer, but a zinc deficiency soon develops. Thus, of all of the trace minerals, storage pools of zinc are the smallest relative to the requirement, and dietary deficiencies are not well buffered (Harland *et al.*, 1975; Emmert and Baker, 1995).

Zinc has a very large number of functions and is an activator or a cofactor of more than 200 enzymes. Often, zinc stabilizes 'zinc finger' structural motifs, which are involved in protein–DNA interactions and are important in gene regulation. Zinc is also an important pigmenting metal in feathers. Clearly, zinc is among the most metabolically active of the trace minerals, and deficiencies impair all metabolic pathways, regulation of gene expression, and cell division. Excess zinc is excreted through pancreatic secretions and bile.

Requirement, deficiency, and toxicity

The zinc requirement of poultry, ducks, Japanese Quail, and Ring-necked Pheasants does not exceed 70 mg kg^{-1} dry matter during any stage of the life cycle (NRC, 1994). The zinc requirement is greater for pheasants and turkeys than for chickens or quail. Bone zinc accumulation and appetite are typically used to determine the zinc requirement of birds.

Zinc deficiencies are relatively common in captive birds fed grain-based diets. Even a moderate deficiency of dietary zinc causes a decrease in appetite. More severe deficiencies cause a sharp drop in food consumption. A zinc

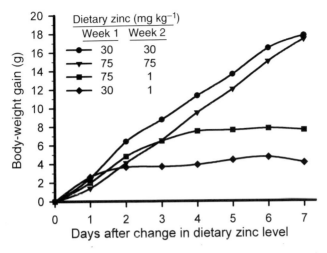

Fig. 10.5. Effect of the level of dietary zinc fed during the first week following hatching on body-weight gain of Japanese Quail chicks fed adequate (30 mg kg^{-1}), surfeit (75 mg kg^{-1}), or deficient (1 mg kg^{-1}) diets during their second week. Body-weight gain is a sensitive indicator of zinc status in young chicks (Harland *et al.*, 1975).

deficiency impairs all important physiological processes, including cell replication and growth, feathering, bone growth, fertility, immunocompetence, electrolyte balance, learning, and behavior. Shortening and thickening of tibiotarsus and tarsometatarsus bones, enlarged hocks, and poor feather pigmentation are diagnostic of a zinc deficiency in growing pheasants and quail. In laying females, a zinc deficiency impairs the number of eggs produced. Chicks from marginally deficient pheasant and quail eggs are weak at hatching and have labored breathing. Severe deficiencies are teratogenic, and embryos die at an early developmental stage with skeletal deformations. Zinc-deficient adults develop a dermatitis on their feet and legs, around the beak, and on their tongue. In ducks, the interdigital webs of the feet exhibit a severe dermatitis, characterized by hyperkeratosis. Following a molt, new feathers are frayed and the feather shaft may contain blisters that weaken their structural integrity (Wight and Dewar, 1976; Cook *et al.*, 1984; McDowell, 1992).

Birds can tolerate relatively high levels of dietary zinc, probably due to proficient regulation of zinc absorption. In chickens, toxicities have not been reported at levels of dietary zinc that are below 1 g kg^{-1}, and levels above 2 g kg^{-1} are usually tolerated by adults. Psittacines and ducks are more susceptible. A zinc toxicity most commonly results from the consumption of nonfood items, such as coins or galvanized steel. Toxicity symptoms include anemia and slow growth in chicks or weight loss in adults. The lining of the gizzard becomes pale-colored and infiltrated by inflammatory cells and may exhibit fissures. In diving ducks, a zinc toxicity causes necrosis of the pancreatic acinar cells and maldigestion, due to the loss of digestive enzymes. In chickens, high dietary zinc (> 2 g kg^{-1}) stops egg laying and induces a molt. Usually, the major problems associated with foods high in zinc are secondary to interactions with other trace minerals. High dietary zinc increases the requirement for selenium, iron, and copper (Bafundo *et al.*, 1984; McCormick and Cunningham, 1984; Wight *et al.*, 1986; Zdziarski *et al.*, 1994).

Dietary sources

The concentration and bioavailability of zinc in foods of plant origin are highly variable. The level of zinc in the vegetative parts of plants ranges from 10 to 300 mg kg^{-1} dry matter, depending on the species, the environment, and the amount of zinc in the soil. Seeds tend to be slightly more consistent in their zinc content, with levels ranging from 10 to 100 mg kg^{-1}. However, the zinc in many seeds has low bioavailability, due to phytic acid and oxalic acid. Most foods of animal origin are a very good source of highly bioavailable zinc.

The molar ratio between zinc and phytic acid is a primary influence on the bioavailability of plant zinc. Molar phytate:zinc ratios of 12–15 are detrimental to zinc absorption. The negative influence of phytic acid on zinc absorption is exacerbated by high dietary calcium (Fordyce *et al.*, 1987).

Zinc is typically supplemented as zinc sulfate, zinc carbonate, zinc oxide, or zinc complexed to amino acids. Galvanized pipe enriches the water supply with zinc and can prevent a deficiency. Galvanized wire found in cages can be a major supply of zinc. Many captive birds will gnaw at their cage and may ingest enough zinc to cause a copper or iron deficiency. New cages are a particularly rich source of zinc, and a variety of psittacines have developed a fatal zinc toxicity within a few weeks of being housed (Howard, 1992).

Manganese

Manganese absorption occurs throughout the small intestine, but the rate is relatively slow and variable. Typically, less than a few percent of dietary manganese is absorbed. Absorption is negatively affected by the level of phytic acid, of fiber, and of other minerals, such as calcium, phosphate, iron, and cobalt. Following absorption, manganese is bound to α_2-macroglobulin and transferrin in the blood. Bone, liver, pancreas, and kidney are the most active tissues at taking up circulating manganese and have the highest concentrations. Manganese is an important activator of many enzymes and is also a component of arginase, pyruvate carboxylase, and manganese superoxide dismutase. During a deficiency, some of the functions of manganese can be accomplished by other minerals, such as magnesium. Other functions, such as activation of glycosyl-transferases, are very sensitive to a deficiency of manganese and are the basis for problems in the synthesis of mucopolysaccharides and glycoproteins in bone and cartilage (Baker and Halpin, 1991; McDowell, 1992; Liu *et al.*, 1994, Henry, 1995).

The amount of manganese that accumulates in a bird's body appears to be regulated by the rate of both absorption and excretion. Bile appears to be the primary excretory route. In growing chickens, body stores of manganese that accumulate during periods of high dietary manganese consumption can supply sufficient manganese to last for about a month of consuming a severely manganese-deficient diet (Baker *et al.*, 1986).

Requirement, deficiency, and toxicity

The manganese requirement of poultry, Ring-necked Pheasants, Japanese Quail, and White Pekin Ducks consuming a grain-based diet is less than 70 mg kg^{-1} dry matter for all stages of the life cycle (NRC, 1994). The requirement for birds fed a purified diet is considerably lower, due to the absence of factors that negatively impact absorption. Presumably, faunivorous birds have manganese requirements that are considerably lower, due to the higher bioavailability of manganese in foods of animal origin. However, manganese in animal prey is found in greatest amounts in the internal organs, such as liver. Peregrine Falcons fed eviscerated pigeons develop perosis, due to a manganese deficiency (Sykes *et al.*, 1982).

A manganese deficiency in growing chicks causes a perosis-like syndrome, characterized by shortened and thickened long bones, malformed tibiometa-tarsal (hock) joints, and weakened cartilage (see Fig. 10.4c; Jensen, 1968; Leach, 1988). The Achilles' tendon may detach from flattened and weakened lateral condyles of the hock joint, causing deformation and inflammation of the hock joint and twisting of the distal tibiotarsus and the proximal tarsometatarsus bones. This type of lameness is common in many species of precocial birds (e.g., Galliformes, ratites, waterfowl) raised in captivity and permitted ad libitum access to diets that are very high in energy, protein, and calcium, but marginally deficient in bioavailable manganese. Among birds in the wild, it has been reported in wild Canada Geese. Sufficient dietary manganese may not always prevent this condition, unless growth is slowed by restricting the amount of food consumed each day. Apparently, manganese partitioned towards the accretion of new skeletal muscle and other tissues takes precedence over its use for the maturation of connective tissue. This situation causes the cartilage in the joints and the connective tissue of the tendons to be inappropriately weak for the size and strength of the skeletal muscles that pull on them. When growth is slowed by caloric restriction, the timing between muscle growth and joint strength is better matched. Although a manganese deficiency is the most common cause for perosis, copper or choline deficiencies may contribute.

In laying birds, a manganese deficiency causes shell thinning, reduced hatchability, and chondrodystrophy of the embryo. Thinning of the eggshell is due to disruptions in the synthesis of the organic matrix. Chickens that hatch from manganese-deficient eggs may have wiry down, skeletal abnormalities, and ataxia and may assume a 'stargazing posture,' with their head bent sharply backwards (Scott, 1982).

Manganese is considered to be one of the least toxic trace minerals, and dietary levels of 1000 mg kg^{-1} are tolerated. Many of the symptoms due to chronic consumption of very high manganese are due to secondary deficiencies of other minerals, particularly iron.

Dietary sources

The vegetative parts of plants are a good source of manganese (60–800 mg kg^{-1} dry matter). Their actual manganese concentration depends on soil and climate conditions. Seeds contain less manganese (10–80 mg kg^{-1}). The concentration of manganese in a food often tells little about the value of that food as a manganese source, due to extreme variability in bioavailability.

Manganese is usually supplemented to diets as manganese sulfate, manganese dioxide, manganese carbonate, or complexes of manganese and methionine (Henry, 1995). Complexed manganese has a greater bioavailability than inorganic forms and is less affected by other dietary factors.

Iodine

Iodine is an integral part of the thyroid hormones triiodothyronine and thyroxine. No other metabolic functions have been described in birds. Most iodine in foods is in an inorganic form, which is readily absorbed through the gastrointestinal tract as iodide (I$-$), probably using the same transport system used for chloride. Organic forms of iodine, such as iodinated amino acids, are also efficiently absorbed. Following absorption, iodine is rapidly taken up by the thyroid gland and converted to organic iodine by combining with tyrosine. Thyroglobulin, a glycoprotein with iodinated tyrosine residues, is the predominant storage form for iodine in the thyroid. Excess iodine is excreted in the urine. Developing follicles in the ovary concentrate iodine, and the amount of iodine in the egg reflects that in the diet.

Goitrogens found in the food can impair the selective concentration of iodine in the thyroid and egg follicle. For example, plants in the *Brassica* genus (e.g., rape, kale, cabbage) contain thiocyanates and goitrin (L-5-vinyl-2-thiooxalidone), which block iodine uptake and thyroxine synthesis, respectively. Other foods, such as soybeans, linseeds, lentils, cassava, peas, and peanuts, also contain goitrogenic substances. Many, but not all, of the effects of goitrogens can be overcome by additional supplementation of iodine (Underwood, 1977).

Requirement, deficiency, and toxicity

The iodine requirement of poultry and Japanese Quail is between 0.3 and 0.4 mg kg^{-1} dry matter for all stages of the life cycle (Stallard and McNabb, 1990; NRC, 1994). However, the requirement is greatly affected by the nature of the diet and is increased by more than twofold by the presence of goitrogenic foods in the diet. Japanese Quail have the capacity to adjust thyroid function to maintain homeostasis over more than a 25-fold range in dietary iodine concentrations. Once iodine stores are replete, it takes more than a year to deplete them in a chicken laying eggs daily (Rogler *et al.*, 1961; McDowell, 1992).

The primary deficiency symptom of an iodine deficiency is an enlargement of the thyroid gland, or goiter (Goodman, 1996). This is because low thyroid hormone levels, due to insufficient dietary iodine, cause the release of high levels of thyroid-stimulating hormone from the pituitary, stimulating the hypertrophy of the thyroid. An amplifying endocrine feedback loop occurs, because the iodine deficiency prevents additional thyroid hormone production, perpetuating the cycle. In Budgerigars, the thyroid increases from 1.5 mg to 150–300 mg. In small birds, the enlarged thyroids put pressure on the esophagus and trachea, causing food regurgitation, impaired crop emptying, and respiratory 'peeps.' Low levels of thyroxine due to an iodine deficiency, result in obesity and the growth of abnormally long, lacy feathers, with conspicuous changes in pigmentation. In the laying female, deficient levels of dietary iodine result in very low egg iodine levels, causing poor hatchability and retarded absorption of the yolk sac. The requirement for maximal egg numbers is about six times less than the requirement for egg hatchability and viability of the hatchlings.

Symptoms of iodine excess are often similar to those of a deficiency (Wheeler and Hoffmann, 1949; Russel, 1977). Toxic levels of iodine also cause an enlarged thyroid, by interfering with thyroid hormone synthesis and inducing thyroid-stimulating hormone. In Little Penguins and Mallards, excess iodine causes ataxia, irritability, and watery droppings, as well as goiter. Iodine becomes toxic at about 100 mg kg^{-1} in growing chickens. Levels in excess of this are commonly found in ocean fish and especially mollusks and shellfish, but apparently are not harmful to the wild birds that eat them.

Dietary sources

Iodine is not distributed uniformly in the environment, and the iodine content of water and avian foods is extremely variable. Plants grown in regions with iodine-deficient soils are deficient in iodine, and animal prey that consumes these plants is low in iodine as well. In general, animal prey contains higher levels of iodine than plants, and the iodine in marine fish and plants is very high.

Iodine is usually supplemented to the diet of captive birds as calcium iodate ($Ca(IO_3)_2$), potassium iodate (KIO_3), or potassium iodide (KI). All of these forms have similar bioavailability (Miller and Ammerman, 1995). Potassium iodide is more volatile than the other forms and may not survive the heat or moisture inherent in pelleting diets, unless it is stabilized with other minerals. Marine vegetation, such as kelp, is sometimes used to supplement diets.

Selenium

A quantitative assessment of selenium nutrition, including required and toxic levels, is complicated by its occurrence in many different chemical forms. The four common valence states of selenium are selenide (-2), elemental selenium (0), selenite ($+4$), and selenate ($+6$). The bioavailability and thus the requirement and toxic levels are highly dependent upon the dietary form. For example, elemental selenium is insoluble in water or lipids and is very poorly absorbed from the intestine. Thus, from a nutritional perspective, discussions on selenium levels or requirements refer to nonelemental selenium. Selenite, selenate, and organic selenides are absorbed with high efficiency in the small intestine, particularly the duodenum. In chickens, there is little homeostatic regulation of selenium absorption and uptake is high even when tissue levels are replete. Selenite, selenate, and selenomethionine absorption mimics that of sulfite, sulfate, and methionine absorption, respectively. Urinary excretion as trimethylselenonium $((CH_3)_3Se^+)$ is the primary route of selenium loss from the bird's body (Humaloja and Mykkanen, 1986).

Selenium is an essential component of at least three enzymes: glutathione peroxidase, phospholipid hydroperoxide glutathione peroxidase, and 5′-iodothyronine deiodinase. The glutathione peroxidase enzymes destroy peroxides as they are formed and are important in defending against the oxidative damage of membranes. Vitamin E is also an important antioxidant and glutathione peroxidase functions in concert with vitamin E to mitigate oxidative damage (see Chapter 11, p. 296). This complementary nature of selenium and vitamin E is the basis for their mutual sparing effect on requirements (Combs and Combs, 1986).

Requirement, deficiency, and toxicity

The selenium requirement of poultry, ducks, and Japanese Quail is 0.2 mg kg^{-1} or less during all stages of the life cycle. High levels of vitamin E partially replace the need for glutathione peroxidase and decrease the selenium requirement. Chickens fed diets high in vitamin E have a selenium requirement that is at least fivefold less than in those fed low vitamin E. Large amounts of selenium are stored in body tissues when intake is high, and these stores buffer transient dietary deficiencies (Thompson and Scott, 1969; NRC, 1994).

Selenium deficiency occurs predominantly in birds consuming grain-based diets in regions where the soil is very low in selenium. Selenium-deficient Japanese Quail show severely depressed growth, with poor feathering, and may develop exudative diathesis if dietary vitamin E is low. Exudative diathesis is caused by oxidative damage to the membranes of capillaries and the leakage of fluids, resulting in a gelatinous edema, which is often visible under the skin of the breast and abdomen. Small hemorrhages cause the accumulation of blood in the fluid, giving it the green-blue color characteristic of a bruise. Growing ducks fed a selenium-deficient diet also develop myopathies in their gizzard, cardiac, and skeletal muscles. The muscular dystrophy that accompanies a selenium and/or vitamin E deficiency is especially prevalent in the pectoralis muscle, where characteristic white striations develop, due to the replacement of muscle fibers with connective tissue. In chickens, an uncomplicated selenium deficiency causes pancreatic fibrosis. Pancreatic dysfunction decreases the absorption of fat and fat-soluble vitamins. The resulting combined deficiency of selenium and vitamin E causes muscular dystrophy and exudative diathesis (Scott *et al.*, 1982; Combs and Combs, 1986).

Selenium becomes toxic to birds at levels (5–20 mg kg^{-1} dry matter) about 50-fold higher than the requirement. Thus, selenium is the most toxic of the trace minerals and has the lowest margin between deficient and excess dietary levels. The precise dietary level that is toxic depends upon the chemical form. Organic selenides are the most toxic. Concentrations of selenium in natural foods found in some regions can be sufficiently high to cause acute or chronic toxicity. For example, irrigation practices in the semiarid Central Valley of California dissolved naturally occurring selenium salts from the soil. When this selenium-laden water was drained from agricultural lands into wetlands, its evapoconcentration resulted in concentrations of over 100 mg kg^{-1} selenium in aquatic vegetation and invertebrates, mostly as the highly toxic organoselenium compounds. Selenium toxicosis became common in aquatic birds in these wetlands and was first observed as poor reproductive success and a very high incidence of grossly deformed embryos and chicks (Ohlendorf *et al.*, 1990).

Excess selenium interferes with sulfur metabolism, due to the formation of sulfur–selenium complexes and the substitution of selenium for sulfur in cysteine metabolism. This diminishes protein synthesis and interferes with protein–gene interactions, causing mutagenesis. The developing embryo is particularly affected by high selenium. Hatchability is typically poor and deformations are common. Chickens and Japanese Quail are more sensitive to the teratogenic effects of selenium toxicity than Mallards, which are more sensitive than Screech Owls and Black-crowned Night Herons. Malformations include legs, toes, wings, beaks, and eyes that are rudimentary or entirely lacking. Disturbances in the normal formation of bone and cartilage may be evident and the down is often wiry. If hatching is successful, symptoms of selenium toxicity include slow growth, pectoral muscle atrophy, impaired immunity, hepatotoxicity, edema, and claw and feather loss. The toxicity of selenium is often reduced by the presence of other trace minerals, such as arsenic, cadmium, and lead, and by high levels of dietary methionine (Hoffman *et al.*, 1992, 1996; Albers *et al.*, 1996).

Dietary sources

The selenium level in foods of plant origin is extremely variable, depending upon the plant species and the level of selenium in the soil and water. The importance of plant species is more striking than for any other essential mineral. Some species accumulate toxic levels of selenium in their tissues, whereas other plants growing in the same soil may have moderate levels. Many regions of the world are selenium-depleted and plants and their seeds have very low levels of selenium (< 0.05 mg kg^{-1} dry matter). The amount of selenium in invertebrate and vertebrate foods is highly dependent upon their position in the food chain and the amount of selenium in the environment. The bioavailability of selenium in foods of animal origin is low relative to that in plants – a situation that is different from that in all other minerals (Cantor and Tarino, 1982; Henry and Ammerman, 1995).

Supplemental selenium is commonly added to diets as sodium selenite (Na$_2$SeO$_3$), sodium selenate (Na$_2$SeO$_4$), or selenomethionine. The amount of selenium added to diets is very small and the margin between the bird's requirement and the toxic level is narrow, so selenium supplementation must be done carefully. Selenium is a potent human carcinogen and a potential environmental hazard, making its supplementation to animal feeds subject to considerable government regulation in many countries.

References

Albers, P.H., Green, D.E. and Sanderson, C.J. (1996) Diagnostic criteria for selenium toxicosis in aquatic birds – dietary exposure, tissue concentrations, and macroscopic effects. *Journal of Wildlife Diseases* 32, 468–485.

Almasri, M.R. (1995) Absorption and endogenous excretion of phosphorus in growing broiler chicks, as influenced by calcium and phosphorus ratios in feed. *British Journal of Nutrition* 74, 407–415.

Aoyagi, S., Baker, D.H. and Wedekind, K.J. (1993) Estimates of copper bioavailability from liver of different animal species and from feed ingredients derived from plants and animals. *Poultry Science* 72, 1746–1755.

Arnold, S.A., Kram, M.A., Hintz, H.F., Evans, H. and Krook, L. (1973) Nutritional secondary hyperthyroidism in the parakeet. *Cornell Veterinarian* 64, 37–46.

Austic, R.E. and Patience, J.F. (1988) Undetermined anion in poultry diets: influences on acid–base balance, metabolism and physiological performance. *Critical Reviews in Poultry Biology* 1, 315–345.

Bafundo, K.W., Baker, D.H. and Fitzgerald, P.R. (1984) The iron–zinc interrelationship in the chick as influenced by *Eimeria acervulina* infection. *Journal of Nutrition* 114, 1306–1312.

Bain, S.D. and Watkins, B.A. (1993) Local modulation of skeletal growth and bone modeling in poultry. *Journal of Nutrition* 123, 317–322.

Baker, D.H. and Ammerman, C.B. (1995a) Copper bioavailability. In: Ammerman, C.B., Baker, D.H. and Lewis, A.J. (eds) *Bioavailability of Nutrients for Animals*. Academic Press, San Diego, pp. 127–156.

Baker, D.H. and Ammerman, C.B. (1995b) Zinc bioavailability. In: Ammerman, C.B., Baker, D.H. and Lewis, A.J. (eds) *Bioavailability of Nutrients for Animals*. Academic Press, San Diego, pp. 367–398.

Baker, D.H. and Halpin, K.M. (1991) Manganese and iron interrelationship in the chick. *Poultry Science* 70, 146–152.

Baker, D.H., Halpin, K.M., Laurin, D.E. and Southern, L.L. (1986) Manganese for poultry – a review. In: *Proceedings Arkansas Nutrition Conference*. Arkansas, Little Rock, pp. 1–6.

Bird, F.H. (1949) Magnesium deficiency in the chick. *Journal of Nutrition* 39, 13–20.

Cade, T.J. (1964) Water and salt balance in granivorous birds. In: Wayner, M.J. (ed.) *Thirst*. Pergamon Press, Oxford, pp. 237–256.

Cain, J.R., Beasom, S.L., Rowland, L.O. and Rowe, L.D. (1982) The effects of varying dietary phosphorus on breeding bobwhites. *Journal of Wildlife Management* 46, 1061–1065.

Campbell, E.G. and Koplin, J.R. (1986) Food consumption, energy, nutrient and mineral balances in a Eurasian kestrel and screech owl. *Comparative Biochemistry and Physiology* 83A, 249–254.

Cantor, A.H. and Tarino, J.Z. (1982) Comparative effects of inorganic and organic dietary sources of selenium on selenium levels and selenium–dependent glutathione per-oxidase activity in blood of young turkeys. *Journal of Nutrition* 112, 2187–2194.

Clunies, M., Emslie, J. and Leeson, S. (1992) Effect of dietary calcium level on medullary bone calcium reserves and shell weight of Leghorn hens. *Poultry Science* 71, 1348–1356.

Combs, G.F. and Combs, S.B. (1986) *The Role of Selenium in Nutrition*. Academic Press, New York.

Cook, M.E., Sunde, M.L., Stahl, J.I. and Hanson, L.E. (1984) Zinc deficiency in pheasant chicks fed practical diets. *Avian Diseases* 28, 1102–1109.

Cork, S.C., Alley, M.R. and Stockdale, P.H.G. (1995) A quantitative assessment of haemosiderosis in wild and captive birds using image analysis. *Avian Pathology* 24, 239–254.

Cosson, R.P. (1989) Relationships between heavy metal and metallothionein-like protein levels in the liver and kidney of two birds: the greater flamingo and the little egret. *Comparative Biochemistry and Physiology* 94C, 243–248.

Dacke, C.G., Arkle, S., Cook, D.J., Wormstone, I.M. and Jones, S. (1993) Medullary bone and avian calcium regulation. *Journal of Experimental Biology* 184, 63–88.

Davis, P.N., Norris, L.C. and Kratzer, F.H. (1962) Iron deficiency studies in chicks using treated isolated soybean protein diets. *Journal of Nutrition* 78, 445–451.

Dierenfeld, E.S., Pini, M.T. and Sheppard, C. (1994) Hemosiderosis and dietary iron in birds. *Journal of Nutrition* 124, 2685S–2686S.

Drent, P.J. and Woldendorp, J.W. (1989) Acid rain and eggshells. *Nature* 339, 431.

Duke, G.E., Jegers, A.A., Loff, G. and Evanson, O.A. (1975) Gastric digestion in some raptors. *Comparative Biochemistry and Physiology* 50A, 649–656.

Earl, K.E. and Clarke, N.R. (1991) The nutrition of the budgerigar (*Melopsittacus undulatus*). *Journal of Nutrition* 121, 186–192.

Edwards, H.M.J. and Nugara, D. (1968) Magnesium requirements of the chick. *Poultry Science* 47, 963–968.

Emmert, J.L. and Baker, D.H. (1995) Zinc stores in chickens delay the onset of zinc deficiency symptoms. *Poultry Science* 74, 1011–1021.

Etches, R.J. (1996) *Reproduction in Poultry.* CAB International, Wallingford, UK.

Fordyce, E.J., Forbes, R.M., Robbins, K.R. and Erdman, J.W. (1987) Phytates by calcium/zinc molar ratios: are they predictive of zinc bioavailability? *Journal of Food Science* 52, 440–448.

Frings, H. and Frings, M. (1959) Observations on salt balance and behavior of Laysan and Black-footed albatrosses in captivity. *Condor* 61, 305–315.

Gailey-Phipps, P. (1982) Survey on nutrition of penguins. *Journal of American Veterinary Medical Association* 181, 1306–1309.

Gardner, E.E., Rogler, J.C. and Parker, H.E. (1960) Magnesium requirement of the chick. *Poultry Science* 39, 1111–1118.

Gilbert, A.B. (1983) Calcium and reproductive function in the hen. *Proceedings Nutrition Society* 42, 195–212.

Gillis, M.B., Edwards, H.M. and Young, R.J. (1962) Studies on the availability of calcium orthophosphates to chickens and turkeys. *Journal of Nutrition* 78, 155–162.

Goodman, G.J. (1996) Metabolic disorders. In: Rosskopf, W. and Woerpel, R. (eds) *Diseases of Cage and Aviary Birds.* Willians & Wilkins, Baltimore, pp. 218–234.

Grau, C.R., Roudybush, T.E. and McGibbon, W.H. (1979) Mineral composition of yolk fractions and whole yolk from eggs of restricted ovulator hens. *Poultry Science* 58, 1143–1148.

Graveland, J. (1996) Avian eggshell formation in calcium-rich and calcium-poor habitats – importance of snail shells and anthropogenic calcium sources. *Canadian Journal of Zoology – Revue Canadienne de Zoologie* 74, 1035–1044.

Graveland, J. and Van Gijzen, T. (1994) Arthropods and seeds are not sufficient as calcium sources for shell formation and skeletal growth in passerines. *Ardea* 82, 299–314.

Guinotte, F., Gautron, J., Nys, Y. and Soumarmon, A. (1995) Calcium solubilization and retention in the gastrointestinal tract in chicks (*Gallus Domesticus*) as a function of gastric acid secretion inhibition and of calcium carbonate particle size. *British Journal of Nutrition* 73, 125–139.

Hallquist, N.A. and Klasing, K.C. (1994) Serotransferrin, ovotransferrin and metallothionein levels during an immune response in chickens. *Comparative Biochemistry and Physiology B – Biochemistry and Molecular Biology* 108, 375–384.

Harland, B.F., Spivey Fox, M.R. and Fry, B.E. (1975) Protection against zinc deficiency by prior excess dietary zinc in young Japanese quail. *Journal of Nutrition* 105, 1509–1518.

Hart, L.E., Ravindran, V. and Young, A. (1992) Accumulation of calcium and phosphorus in pigeon (*Columba livia*) embryos. *Journal of Comparative Physiology B – Biochemical Systemic and Environmental Physiology* 162, 535–538.

Hempe, J.M. and Cousins, R.J. (1992) Cysteine-rich intestinal protein and intestinal metallothionein – an inverse relationship as a conceptual model for zinc absorption in rats. *Journal of Nutrition* 122, 89–95.

Henry, P.R. (1995) Manganese bioavailability. In: Ammerman, C.B., Baker, D.H. and Lewis, A.J. (eds) *Bioavailability of Nutrients for Animals*. Academic Press, San Diego, pp. 239–256.

Henry, P.R. and Ammerman, C.B. (1995) Selenium bioavailability. In: Ammerman, C.B., Baker, D.H. and Lewis, A.J. (eds) *Bioavailability of Nutrients for Animals*. Academic Press, San Diego, pp. 303–336.

Henry, P.R. and Miller, E.R. (1995) Iron bioavailability. In: Ammerman, C.B., Baker, D.H. and Lewis, A.J. (eds) *Bioavailability of Nutrients for Animals*. Academic Press, San Diego, pp. 169–199.

Hoffman, D.J., Sanderson, C.J., Lecaptain, L.J., Cromartie, E. and Pendleton, G.W. (1992) Interactive effects of arsenate, selenium, and dietary protein on survival, growth, and physiology in mallard ducklings. *Archives of Environmental Contamination and Toxicology* 22, 55–62.

Hoffman, D.J., Heinz, G.H., Lecaptain, L.J., Eisemann, J.D. and Pendleton, G.W. (1996) Toxicity and oxidative stress of different forms of organic selenium and dietary protein in mallard ducklings. *Archives of Environmental Contamination and Toxicology* 31, 120–127.

Houston, D.C., Donnan, D. and Jones, P.J. (1995) The source of the nutrients required for egg production in zebra finches *Poephila guttata*. *Journal of Zoology* 235, 469–483.

Howard, B.R. (1992) Health risks of housing small psittacines in galvanized wire mesh cages. *Journal of American Veterinary Medical Association* 200, 1667–1674.

Humaloja, T. and Mykkanen, H.M. (1986) Intestinal absorption of ^{75}Se-labeled sodium selenite and selenomethionine in chicks: effect of time, segment, selenium concentration and method of measurement. *Journal of Nutrition* 116, 142–148.

Hurwitz, S. and Pines, M. (1991) Regulation of bone growth. In: Pang, P.K.T. and Schreibaum, M. (eds) *Vertebrate Endocrinology: Fundamentals and Biomedical Implications*. Academic Press, New York, pp. 163–189.

Hurwitz, S., Plavnik, I., Shapiro, A., Wax, E., Talpaz, H. and Bar, A. (1995) Calcium metabolism and requirements of chickens are affected by growth. *Journal of Nutrition* 125, 2679–2686.

Jakubas, W.J., Guglielmo, C.G., Vispo, C. and Karasov, W.H. (1995) Sodium balance in ruffed grouse as influenced by sodium levels and plant secondary metabolites in quaking aspen. *Canadian Journal of Zoology – Revue Canadienne de Zoologie* 73, 1106–1114.

Jensen, W.I. (1968) Perosis in Canada geese, *Branta canadensis*. *Bulletin of the Wildlife Disease Association* 4, 95–99.

Johnson, A.L. (1986) Reproduction in the male. In: Sturkie, P.D. (ed.) *Avian Physiology*. Springer-Verlag, New York, pp. 432–451.

Johnston, J.J. and Bildstein, K.L. (1990) Dietary salt as a physiological constraint in white ibis breeding in an estuary. *Physiological Reviews* 63, 190–207.

Keshavarz, K. (1994) Laying hens respond differently to high dietary levels of phosphorus in monobasic and dibasic calcium phosphate. *Poultry Science* 73, 687–703.

Kim, Y.S., Sun, S.S. and Myng, K.H. (1985) A comparison of true available calcium with apparent available calcium values using 6 calcium supplements in breeding Japanese quail. *Korean Journal of Animal Science* 27, 297–309.

Kincaid, A.L. and Stoskopf, M.K. (1987) Passerine dietary iron overload syndrome. *Zoo Biology* 6, 79–88.

Koh, T.S., Peng, R.K. and Klasing, K.C. (1996) Dietary copper level affects copper metabolism during lipopolysaccharide-induced immunological stress in chicks. *Poultry Science* 75, 867–872.

Korschgen, L.J., Chambers, G.D. and Sadler, K.C. (1965) Digestion rate of limestone force-fed to pheasants. *Journal of Wildlife Management* 29, 820–823.

Kratzer, F.H. and Vohra, P. (1986) *Chelates in Animal Nutrition.* CRC Press, Boca Raton.

Krementz, D.G. and Ankney, C.D. (1995) Changes in total body calcium and diet of breeding house sparrows. *Journal of Avian Biology* 26, 162–167.

Leach, R.M. (1988) The role of trace elements in the development of cartilage matrix. In: Hurley, L., Keen, C.L., Lonnerdal, B. and Rucker, R.B. (eds) *Trace Elements in Man and Animals,* Vol. 6. Plenum Press, New York, pp. 267–271.

Leach, R.M. and Gay, C.V. (1987) Role of epiphyseal cartilage in endochondral bone formation. *Journal of Nutrition* 117, 784–790.

Leach, R.M. and Nesheim, M.C. (1963) Studies on chloride deficiency in chicks. *Journal of Nutrition* 81, 193–199.

Leclercq, B., Decarville, H. and Guy, G. (1990) Calcium requirement of male Muscovy ducklings. *British Poultry Science* 31, 331–337.

Liu, A.C.H., Heinrichs, B.S. and Leach, R.M. (1994) Influence of manganese deficiency on the characteristics of proteoglycans of avian epiphyseal growth plate cartilage. *Poultry Science* 73, 663–669.

Long, S. (1982) Acid–base balance and urinary acidification in birds. *Comparative Biochemistry and Physiology* 71A, 519–526.

McCormick, C.C. and Cunnigham, D.L. (1984) High dietary zinc and fasting as methods of forced resting: a performance comparison. *Poultry Science* 63, 1201–1206.

McDowell, L.R. (1992) *Minerals in Animal and Human Nutrition.* Academic Press, New York.

MacLean, S.F. (1973) Lemming bones as a source of calcium for arctic sandpipers (*Caldris* spp.). *Ibis* 116, 552–557.

McWard, G.W. (1967) Magnesium tolerance of the growing and laying chicken. *British Poultry Science* 8, 91–97.

Miller, E.R. and Ammerman, C.B. (1995) Iodine bioavailability. In: Ammerman, C.B., Baker, D.H. and Lewis, A.J. (eds) *Bioavailability of Nutrients for Animals.* Academic Press, San Diego, pp. 157–168.

Miller, S.C. (1977) Osteoclast cell surface changes during egg-laying cycle in Japanese quail. *Journal of Cell Biology* 75, 104–118.

Mitchell, R.D. and Edwards, H.M. (1996) Effects of phytase and 1,25-dihydroxy-cholecalciferol on phytate utilization and the quantitative requirement for calcium and phosphorus in young broiler chickens. *Poultry Science* 75, 95–110.

Mongin, P. (1989) Recent advances in dietary anion–cation balance in poultry. In: Haresign, W. and Cole, D.G.A. (eds) *Recent Developments in Poultry Nutrition.* Butterworths, London, pp. 94–104.

Morck, T.A. and Austic, R.E. (1981) Iron requirement of white Leghorn hens. *Poultry Science* 60, 1497–1501.

Morgan, E.H. (1975) Plasma iron transport during egg laying and after oestrogen administration in domestic fowl. *Quarterly Journal of Experimental Physiology* 60, 233–247.

Nielson, F.H. (1986) Other elements. In: Mertz, W. (ed.) *Trace Elements in Human and Animal Nutrition*, Vol. 2. Academic Press, Orlando, Florida, pp. 313–343.

Norman, A.W. and Hurwitz, S. (1993) The role of the vitamin D endocrine system in avian bone biology. *Journal of Nutrition* 123, 310–316.

Norris, L.C., Kratzer, F.H., Lin, H.J., Hellewell, A.B. and Belhan, J.R. (1972) Effect of quantity of dietary calcium on maintenance of bone integrity in mature white Leghorn male chickens. *Journal of Nutrition* 102, 1085–1091.

Noy, Y., Frisch, Y., Rand, N. and Sklan, D. (1994) Trace mineral requirements in turkeys. *World's Poultry Science Journal* 50, 253–268.

NRC (1994) *Nutrient Requirements of Poultry*. National Academy Press, Washington, DC.

Ohlendorf, H.M., Hothem, R.L., Bunck, C.M. and Marois, K.C. (1990) Bioaccumulation of selenium in birds at Kesterson Reservoir, California. *Archives of Environmental Contamination and Toxicology* 19, 495–507.

Packard, M.J. (1994) Mobilization of shell calcium by the chick chorioallantoic membrane *in vitro*. *Journal of Experimental Biology* 190, 141–153.

Packard, M.J. and Packard, G.C. (1991) Patterns of mobilization of calcium, magnesium, and phosphorus by embryonic yellow-headed blackbirds (*Xanthocephalus xanthocephalus*). *Journal of Comparative Physiology B – Biochemical Systemic and Environmental Physiology* 160, 649–654.

Parsons, A.H. and Combs, G.F.J. (1981) Blood ionized calcium cycles in the chicken. *Poultry Science* 60, 1520–1524.

Pimentel, J.L., Greger, J.L., Cook, M.E. and Stahl, J.L. (1992) Iron metabolism in chicks fed various levels of zinc and copper. *Journal of Nutritional Biochemistry* 3, 140–145.

Prange, H.D., Anderson, J.F. and Rahn, H. (1979) Scaling of skeletal mass to body mass in birds and mammals. *American Naturalist* 113, 103–122.

Randall, M.G. (1981) Nutritionally induced hypocalcemic tetany in an Amazon parrot. *Journal of American Veterinary Medical Association* 179, 1277–1278.

Rao, S.K., Roland, D.A. and Gordon, R.W. (1995) A method to determine and factors that influence *in vivo* solubilization of phosphates in commercial Leghorn hens. *Poultry Science* 74, 1644–1649.

Ravindran, V., Bryden, W.L. and Kornegay, E.T. (1995) Phytates: occurrence, bioavailability and implications in poultry nutrition. *Poultry and Avian Biology Reviews* 6, 125–143.

Richards, M.P. and Packard, M.J. (1996) Mineral metabolism in avian embryos. *Poultry and Avian Biology Reviews* 7, 143–161.

Richardson, P.R.K., Mundy, P.J. and Plug, I. (1986) Bone crushing carnivores and their significance to osteodystrophy in griffon vulture chicks. *Journal of Zoology* 210, 23–43.

Robbins, C.T. (1993) *Wildlife Feeding and Nutrition*, 2nd edn. Academic Press, San Diego.

Rogler, J.C., Parker, H.E., Andrews, F.N. and Carrick, S.W. (1961) The iodine requirements of the breeding hen. *Poultry Science* 40, 1554–1561.

Romanoff, A.L. and Romanoff, A.J. (1949) *The Avian Egg*. John Wiley and Sons, New York.

Roudybush, T. (1996) Nutrition. In: Rosskopf, W. and Woerpel, R. (eds) *Diseases of Cage and Aviary Birds*. Williams & Wilkins, Baltimore, pp. 218–234.

Rowland, L.O., Sloain, D.R., Fry, J.L. and Harmes, R.H. (1973) Calcium requirement for bone maintenance of aged non-laying hens. *Poultry Science* 52, 1415–1418.

Rucker, R.B., Parker, H.E. and Rogler, J.C. (1968) Utilization of calcium and phosphorus from hydrous and anhydrous dicalcium phosphates. *Journal of Nutrition* 96, 513–519.

Russel, W.C. (1977) Iodine-induced goiter in penguins. *Journal of the American Veterinary Medical Association* 171, 959–960.

Saiz, M.P., Marti, M.T., Mitjavila, M.T. and Planas, J. (1993) Iron absorption by small intestine of chickens. *Biological Trace Element Research* 36, 7–14.

Scott, M.L., van Tienhoven, A., Holm, E.R. and Reynolds, R.E. (1960) Studies on the sodium, chlorine and iodine requirements of young pheasants and quail. *Journal of Nutrition* 71, 282–288.

Scott, M.L., Nesheim, M.C. and Young, R.J. (1982) *Nutrition of the Chicken*, 3rd edn. M.L. Scott and Associates, Ithaca, New York.

Shafey, T.M. (1993) Calcium tolerance of growing chickens – effect of ratio of dietary calcium to available phosphorus. *World's Poultry Science Journal* 49, 5–18.

Soares, J.H. (1995) Calcium bioavailability. In: Ammerman, C.B., Baker, D.H. and Lewis, A.J. (eds) *Bioavailability of Nutrients for Animals*. Academic Press, San Diego, pp. 95–118.

Spick, G., Coddeville, B. and Montreuil, J. (1988) Comparative study of the primary structures of sero-, lacto- and ovotransferrin glycans from different species. *Biochimie* 70, 1459–1469.

Stallard, L.C. and McNabb, F.M.A. (1990) The effects of different iodide availabilities on thyroid function during development in Japanese quail. *Domestic Animal Endocrinology* 2, 239–250.

Stillmak, S.J. and Sunde, M.L. (1971) The use of high magnesium limestone in the diet of the laying hen. *Poultry Science* 50, 553–560.

Sullivan, T.W. and Douglas, J.H. (1994) Levels of various elements of concern in feed phosphates of domestic and foreign origin. *Poultry Science* 73, 520–528.

Sykes, G., Hardaswick, V. and Heck, W. (1982) Nutritional deficiency and perosis in peregrine falcons. *Hawk Chalk* 21, 33–36.

Tao, S., Fry, B.E. and Spivey-Fox, M.R. (1983) Magnesium stores and anemia in young Japanese quail. *Journal of Nutrition* 113, 1195–1203.

Teeter, R.G., Smith, M.O., Owens, F.N., Arp, S.C., Sangiah, S. and Breazile, J.E. (1985) Chronic heat stress and respiratory alkalosis: occurrence and treatment in broiler chicks. *Poultry Science* 64, 1060–1064.

Thompson, J.N. and Scott, M.L. (1969) The role of selenium in the nutrition of the chicken. *Journal of Nutrition* 97, 335–343.

Underwood, E.J. (1977) *Trace Elements in Human and Animal Nutrition*. Academic Press, New York.

Wages, D.P., Ficken, M.D., Cook, M.E. and Mitchell, J. (1995) Salt toxicosis in commercial turkeys. *Avian Diseases* 39, 158–161.

Wallach, J.D. and Fleig, G.M. (1970) Cramps and fits in carnivorous birds. *London Zoological Society* 10, 3–4.

Ward, R.J., Smith, T., Henderson, G.M. and Peters, T.J. (1991) Investigation of the aetiology of haemosiderosis in the starling (*Sturnus vulgaris*). *Avian Pathology* 20, 225–232.

Wheeler, R.S. and Hoffmann, E. (1949) Goitrogenic action of iodide and the etiology of goiters in chicks from thyroprotein fed hens. *Proceedings Society Experimental Medicine* 72, 250–254.

Wideman, R.F. (1987) Renal regulation of avian calcium and phosphorus metabolism. *Journal of Nutrition* 117, 808–815.

Wight, P.A.L. and Dewar, W.A. (1976) The histopathology of a zinc deficiency in ducks. *Journal of Pathology* 120, 183–191.

Wight, P.A.L., Dewar, W.A. and Saunderson, C.L. (1986) Zinc toxicity in the fowl: ultrastructural pathology and relationship to selenium, lead, and copper. *Avian Pathology* 15, 23–38.

Wilson, H.R., Persons, H.N., Rowland, L.O. and Harmes, R.H. (1969) Reproduction in white Leghorn males fed various levels of dietary calcium. *Poultry Science* 49, 798–801.

Wilson, S. and Duff, S.R I. (1991) Effects of vitamin or mineral deficiency on the morphology of medullary bone in laying hens. *Research in Veterinary Science* 50, 216–221.

Zdziarski, J.M., Mattix, M., Bush, M.R. and Montali, R.J. (1994) Zinc toxicosis in diving ducks. *Journal of Zoo and Wildlife Medicine* 25, 438–445.

CHAPTER 11
Vitamins

The term vitamin refers to a collection of organic molecules that prevent pathological symptoms when present at very low levels in the diet. Vitamins consist of a mixed group of compounds and are not chemically related to each other, as are other nutrient categories, such as amino acids, fats, or carbohydrates. Most of the vitamins function as cofactors for enzymes, while a few function as hormones or antioxidants.

Birds are unable to synthesize most of the vitamins because they lack the necessary metabolic pathways. A few vitamins may be synthesized by birds, but at rates too slow to meet a bird's requirement under all physiological states. This latter group includes niacin, vitamin D, choline, and, in many species, vitamin C. Some vitamins that cannot be synthesized by the bird's metabolic capabilities are synthesized in nutritionally relevant amounts by the microflora of the gastro-intestinal tract, especially in those birds that have functional ceca. The remainder of the vitamin needs of birds must come from their food.

Vitamins in avian foods often exist as precursor compounds or as coenzymes that are bound to proteins. Digestive processes are required to release vitamins from macromolecular complexes and to convert them to usable chemical forms. Foods vary widely in the specific chemical forms (vitamers) in which vitamins are present (Table 11.1). Often, the various vitamers differ in their nutritional value to the bird. Additionally, some foods contain provitamins, which are compounds that are structurally related to a vitamin but require extensive metabolic transformations to give an active vitamin. For example the provitamin A compound, β-carotene, must undergo extensive hydrolysis in its metabolic conversion to vitamin A. The standardized unit of activity, known as the international unit (IU), is commonly used to describe the *in vivo* biological activity of individual vitamers and provitamins (Table 11.1). The growing chicken has traditionally been used to determine the IU content of foods for purposes of human, animal, and poultry nutrition.

Chickens were utilized for much of the original research and observations that led to the discovery and characterization of the vitamins. In fact, two Nobel prizes have been awarded for vitamin research using the chicken: one to Christian Eijkman in 1929 for discovering an antineuritic vitamin (thiamin) and one to Henrik Dam in 1943 for discovering an antihemorrhagic vitamin (K). We have a very clear picture of the requirements, interrelationships, and food values of the vitamins for chickens and, to a lesser extent, for turkeys, quail, and domestic ducks. Unfortunately, little is known about vitamin nutrition of

Table 11.1. Chemical forms of vitamins found in foods and their bioavailability in chickens (from NRC, 1994; Baker, 1995).

Vitamin	Vitamer form	Amount	Activity
Vitamin A	Retinol	1 µg	3.33 IU
	Retinol acetate	1 µg	2.91 IU
	Retinol palmitate	1 µg	1.82 IU
	β-Carotene	1 µg	1.67 IU
Vitamin D	Cholecalciferol (D_3)	1 µg	40 IU
	Ergocalciferol (D_2)	1 µg	≈4 IU
Vitamin E	DL-α-Tocopheryl acetate	1 mg	1.00 IU
	D-α-Tocopheryl acetate	1 mg	1.36 IU
	D-α-Tocopherol	1 mg	1.49 IU
	DL-α-Tocopherol	1 mg	1.10 IU
	D-γ-Tocopherol	1 mg	0.07 IU
	Tocotrienols	1 mg	1–30 IU
Vitamin K	Phylloquinone (K_1)		100%
	Menaquinone-4 (K_2)		100%
	Menadione (K_3)		60–140%
Vitamin B_1	Thiamin		100%
Vitamin B_2	Riboflavin		100%
Niacin	Nicotinic acid		100%
	Nicotinamide		125%
Vitamin B_6	Pyridoxal		100%
	Pyridoxol		100%
	Pyridoxamine		100%
Pantothenic acid	D-Pantothenate		100%
	DL-Pantothenate		50%
Biotin	D-Biotin		100%
	DL-Biotin		50%
Folic acid	Folic acid		100%
	Polyglutamyl folacins		?
Vitamin B_{12}	Cyanocobalamin	1 µg	1 USP
Choline	Choline		100%
Vitamin C	Ascorbic acid		100%
	Dehydroascorbic acid		100%

nondomestic species, beyond the symptoms of severe deficiencies. It is interesting that the White-headed Munia (a small passerine known as the Rice Bird in its native Java) played a critical role in the discovery of thiamin because of its sensitivity to dietary deficiency. Clearly, important research remains to be done in order to characterize the vitamin nutrition of avian species across diverse nutritional strategies, including faunivores, herbivores, frugivores, and nectarivores.

Nomenclature and Classification

Vitamins are typically divided into two categories: those that are soluble in water – water-soluble vitamins – and those soluble in lipids or lipid solvent – fat-soluble vitamins. Nomenclature of the vitamins can be somewhat confusing, because they were given letter names (e.g., vitamin B_1) prior to discovery of their chemical structures and then, at a later date, were renamed according to their chemical identity (e.g. thiamin). Over time, the chemical names have replaced the letter names in usage for many, but not all, of the vitamins. The Committee on Nomenclature of the American Institute of Nutrition provides up-to-date rules for naming the vitamins and expressing vitamin activity (AIN, 1990).

The fat-soluble vitamins, A, D, E, and K, are usually found in the diet in association with lipids and are absorbed from lipid micelles by the same pathways as fatty acids and cholesterol. Dietary and physiological conditions that optimize the digestion and absorption of dietary lipids favor the absorption of fat-soluble vitamins. The amount of dietary fat required for adequate absorption of fat-soluble vitamins appears to be about 2% of dry matter. This level is found in most avian foods and lower values would typically only be found in artificial diets made with lipid-extracted ingredients. Fat-soluble vitamins are transported from the intestines to the tissues in portomicrons. Hepatic cells internalize portomicrons and their remnants, taking up the fat-soluble vitamins in the process. Tissues that express lipoprotein lipase (e.g., adipose tissue, heart, skeletal muscle) may also take up vitamins from the portomicrons. Fat-soluble vitamins are stored relatively well and birds with replete stores can withstand long periods of depletion before deficiency symptoms manifest. Excesses are excreted in the bile and lost in the feces, but levels of vitamins A and D can build up to toxic levels.

The water-soluble vitamins are absorbed by independent pathways and are transported between tissues either in the free form or complexed to proteins. Except for vitamin B_{12} and choline, they are not stored in appreciable amounts and a fairly constant dietary supply is needed. However, during a general food deficit, tissue catabolism results in the liberation of most of the vitamins from muscle and other tissues, markedly delaying deficiency symptoms. All of the water-soluble vitamins act as cofactors for enzymes and vitamin-dependent enzymes are found in every metabolic pathway. Although the specific functions of the water-soluble vitamins in metabolism are biochemically distinct, the deficiency symptoms are often similar. Usually, the overall growth rate of young birds is impaired and tissues that turn over rapidly are most affected, including the growth plate of bones, the epithelial surfaces of skin, feather follicles, and hemopoietic and leukopoietic tissues. Common symptoms that arise from defects in these tissues are perosis, dermatitis, poor feathering, anemia, and increased susceptibility to infectious diseases. Water-soluble vitamins are excreted mostly in urine and are relatively nontoxic.

Vitamin Requirements

The minimal vitamin requirements for growth and egg production of chickens, turkeys, domestic ducks, and Japanese Quail have been determined experimentally (Table 11.2). Vitamin requirements at maintenance are almost completely unknown, but are probably considerably lower than levels needed for growth or egg production. The values in Table 11.2 are based on maximizing rates of gain and egg production and on the prevention of specific deficiency pathologies. Higher levels are usually advised to optimize vitamin stores and to overcome inefficiencies in absorption or metabolism due to other dietary constituents or the physiological state of the bird. The requirements for maximal hatchability of eggs are usually greater than for maximum numbers or sizes of eggs, and the requirements to optimize stores in the hatchlings are even higher. Thus, prudent diets contain a generous 'margin of safety.'

Table 11.2. Vitamin requirements of birds (from NRC, 1994).[*]

	Chicken		Turkey		Pekin Duck		Japanese Quail	
Vitamin	Growth	Eggs	Growth	Eggs	Growth	Eggs	Growth	Eggs
A (IU)	1500	3000	5000	5000	2500	4000	1650	3300
D (IU)	200	300	1100	1100	400	900	750	900
E (IU)	10	10	12	25	10	10	12	25
K (IU)	0.5	1.0	1.75	1.0	0.5	0.5	1	1
Riboflavin (mg kg^{-1} diet)	3.6	3.6	4.0	4.0	4.0	4.0	4	4
Pantothenic acid (mg kg^{-1} diet)	10	7.0	10.0	16.0	11.0	11.0	10	15
Niacin (mg kg^{-1} diet)	27	10.0	60.0	40.0	55	55	40	20
B$_{12}$ (mg kg^{-1} diet)	0.009	0.08	0.003	0.003	?	?	0.003	0.003
Choline (mg kg^{-1} diet)	1300	1050	1600	1000	?	?	2000	1500
Biotin (mg kg^{-1} diet)	0.15	0.10	0.25	0.20	?	?	0.3	0.15
Folate (mg kg^{-1} diet)	0.55	0.35	1.0	1.0	?	?	1	1
Thiamin (mg kg^{-1} diet)	1.0	0.7	2.0	2.0	?	?	2	2
B$_6$ (mg kg^{-1} diet)	3.0	4.5	4.5	4.0	2.5	3.0	3	3

[*]Chicken values are for white-egg-laying strains of Leghorns. Values for growth are for the starter diet. Values for egg production are for breeding hens and optimized vitamin stores in the hatchling.

Although research with nondomestic species is limited, it is apparent that vitamin requirements of captive game birds and waterfowl are similar to or somewhat higher than those of chickens. The domestic turkey also has a slightly higher requirement for most of the vitamins than the chicken, and requirement levels for turkeys are commonly used as a reference point for captive precocial birds of all species. Inference of the vitamin requirements of altricial birds from those determined for poultry does not currently have a sound experimental basis. It is probable that there are important qualitative and quantitative differences between the vitamin requirements of domestic birds and those of distantly related species, especially those that have adopted very different nutritional strategies.

Vitamins are required at dietary concentrations in the same general range as trace minerals. Unlike their mineral counterparts, the physiology of vitamin digestion is orientated towards high rates of absorption. At dietary levels near the requirement, most of the vitamins are absorbed with an efficiency of more than 50%, although biotin and folate are absorbed less efficiently.

Research in chickens indicates that a variety of factors can quantitatively impact the requirement for various vitamins, including the extent of microbial synthesis, environmental stresses, diseases, and various dietary factors. Nutritionally relevant amounts of the B vitamins may be synthesized by microbes in the ceca and rectum, but absorption is not very efficient (Coates *et al.*, 1968). The very high concentration of vitamin K and most of the water-soluble vitamins in cecotropes can make an important dietary contribution, even if they are only occasionally consumed. In fact, deficiencies of several vitamins are rare in poultry if they have access to their feces but are readily induced when husbandry conditions prevent coprophagy. Species with short gastrointestinal tracts (e.g., Passeriformes) or fast rates of passage (e.g., many frugivores) probably gain little vitamin nutrition from intestinal synthesis, but this has not been adequately investigated.

The extent to which stress contributes to increased vitamin requirements is controversial. Because of the catalytic role of most vitamins, they are not used up by metabolic reactions, and requirements do not increase measurably when the metabolic rate increases due to activity or exposure to the cold. In fact, conditions that cause a higher level of food intake decrease the vitamin concentration required in the diet. However, stresses associated with infection impair absorption, especially of the fat-soluble vitamins, and increase the requirement of some vitamins. Heat stress decreases food intake and may increase endogenous losses of some vitamins, increasing the vitamin requirements.

The lack of chemical similarity between the vitamins limits the potential for nutritionally relevant interactions between vitamins, but two important interactions have been characterized: (i) choline, B_{12}, and folic acid interact in the metabolism of methyl groups, and dietary levels of one affect the requirement of the others; and (ii) the fat-soluble vitamins interact at the level of absorption and large excesses of one increase the requirement for the others. For example, high levels of vitamin E compete with vitamin A for absorption and increase its dietary requirement. Similarly, high levels of vitamin A increase the vitamin E requirement by decreasing its absorption (Abawi and Sullivan, 1989).

The distribution of vitamins in foods consumed by birds is a major factor that determines the probability of deficiencies across species with widely varying nutritional strategies (Table 5.2). Among the fat-soluble vitamins, vitamins A and D are most prevalent in foods of animal origin, whereas vitamins E and K are found in considerably higher concentrations in plants. The water-soluble vitamins are widely distributed, but foods of animal origin generally have higher levels of readily bioavailable forms, with the exception of vitamin C. Vitamins A, D, and E are the vitamins most likely to be deficient in the diets of domestic birds, which are granivores or omnivores. Diets based on domestic grains and soybean meal, but lacking a source of animal protein and vegetation, must also be supplemented with riboflavin, pantothenic acid, niacin, biotin, B_{12}, and choline. Vitamins A, E, and thiamin are most likely to be deficient in captive faunivores, especially if eviscerated prey items are fed (Graham and Halliwell, 1986; Lowenstine, 1986).

Vitamin Nutrition of the Embryo

Most vitamins are transported into the egg in association with specific binding proteins (Table 11.3). Vitamins bound for the yolk are usually complexed to binding proteins that were synthesized by the liver and are taken up by receptors in the follicle. Vitamins in the albumen are complexed to vitamin-binding

Table 11.3. Transporter responsible for the delivery of vitamins to the albumen and yolk of chicken eggs (see text for references).

Vitamin	Transporter	Destination
Retinol (A)	Retinol-binding protein, transthyretin complex	Yolk
Vitamin C	None	None
Vitamin D_3	Cholecalciferol-binding protein	Yolk
α-Tocopherol (E)	VLDLy	Yolk
Vitamin K	VLDLy	Yolk
Thiamin	Thiamin-binding protein	Yolk and albumen
Riboflavin	Riboflavin-binding protein	Yolk and albumen
Niacin	None	Albumen
Pantothenic acid	?	Yolk
Pyridoxal (B_6)	None	Yolk (75%), white (25%)
Vitamin B_{12}	Transcobalamin	Yolk
Biotin	Biotin-binding proteins, avidin	Yolk (90%), white (10%)
Folacin	?	Yolk
Vitamin B_{12}	B_{12}-binding protein	Yolk
Choline	VLDLy	Yolk
Vitamin C	None	None

VLDLy, yolk-targeted very-low-density lipoprotein.

proteins produced in the oviduct. The form of each vitamin that is best transferred to the egg is dependent upon binding affinities to the binding proteins. For example, retinol and 25-hydroxy vitamin D_3 ($25(OH)D_3$) bind with high affinities to their respective binding proteins and are supplied to the egg, whereas retinoic acid and 1,25-dihydroxy vitamin D_3 ($1,25(OH)_2D_3$) have low affinity for the binding proteins and are of minimal value in provisioning the egg. The level of deposition of most of the vitamins into the egg directly reflects the amount in the diet (see Fig. 11.3). Marginal deficiencies of vitamins do not decrease egg production but are likely to influence embryonic development (Naber and Squires, 1993a,b; Jiang *et al.*, 1994; Wilson, 1997).

The transfer of vitamins from yolk and albumen stores to the developing embryo is not very well characterized (Gaal *et al.*, 1995). Fat-soluble vitamins are probably included in triglyceride-rich lipoproteins synthesized by the yolk-sac membrane and are released into the embryonic circulation. However, fat-soluble vitamins are found in association with differing yolk components and are transferred to the embryo at different stages. Vitamin A resides in the yolk's aqueous phase and is transferred throughout embryogenesis, whereas vitamin E resides in the lipid-rich fraction of the yolk and is mobilized during the last few days of incubation. Presumably, the transfer of water-soluble vitamins from the yolk and albumen involves transport proteins produced by the embryo or its embryonic membranes.

Stability and Supplementation

The vitamins are the least stable of the nutrients. Stored foods or supplements lose their vitamin activity over time, diminishing their nutritional value. Factors that decrease vitamin stability include high temperatures, level and form of dietary trace minerals, exposure to ultraviolet (UV) light, low antioxidant levels, and high moisture (Table 11.4). In general, water-soluble vitamins are less susceptible to destruction than the fat-soluble vitamins, but there are exceptions. Vitamin C, for example, is almost completely lost when food is stored for 6 months. Vitamins A, D, and E are very sensitive to oxidation, which can be catalyzed by trace minerals or by peroxides originating from unsaturated fatty acids. Of the trace minerals, iron and copper have high redox potentials and cause the most damage. Typically, commercial feeds are supplemented with chemical forms of vitamins that maximize stability; however, manufacturing processes (pelleting or extrusion) may still cause high losses of vitamins C, K, and thiamin.

Most commercial avian feeds are supplemented with all of the vitamins, with the frequent exception of vitamin C, in order to rectify losses due to processing and storage and to account for the high variability in vitamin levels in feed ingredients. The quantities supplemented are partly dependent upon the bird's requirements, but are also contingent upon the stability of each vitamin during the feed manufacturing process and the length and conditions under which the feed will be stored (Fig. 11.1). From an economic viewpoint, vitamins E and A are

Table 11.4. Factors that impinge upon the stability of vitamins in foods.*

Vitamin	Stability factor						Typical losses (% month^{-1})
	Moisture	Reduction	Oxidation	Minerals	Heat	Light	
A	++	–	++	++	+	+	10
D	++	–	++	++	+	+	8
E	–	–	+	++	–	–	40†
K	+++	+	–	+++	+	++	17
Thiamin HCl	++	++	++	+	++	–	11
Riboflavin	–	+	–	–	–	+	3
Pyridoxine	–	–	–	+	–	++	4
B$_{12}$	–	++	+	+	+	++	1
Pantothenate	++	–	–	–	+	–	2
Folate	++	+	+	++	+	+	5
Biotin	–	–	++	–	–	–	4
Niacin	–	–	–	–	–	–	5
Ascorbic acid	++	–	+++	+++	–	+	30
Choline	+++	–	–	–	–	–	1

*Data for vitamins A, D, and E are for protected beadlets; unprotected forms are considerably more sensitive. The rate of loss is for commercial feeds stored under normal conditions.
†Alcohol form.
+, Moderately sensitive; ++, sensitive; +++, very sensitive; –, resistant, HCl, hydrogen chloride.

the most expensive; however, vitamin fortification accounts for less than 2% of the total costs of commercial poultry feeds. Because of their inexpensive nature, susceptibility to loss, and general low toxicity, vitamins are usually supplemented with a very large margin of safety, often in the range of two to ten times the requirement. Even higher levels are commonly used for nondomestic species for which requirements are unknown.

In the manufacture of commercial diets, vitamin premixes, containing each of the vitamins plus a carrier, are necessary because several of the vitamins are supplemented in very small amounts. For example, supplementation of vitamin B$_{12}$ to meet the chicken's entire requirement necessitates the addition of only 0.01 mg kg^{-1} diet, which is 1 part per 100,000,000, or 1 g 100 t^{-1}. The premix permits homogeneous distribution of these trace vitamin levels in the feed.

Vitamin A

From a nutritional perspective, vitamin A is the most challenging of the vitamins, because it is the most likely vitamin to be deficient in either captive or wild birds and is present at extremely variable levels in foods. Vitamin A refers to all noncarotenoid β-ionone derivatives that have biological activity similar to all-*trans*-retinol. Plants do not contain vitamin A itself, but have a variety of

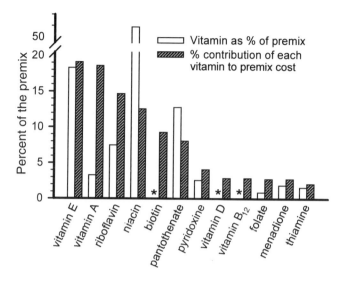

Fig. 11.1. The percent of each vitamin added to the diet of commercial feeds typically fed to broiler chickens between 23 and 45 days of age, and the relative cost of adding each of these vitamins. Inclusion rates designated with ˙ are: biotin, 0.08%; vitamin D, 0.08%; vitamin B_{12}, 0.01%.

carotenoids that are precursors to vitamin A. In animal tissues, vitamin A exists predominantly as retinal, retinol, retinaldehyde, retinoic acid, and retinyl esters.

Metabolism

Vitamin A in most foods of animal origin is in the form of retinyl esters. These are hydrolyzed in the lumen of the small intestine by retinyl ester hydrolase and the free forms of vitamin A enter micelles. Following absorption, vitamin A is reesterified within the intestinal epithelium. Some carotenoids consumed by birds can be cleaved by 15,15′-carotenoid dioxygenase and other hydrolases in the intestinal epithelium to retinaldehyde, which is then reduced to give one molecule of retinol per molecule of carotenoid (Fig. 11.2). Of the 500 carotenoids that exist in nature, about 50 are thought to be converted to retinol, with variable efficiency. Species differences in the enzymatic capacity to convert β-carotene and other carotenoids into vitamin A are not well studied, but are probably extensive. Relative to chickens, use of β-carotene is inefficient in Northern Bobwhite Quail and domestic ducks. Mammalian carnivores (e.g., cats and minks) are unable to convert β-carotene into retinol, but it is not known if the same is true for avian carnivores (Harper *et al.*, 1952; Scott and Dean, 1991).

Retinol and retinyl esters absorbed from the food are transported by portomicrons to the liver and stored mostly as retinyl-palmitate. The liver has the capacity to store large amounts of retinyl-palmitate and to provision other tissues as needed. The liver synthesizes retinol-binding protein, which transports retinol from storage pools to the peripheral tissues. Cells that need vitamin A express a receptor for retinol-binding protein and internalize retinol. Vitamin A

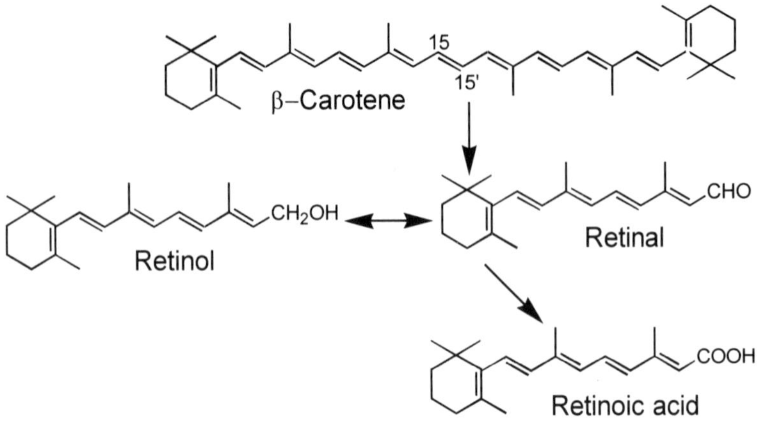

Fig. 11.2. β-Carotene and many other carotenoids can be enzymatically degraded to retinal in the intestinal epithelium. Retinal can be used to synthesize retinol and retinoic acid, but dietary retinoic acid cannot be converted to retinal.

has two very different metabolic functions in cells: the hormone-like regulatory actions of retinoic acid and the photoreception actions of retinal.

The classic photoreception function of retinal in the rods of the eyes is due to the capacity of this molecule to resonate between isomeric forms when struck by light photons. This transformation engages a messenger system that depolarizes the cell membrane of the rods, initiating neural transmission to the brain, where light is perceived. This vitamin A-dependent photoreception is only useful to the bird when light is dim and the rods of the eye are utilized. In conditions of bright light, this system is inactivated and cones, which do not utilize vitamin A, are active.

The hormonal functions of vitamin A are due to the interactions of retinol and retinoic acid with cytoplasmic and nuclear-binding proteins (Halevy *et al.*, 1994; Chambon, 1996). In particular, all-*trans*-retinoic acid and 9-*cis*-retinoic acid bind to specific nuclear receptors and induce the expression of genes that regulate cell replication, differentiation, and preprogrammed cell death. For example, retinoic acid induces the differentiation of epithelial basal cells into the cubodial, columnar, and goblet cells characteristic of the soft moist epithelia of the respiratory and gastrointestinal tracts. In the absence of vitamin A, the basal cells of the respiratory tract, mouth, esophagus, cloaca, ureters, bursa of Fabricius, vaginal region of the oviduct, and the conjunctiva of the eye differentiate into keratinizing cells, which form a hard dry squamous epithelium characteristic of the skin. Retinoic acid is also a critical regulatory molecule controlling cell differentiation and developmental events in the embryo, including the expression of *Hox* genes along the axis of the early-stage embryo and the induction of digit formation. Additionally, the retinoic acid-dependent regulation of the commitment of differentiating stem cells along specific paths is important in bone modeling, spermatogenesis, and leukopoiesis.

The established minimum requirement for vitamin A (see Table 11.2) is adequate to prevent deficiency symptoms, but levels several-fold higher are commonly recommended in order to: account for variability in stores originating from the egg; buffer malabsorption problems due to intestinal parasites or other diseases; build adequate tissue stores; and compensate for the high instability of the vitamin.

Retinol, retinyl esters, and retinal can be interconverted and can be used to synthesize retinoic acid (Fig. 11.2). Thus, the vitamin A requirement can be met by either retinol or retinal. Retinoic acid cannot be converted to retinol or retinal, so it does not support the photoreception function of vitamin A. Additionally, retinoic acid is not transferred to the egg. Hens fed diets with retinoic acid as the sole source of vitamin A lay eggs, but the embryos do not develop. Retinol is the primary form of vitamin A transferred to the egg because it is the primary form of vitamin A that binds to the retinol-binding protein (Thompson *et al.*, 1965; Squires and Naber, 1993; Vieira *et al.*, 1995).

Deficiencies

Chicks from hens consuming diets high in vitamin A have sufficient stores from the yolk to buffer 2–3 months of a diet devoid in vitamin A. Chicks hatched from eggs laid by hens consuming diets marginally adequate in vitamin A have low stores and may begin to express deficiency symptoms soon after hatching. Deficiency symptoms include anorexia, poor growth, ruffled plumage, weakness, poor coordination, and lack of carotenoid pigmentation in the skin and feathers. Deficient chicks are very susceptible to infections, due to inadequate antibody production because of keratinization of the bursa and poor differentiation of B lymphocytes. Responses of T lymphocytes are also impaired. Exacerbating the situation, infectious challenges are more common due to loss of integrity of the mucous epithelia. Birds that are successful in thwarting an infectious challenge often have an impaired ability to repair damaged tissues, making secondary infections more probable. Death from infections may occur before gross symptoms that are diagnostic of vitamin A deficiency are evident (Scott *et al.*, 1982; Tengerdy *et al.*, 1990; Rombout *et al.*, 1992; Sklan *et al.*, 1994).

The content of vitamin A in eggs varies directly with the concentration in the hen's diet and is an excellent indication of the dietary adequacy of the hen (Fig. 11.3). Carotenoid concentrations in the egg also vary directly with dietary levels, but carotenoids, including β-carotene, are not absorbed efficiently from the yolk. Normal stores of vitamin A in the liver of a laying chicken can supply sufficient retinol to the egg for at least 3 months of consumption of a diet devoid of vitamin A before embryonic pathology is frequent. Embryos from vitamin A-deficient Japanese Quail or turkeys usually die at about 48 h of incubation from failure of the circulatory system to develop. Abnormalities of the kidneys, eyes, and skeleton are apparent in less severe deficiencies. Problems with embryonic development manifest themselves before the hen shows symptoms of a vitamin A deficiency (Heine *et al.*, 1985; Haq and Bailey, 1996).

In nonlaying adults, liver stores of vitamin A can buffer a dietary deficiency

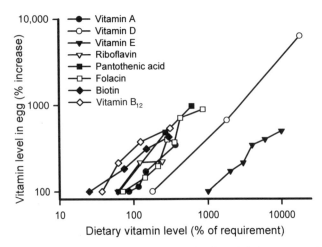

Fig. 11.3. The relationship between dietary vitamin levels and their concentrations in the eggs of chickens. Values are calculated from data summarized by Naber (1993).

for many months (Austic and Scott, 1991; Honour *et al.*, 1995a,b). For example, Mallard ducks with liver stores > 600 μg g^{-1} do not show signs of deficiency until 5 months of consuming a diet virtually devoid in vitamin A. In adult chickens, turkeys, and Mallards, the first outward symptoms of a vitamin A deficiency appear in the esophagus, where the normal mucous epithelium is replaced with stratified squamous epithelium. This condition, called squamous metaplasia, blocks the ducts of mucous glands and necrosis of their contents causes the development of small white pustules, which are easily visible to the naked eye. Squamous metaplasia also develops in the mouth, salivary glands, and nares (e.g., in parrots). Infections of these surfaces are common. Hyperkeratinization on the balls of the feet may lead to infections and inflammation, giving a syndrome known as bumblefoot. The mucous epithelium of the intestines does not become keratinized, but the number of goblet cells in the intestine decreases, reducing mucous secretion and causing enteritis and increased infections. In some species (e.g., psittacines, raptors), inappropriate differentiation of the kidney tubules causes diminished uric acid excretion, the crystallization of urates in soft tissues, and gout (Graham and Halliwell, 1986).

Vitamin A deficiency impairs the function of the rods in the eyes, causing night blindness, but is reversible with just a few days of vitamin A supplementation, if caught at the early stages. More severe deficiencies result in keratinization of the conjunctiva and inadequate lubrication of the cornea. Subsequent abrasions and infections cause irreversible loss of sight. This condition is known as xerophthalmia and is characterized by the accumulation of white inflammatory fluids in the eye and a discharge of exudates, which often causes the eyelids to crust together. A sticky exudate is also discharged from the nares if they become infected.

In some populations of birds living in the wild, the liver stores of vitamin A vary seasonally, as intakes and requirements fluctuate (Nestler *et al.*, 1949; Wobeser and Kost, 1992). A deficiency of vitamin A has been documented in populations of game birds and waterfowl. For example, wild ducks consuming agricultural grains, such as wheat or barley, may develop a vitamin A deficiency. This situation is most likely in the winter, when vegetation is unavailable and the low levels of β-carotene found in these grains is diminished by weathering in the fields. Among the domestic grains, corn is a rich source of β-carotene and protects populations that consume it.

Toxicity

Superficially, many of the symptoms of a vitamin A toxicity are somewhat similar to those of a deficiency. High tissue levels of retinoic acid disrupt the differentiation of mucous epithelial cells, causing hyperplasia. This condition leads to necrotic discharges and pustules in the mouth, nares, and eyes. Altered differentiation of osteoblasts causes disruption of the growth plate of long bones. T-lymphocyte function also declines when retinoic acid levels are high. Because retinoic acid is the vitamer that causes these problems, dietary retinoic acid is considerably more toxic than retinol or retinal. Retinal is more toxic than retinol, because it is more easily converted to retinoic acid. The enzymatic conversion of dietary carotenoids to vitamin A is regulated roughly according to need, and consumption of high amounts of carotenoids does not usually cause a toxicity (Scott *et al.*, 1982; Halevy *et al.*, 1994).

Dietary sources

The fat and liver, but not muscles, of vertebrate prey are a rich source of vitamin A (Dierenfeld *et al.*, 1991). Those plants and insects that are high in carotenoids are also a good source of vitamin A activity. β-Carotene is the most prevalent carotenoid in most avian foods and has the highest vitamin A bioavailability, approaching 50% of retinol in chickens, but may be as low as 25% in Northern Bobwhite Quail. In chickens, α-carotene and γ-carotene have about half of the bioactivity of β-carotene, and canthaxanthin and zeaxanthin are considerably less active. In fact, high levels of canthaxanthin may competitively inhibit vitamin A absorption or metabolism (Nestler *et al.*, 1948; Sklan *et al.*, 1989; Hencken, 1992).

Vitamin A esters are more stable in avian foods than retinol, so retinyl-acetate and retinyl-palmitate are typically the forms used for supplementation of avian diets. Often, they are manufactured as coated spheres containing an antioxidant, such as ethoxyquin or butylated hydroxytoluene (BHT). These coated beads are relatively stable during the pelleting, extrusion, or storage of feed. Although dietary carotenoids have considerably lower vitamin A activity than retinol, they are a very safe way to supplement diets, as they have the lowest potential toxicity.

Vitamin D

Vitamin D activity is found in a group of related sterols, including cholecalciferol (vitamin D_3), ergocalciferol (vitamin D_2), and their metabolites. Birds can synthesize cholecalciferol from cholesterol when they receive adequate sunlight. Hence the requirement for vitamin D can be met either by endogenous synthesis or by a dietary supply. In birds kept in captivity, endogenous synthesis is often insufficient to completely negate the need for a dietary source. Vitamin D is a required component of the endocrine system of birds and regulates calcium and phosphorus homeostasis, bone mineralization, and eggshell formation. As such, vitamin D is considered to be a hormone as well as a vitamin.

Metabolism

Endogenous synthesis of vitamin D_3 starts with the synthesis of 7-dehydrocholesterol from cholesterol in the epithelial cells of the skin (Fig. 11.4). This provitamin can be converted to cholecalciferol by photoisomerization induced by solar radiation. Ultraviolet light, with wavelengths between 285 and 315 nm, is most effective. Studies in poultry indicate that 11–30 min of strong

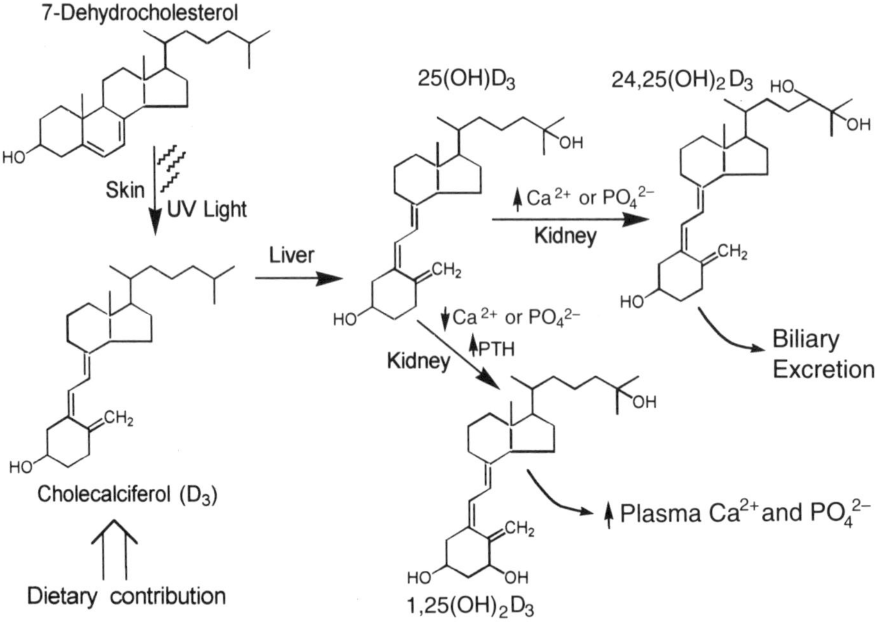

Fig. 11.4. The conversion of 7-dehydrocholesterol to cholecalciferol (vitamin D_3) occurs in the skin of birds and requires UV light. The hydroxylation of 25(OH)D_3 to 1,25(OH)$_2D_3$ occurs when plasma levels of either calcium ion (Ca^{2+}) or inorganic phosphate (PO$_4^{2-}$) are low or levels of parathyroid hormone (PTH) are high. The hydroxylation of 25(OH)D_3 to 1,24(OH)$_2D_3$ occurs when plasma levels of Ca^{2+} or PO$_4^{2-}$ are high and facilitates the excretion of vitamin D.

sunshine each day prevents a vitamin D deficiency in growing chicks. Sunlight is most potent in the tropics, at high altitudes, at midday, and during the summer. 7-Dehydrocholesterol synthesis is especially active in featherless areas, such as the lower legs, feet, and areas of the head, such as combs, wattles, and around the eyes. In addition to feathers, skin pigments block the penetration of UV light, and the effectiveness of vitamin D synthesis probably varies between species, depending upon the amount of exposed skin and its pigmentation. Contrary to popular belief, the uropygial gland of wild and domestic birds does not contain high concentrations of 7-dehydrocholesterol and birds do not ingest nutritionally relevant amounts of vitamin D during the preening process (Hart *et al.*, 1923; Jacob and Ziswiler, 1982; Bernard *et al.*, 1989).

Cholecalciferol, whether of dietary origin or synthesized endogenously, has little metabolic activity unless it is activated by a series of two hydroxylation reactions to the highly potent $1,25(OH)_2D_3$. The 25-hydroxylation occurs first and is catalyzed by a hepatic enzyme. This step is not highly regulated and most cholecalciferol is quickly converted to $25(OH)D_3$. This more polar metabolite is the main circulating and storage form of vitamin D. The subsequent 1-hydroxylation of $25(OH)D_3$ occurs in the kidney and is highly regulated by parathyroid hormone and calcium concentrations. 1,25-Dihydroxy vitamin D_3 is 500–1000-fold more active than its precursor, $25(OH)D_3$. Over 30 other hydroxylation metabolites can be produced, but $1,25(OH)_2D_3$ is the primary active form. Vitamin D_3 and its metabolites circulate in the blood bound to vitamin D-binding protein. This protein delivers vitamin D to tissues, acts to conserve the vitamin during deficiencies, and buffers against toxicities. The affinity of vitamin D-binding protein is greatest for $25(OH)D_3$ followed by D_3 and then $1,25(OH)_2D_3$. A unique vitamin D-binding protein is synthesized in laying females. This binding protein has a higher affinity for D_3 than for $25(OH)D_3$ and forms a complex with phosvitin to deliver D_3 to the follicle for deposition into the egg yolk. In penguins, the production of this second calcium-binding protein appears to be induced by courtship behavior (Fraser and Emtage, 1976; Griffiths and Fairney, 1988; Norman and Hurwitz, 1993).

When dietary calcium is deficient, high levels of parathyroid hormone activate the renal 1-hydroxylase enzyme and increase the circulating levels of $1,25(OH)_2D_3$ (Fig. 11.4). These two hormones act in concert to mediate a variety of actions that increase plasma calcium concentrations, including increased calcium absorption from the small intestine, calcium mobilization from bone, and increased reabsorption of calcium from the kidney tubules. During a phosphorus deficiency, growth hormone mediates an increase in 1-hydroxylase activity, causing high $1,25(OH)_2D_3$ in the absence of high levels of parathyroid hormone. This hormone combination causes mobilization of calcium and phosphorus from the bone, but the calcium is excreted, so that normal plasma calcium and phosphorus levels are maintained. The tight regulation of plasma calcium and phosphorus levels requires stringent regulation of the amount of $1,25(OH)_2D_3$ synthesized, and the regulation of $1,25(OH)_2D_3$ catabolism via 24-hydroxylation is an important control point. Although the classic effect of the vitamin D endocrine system on calcium and phosphorus metabolism is best characterized, this regulatory system also influences cell differentiation in the immune system, bone, intestine, and skin. In fact, cells of the skin (keratinocytes) and the immune system (activated macrophages) can activate $25(OH)D_3$ to $1,25(OH)_2D_3$. In general, $1,25(OH)_2D_3$ has immunosuppressive actions, particularly on T-helper lymphocytes of the subset-1 class. It also mediates the differentiation of monocytes into osteoclasts (Walters, 1992; Norman and Hurwitz, 1993; Soares, 1995).

Vitamin D_3 is present in the yolk of the egg and is taken up by the embryo throughout development. The vitamin D endocrine system becomes competent in the chicken embryo at 6–8 days of incubation, with activating hydroxylation occurring in the mesonephric kidney. Beginning at this early stage, $1,25(OH)_2D_3$ regulates calcium homeostasis by activating the uptake of calcium by the yolk-sac membrane. As the calcium needs of the embryo increase during the last half of development, $1,25(OH)_2D_3$ mediates the uptake of calcium from the shell via the chorioallantoic membrane (Narbaitz, 1987; Clark *et al.*, 1990).

Both genomic and nongenomic mechanisms are involved in the endocrine-like action of $1,25(OH)_2D_3$. 1,25-Dihydroxy vitamin D_3 can traverse the cellular membranes and bind to receptors in the nucleus. Receptors are present in almost all tissues of the body and, to date, over 100 genes are known to be regulated by $1,25(OH)_2D_3$. These include calbindin, which is important in calcium absorption and eggshell mineralization, and osteocalcin, which is important in bone mineralization. The opening of calcium channels in the cell membrane is an important nongenomic action of $1,25(OH)_2D_3$, which initiates a variety of immediate regulatory actions in cells (Norman and Hurwitz, 1993).

Unlike the other fat-soluble vitamins, vitamin D is not preferentially stored in the liver but is distributed relatively evenly among the various tissues, where it resides in lipid components. In replete birds, the persistence of vitamin D during periods of deficiency is due to large amounts in the blood and the slow release from adipose tissue and skin (Norman, 1990).

Vitamin D_2

Ergosterol is a provitamin sterol synthesized by plants, fungi, molds, lichens, and some invertebrates (e.g., snails and worms). Ergosterol does not exhibit vitamin D activity unless it is converted to ergocalciferol (vitamin D_2) prior to consumption. In plants, most of this conversion occurs on the surface of the leaves by the action of UV light (285–315 nm), which brings about a photoisomerization similar to that which produces vitamin D_3. This process is most active as plants perish and their protective pigments fade. Young, fresh, green plants have low levels of vitamin D_2. Following consumption, vitamin D_2 is hydroxylated to $1,25(OH)_2D_2$ by the same enzymes as vitamin D_3 and has hormonal activity in cells. However, in all birds tested to date, vitamin D_2 does not bind to the plasma vitamin D-binding protein with sufficient affinity to prevent its rapid conjugation and excretion in the bile (Table 11.5). This metabolic loss results in a bioavailability of only about 7–10% of that for cholecalciferol in poultry. Unless nutrition experiments prove otherwise, it should be assumed that all birds utilize vitamin D_2 with an efficacy of less than 10% of vitamin D_3 (Chen and Bosmann, 1964; Hay and Watson, 1977; LeVan *et al.*, 1981).

Table 11.5. Species in which dietary vitamin D_2 is thought to be poorly utilized.*

Anseriformes	Pelecaniformes
Grey-lag Goose	Dalmation Pelican
Barnacle Goose	Piciformes
Ciconiiformes	Chestnut-mandibled Toucan
Saddle-billed Stork	Psittaciformes
Coraciformes	Salmon-crested Cockatoo
Laughing Kookaburra	Ratites
Falconiformes	Ostrich
Imperial Eagle	Greater Rhea
Galliformes	Strigiformes
Domestic chicken	Tawny owl
Common Pheasant	Nduk Eagle Owl
Copper Pheasant	Buffy Fish Owl
Gruiformes	
Brolga Crane	
White-naped Crane	

*Based on the relative affinity of the vitamin D-binding protein for 25-hydroxy vitamin D_3 versus 25-hydroxy vitamin D_2 (Hay and Watson, 1977). Other species have not been examined and should be assumed to have similarly poor utilization until proved otherwise.

Requirement

Birds do not have a requirement for vitamin D if they receive adequate sunlight; however, the established requirement is based on the assumption that no endogenous synthesis occurs (see Table 11.2). In addition to insufficient sunlight, the vitamin D requirement is increased when the dietary level of calcium is low, when the calcium to phosphorus ratio is low or high, and when dietary phosphorus is present as phytic acid. In poultry, levels of vitamin D_3 well above the requirement or, in some cases, the use of more active metabolites, such as $25(OH)_2D_3$ or $1,25(OH)_2D_3$, is effective at treating or preventing bone diseases, such as tibial dyschondroplasia or osteomalacia (Edwards, 1990; Xu *et al.*, 1997).

Deficiency

Vitamin D deficiencies are common among captive birds kept indoors (Bernard *et al.*, 1989). Artificial lights typically do not produce sufficient UV light in the wavelength range of 285–315 nm to permit endogenous synthesis of vitamin D. For example, cool white fluorescent lights produce little measurable radiation below 400 nm. Most types of glass and plastic transmit little radiation below 330 nm, so light received through windows is not a useful adjunct. Vitamin D deficiencies are probably rare in free-living birds, due to their exposure to sunlight. Presumably, the faunivorous diet of nocturnal birds (e.g., owls) provides adequate vitamin D, but the development of a vitamin D-independent endocrine system, such as that of the naked mole rat, cannot be excluded. Birds that live at polar latitudes experience long periods of low sunlight but consume diets with very high levels of vitamin D (e.g., plankton, fish).

The symptoms of a vitamin D deficiency are usually similar to those of a calcium deficiency. In young growing chickens fed a vitamin D-deficient diet, early symptoms are slow growth and an awkward gait. There is a tendency for the bird to rest frequently in a squatting position and to display apparent pain when walking. As the deficiency advances, rickets become evident, ribs develop beading at their juncture with the spinal column, and long bones are easily bent, due to insufficient calcification. The epiphyseal plate of long bones becomes wide and degenerative, due to the failure of cartilage-producing cells to mature, leading to their accumulation rather than replacement by osteoblasts. Non-domestic species of ducks and geese raised in captivity exhibit an outward twisting of the metacarpals ('slipped wing') at the time that they grow their primary feathers, when vitamin D is deficient. When yolk stores are low, a vitamin D deficiency becomes evident by the fifth day posthatch in Ring-necked Pheasants (Millar *et al.*, 1977; Scott *et al.*, 1982; Kear, 1986).

Vitamin D-deficient hens lay eggs with thin shells, exhibit hypertrophy of the parathyroid gland, and develop osteomalacia. The high levels of parathyroid hormone that accompany a vitamin D deficiency cause osteodystrophia fibrosa, characterized by demineralization of the medullary bone and infiltration of fibrous connective tissue. Replete laying hens have sufficient stores of vitamin D to buffer about 4 weeks of a severe dietary deficit. Embryos from eggs with inadequate vitamin D_3 have impaired calcium transport from the eggshell via the chorioallantoic membrane and bone calcification is impaired. In severe incidences, chicks may die at the end of incubation or they may be weak and unable to pip the shell. Deficient embryos often have a malformed upper mandible or incomplete formation of the beak (Shen *et al.*, 1981; Narbaitz, 1987; Wilson and Duff, 1991).

Toxicity

Excessive intake of vitamin D_3 or its metabolites causes disruptions in calcium and phosphorus metabolism (Brue, 1994; Soares, 1995). The relative toxicity of the vitamers follows the same pattern as their bioactivity: $D_2 < D_3 < 25(OH)D_3 < 1,25(OH)_2D_3$. Although vitamin D_3 and $25(OH)D_3$ have little metabolic activity themselves, their affinity for the vitamin D-transport protein causes the displacement of vitamin $1,25(OH)_2D_3$, which is then free to activate calcium mobilization. Elevated rates of calcium absorption and mobilization from bone cause abnormally high blood calcium levels, resulting in soft-tissue calcification, cellular degeneration, and inflammation. Affected tissues may include joints, synovial membranes, kidneys, myocardium, blood vessels, and pancreas. Kidney-tubule calcification often results in a fatal buildup of excretory products. Vitamin D toxicity is exacerbated by high dietary levels of either calcium or phosphorus, especially in the growing chick. Although hens are more resistant to vitamin D toxicity than growing chicks, toxic levels of vitamin D may be transferred into the egg. Vitamin D toxicosis occurs in the late stages of embryonic development, due to excessive mobilization of shell calcium. There appear to be considerable species differences in the level of vitamin D that causes a toxicity. A vitamin D toxicity has been induced in several species of macaw parrots at levels that are considerably lower than the levels tolerated by most other species (1,000,000 IU vitamin D_3 kg^{-1} dry matter).

Both phytoplankton and zooplankton are rich in vitamin D_3 and, consequently, marine fish are also an excellent source of vitamin D_3. Other vertebrate prey may have marginally low vitamin D_3 levels, relative to the requirement of birds that are not exposed to sunlight. Most plants contain very low levels of bioavailable vitamin D, and seeds and fruits are particularly deficient. Thus, natural diets of florivorous birds are low in vitamin D and they are apparently dependent upon endogenous synthesis. Grain-based diets fed to captive birds must be supplemented with vitamin D when husbandry conditions cause inadequate sunlight for endogenous synthesis.

Cholecalciferol is typically used for supplementing avian diets. Some poultry feeds are supplemented with $25(OH)D_3$ to aid in the prevention of bone abnormalities. 1,25-Dihydroxy vitamin D_3 is not widely used as a nutritional supplement, for two reasons: it does not support normal embryonic development; and it bypasses the important 1-hydroxylase control step, resulting in a high probability of a toxicity. As vitamin D is unstable and decomposes during the manufacture and storage of feeds, it is usually supplemented at levels of three to ten times the requirement.

The poor bioavailability of vitamin D_2 makes it a poor choice for supplementing birds. Additionally, the vitamin D levels reported for foods consumed by humans and livestock are calculated based on equivalent bioactivity of vitamin D_2 and D_3. Although appropriate for those mammalian species, the values should not be considered accurate for birds. This is particularly relevant for the vitamin D in marine fish, much of which is present as vitamin D_2. Some foods produced for human consumption are supplemented with vitamin D_2, and this contribution is of little value to birds.

Vitamin E

The term vitamin E refers to two groups of compounds that have antioxidant activity in cellular membranes: α-, β-, γ-, and δ-tocopherols and α-, β-, γ-, and δ-tocotrienols. The most biologically active form is D-α-tocopherol and this compound is used as the standard for description of dietary requirements and food levels. The bioactivity of other forms is not well characterized in birds, but in those species that have been studied only α-tocopherol is incorporated into tissues. Dietary γ- and δ-tocopherols are well absorbed and circulate in blood lipoproteins, but they are not integrated into cellular membranes or triglyceride stores (Jakobsen *et al.*, 1995).

Most dietary vitamin E is absorbed in the free alcohol form and is transported by plasma lipoproteins. There are no known specific transport proteins or important metabolic interconversions of vitamin E. Vitamin E is integrated into the lipid bilayer of cellular membranes and defends them from oxidative damage, by terminating the chain reactions of reactive oxygen intermediates and polyunsaturated fatty acids. Because α-tocopherol can compete for free radicals much faster than polyunsaturated fatty acids, small amounts of the vitamin are able to defend relatively large amounts of membrane lipid. Several enzyme systems aid vitamin E in protecting cells against oxidation, including superoxide dismutase, glutathione peroxidase, and catalase. These enzymes utilize zinc, copper, manganese, selenium, and iron as cofactors, and deficiencies of these trace minerals can increase the demand for vitamin E. Vitamin E is unique among these antioxidant systems in that it is preferentially sequestered in cellular membranes and not the cytoplasm.

Tissue α-tocopherol content increases proportionally with dietary intake (Fig. 11.5). Unlike most other vitamins, there is no saturation threshold, and storage levels can become very high. The hepatic pool is particularly labile and is mobilized during short periods of dietary deficit. The vitamin E in adipose tissue is not a reliable storage site, unless this tissue is being catabolized due to an energy deficit. Vitamin E homeostasis is extremely variable across species. For example, 11 different species of raptors fed diets with high vitamin E levels exhibit a 44-fold difference in serum α-tocopherol levels (Calle *et al.,*

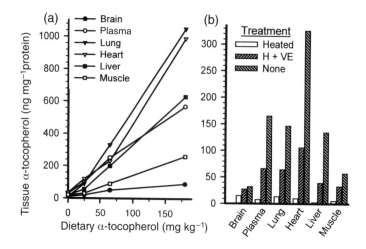

Fig. 11.5. The amount of vitamin E in the tissues is dependent upon the dietary concentration of vitamin E and the amount of oxidized fat in the diet. (a) Chickens were fed diets with 0, 25, 65, and 180 mg α-tocopherol kg⁻¹ from hatching to 24 days of age (Sheehy *et al.,* 1991). (b) Chickens were fed diets containing sunflower oil that received the following treatments (Sheehy *et al.,* 1994):

1. None: the oil was used fresh and contained 30 mg kg⁻¹ vitamin E.
2. Heated: the oil was heated to 120°C for 11 h to induce oxidative rancidity. Virtually all of the vitamin E was destroyed by this process.
3. H + VE: heated sunflower oil to which vitamin E was added to a level of 30 mg kg⁻¹.

1989; Hassan and Hakkarainen, 1990; Mainka *et al.*, 1992).

Vitamin E influences a variety of physiological processes via modulation of eicosanoid metabolism. The conversion of arachidonic acid into prostaglandins and thromboxanes involves a number of oxidative steps, which can be modified by the level of vitamin E in membranes. For example, high levels of vitamin E decrease the amount of prostaglandin E_2 that chicken macrophages synthesize from arachidonic acid during an immune response. The immunomodulatory actions arising from altered prostaglandin levels may be beneficial in some types of disease challenges (Lawrence *et al.*, 1985; McIlroy *et al.*, 1993; Romach *et al.*, 1993; Panda and Rao, 1994; Fuhrmann and Sallmann, 1995).

Requirement

The established minimum requirement for vitamin E (see Table 11.2) is a useful starting point for nutritionists, but a wide variety of dietary factors may markedly increase the requirement, including: high levels of polyunsaturated fats; high levels of vitamin A or β-carotene; low levels of other dietary antioxidants; rancid dietary fats; and a selenium deficiency. Feeding diets with vitamin E at the level required by chickens (10 IU kg^{-1}) commonly results in a deficiency in many species of birds, but the actual requirement for nonpoultry species is unknown. It is common practice to provide 100–250 IU kg^{-1} vitamin E, or more, to captive populations of faunivores. In the production of poultry for human consumption, dietary levels around ten times the requirement are commonly fed to prolong the shelf life of meat by preventing oxidative rancidity (Calle *et al.*, 1989; Mainka *et al.*, 1992).

The type, quantity, and quality of fat in the diet has a marked impact on the vitamin E requirement (Vericel *et al.*, 1991; Applegate and Sell, 1996). The requirement increases linearly with the amount of polyunsaturated fatty acids in the diet. Of the polyunsaturated fatty acids, linoleic acid is the most active in increasing the vitamin E requirement, because it not only increases the potential oxidative load in triglycerides, but also increases the production of eicosanoids, which mediate some of the pathology induced by a deficiency. Vitamin E nutrition is also compromised by the oxidative rancidity of dietary fats prior to their consumption, through two synergistic mechanisms: the vitamin E in the diet is depleted as the fat goes rancid; and oxidation products, which result from peroxidation of polyunsaturated fatty acids, deplete α-tocopherol in the tissues of the bird (Fig. 11.5). Consequently, rancid dietary fats increase the vitamin E requirement by several-fold over that predicted by their polyunsaturated fatty acid content.

Piscivorous birds typically consume high levels of polyunsaturated fatty acids and probably have a much higher vitamin E requirement than domestic birds, but the precise level awaits investigation. The practice of freezing, thawing, eviscerating, and storing fish for long periods of time prior to feeding depletes the vitamin E and contributes to the problem. Wild birds consuming dead rancid fish also have a high requirement and are susceptible to deficiencies (Nichols and Montali, 1987).

Synthetic antioxidants, such as butylated hydroxytoluene (BHT) butylated hydroxyanisole (BHA), or ethoxyquin, protect dietary vitamin E and can perform some of the functions of vitamin E in tissues, decreasing the dietary requirement. Conversely, a selenium deficiency impairs the glutathione peroxidase anti-oxidant system and increases the need for vitamin E. High dietary levels of vitamin A or β-carotene also increase the vitamin E requirement, by decreasing the absorption and tissue deposition of vitamin E (Kim and Combs, 1993; Jiang *et al.*, 1994; Pellett *et al.*, 1994).

Deficiency and toxicity

Most of the diagnostic symptoms of a vitamin E deficiency are manifestations of cellular-membrane dysfunctions, resulting from the oxidative degradation of polyunsaturated fatty acids. Depending upon the species, symptoms may include encephalomalacia, exudative diathesis, muscular dystrophy, gizzard myopathies, and increased fragility of red blood cells. Polyunsaturated fatty acids in storage fat also become oxidized. In piscivorous birds, the adipose tissue may become necrotic and granulomatous. Encephalomalacia (also called 'crazy chick disease') occurs under severe deficiency conditions and is characterized by lesions in the membranes of blood vessels in the cerebellum. These lesions cause hemorrhages and edema in the brain, resulting in ataxia and torsion of the neck. Repro-ductively active males that are vitamin E-deficient have reduced fertility, due to abnormal sperm and lower sperm production. Eggs from deficient females are normal in appearance but have low hatchability, due to late embryonic mortality, characterized by pathologies similar to those seen in adults. A dietary vitamin E deficiency does not cause slower growth, changes in body composition, or numbers of eggs produced, unless it is very severe and long in duration. Mild symptoms of a vitamin E deficiency are usually reversible upon supplementation (Jensen, 1968; Kling and Soares, 1980; Nichols and Montali, 1987).

Vitamin E has a low level of toxicity for poultry, and levels greater than 100 times the requirement are well tolerated. High levels of dietary vitamin E may cause a deficiency of other fat-soluble vitamins, and the toxicity symptoms usually resemble the deficiency symptoms of vitamin A or K. For example, Pink-backed Pelicans fed fish supplemented with high levels of vitamin E (> 1000 IU kg^{-1}) for 1–2 years develop hemorrhagic problems symptomatic of a vitamin K deficiency (Calle *et al.*, 1989; Nichols *et al.*, 1989).

Dietary sources

Vitamin E is synthesized only by plants and is found at highest levels in foods of plant origin. α-Tocopherol is the primary form of vitamin E found in the chloroplasts of plants and is located mostly in leaves. γ-Tocopherol is found in high concentrations in the storage lipids of plants and is prevalent in many grains and vegetable oils. α-Tocopherol predominates in foods of animal origin. In general, the vitamin E value of foods is related to their fat content, and at a given fat content, foods of plant origin are considerably higher in vitamin E than foods of animal origin. However, the amount present in foods of animal origin is dependent upon the species and the amount of vitamin E in their diet. The vitamin E concentration in the tissues of domestic chicks, mice, rats, and quail fed to captive faunivores is often low and may result in deficiencies. For example, Peregrine Falcons fed quail chicks raised on a typical commercial diet (70 IU kg^{-1} vitamin E) have abnormally low circulating vitamin E levels and poor hatchability of their eggs. If the falcons are fed quail that were raised on a high-vitamin E diet (220 IU kg^{-1}), these problems are eliminated and the falcon's circulating vitamin E levels are similar to those of their counterparts living in the wild (Dierenfeld *et al.*, 1989).

The bioactivities of tocopherols for the prevention of myopathy in the chicken are: α-tocopherol = 100; β-tocopherol = 12; γ-tocopherol = 5. Although γ-tocopherol is absorbed about as well as α-tocopherol, its bioavailability is low, because it is not integrated into cellular membranes and is quickly excreted. Tocotrienols also have low bioavailability in birds (1–20%, depending upon the isomer). Thus, the vitamin E content of foods used in avian nutrition should be based on the α-tocopherol content. In some foods, other forms of tocopherol are present at up to fourfold the level of α-tocopherol, and total tocopherol values have little nutritional applicability (McDowell, 1989).

Most vitamin E in natural foods is in the free form and is not very stable (Schneider, 1986). During processing, bioactivity is lost with increasing heat, moisture, unsaturated fatty acids, alkalinity, and trace minerals. Organic acids and finely ground limestone or magnesium oxide (MgO) in a diet rapidly destroy vitamin E. Esterification of the reactive hydroxyl group of vitamin E to an organic acid, such as acetate, markedly improves its stability. This ester linkage is hydrolyzed by the intestinal epithelial cells of birds to give the active vitamin. D,L-α-Tocopheryl acetate is commercially available and is the form of vitamin E most commonly supplemented to diets. The L-form of tocopherol or tocopheryl esters are considerably less bioavailable than the D form, but cost considerations usually prohibit the use of pure D-α-tocopheryl. D,L-α-Tocopheryl acetate is a viscous oil, but it may be adsorbed to silica, to give a dry powder, or to starch or sugars, to form a water-dispersible powder.

Vitamin K

Vitamin K refers to a group of compounds related to menadione (2-methyl-1,4-naphthoquinone; vitamin K$_3$) that exhibit antihemorrhagic activity. Menadione does not occur naturally, but is the common synthetic form of vitamin K. The two natural sources of vitamin K are the phylloquinones (vitamin K$_1$), synthesized by plants, and the menaquinones (vitamin K$_2$), synthesized by bacteria such as *Actinomyces*. The number of isoprenoid groups attached to naturally occurring menaquinones is variable and influences the absorption and, consequently, the bioavailability of the vitamin. Menaquinones with three to five isoprenoid groups have maximal vitamin K activity.

Vitamin K is transported between tissues via lipoproteins and is taken up by tissues during lipoprotein metabolism. Vitamin K functions as a cofactor of a hepatic microsomal carboxylase that catalyzes the posttranslational carboxylation of specific glutamate residues in proteins to γ-carboxyglutamate. This carboxylation is important for the function of at least 12 proteins, including osteocalcin, clotting factors VII, IX, and X, and prothrombin. Both phylloquinone and menaquinone can act as cofactors for γ-glutamylcarboxylase with similar efficacy (Suttie, 1980).

The storage of vitamin K is minor and its tissue turnover is relatively rapid compared to the other fat-soluble vitamins. Dietary menadione is lost particularly fast, because it lacks the isoprene units that apparently prevent its excretion following absorption. However, phylloquinone and menaquinone are lost faster in chickens than in mammals, because they are not recycled efficiently following their catalytic role with γ-glutamylcarboxylase. The kinetic characteristics of the vitamin K epoxide reductase responsible for this recycling account for the higher vitamin K requirement in chickens than in laboratory mammals or humans (Will *et al.*, 1992).

Requirement, deficiency, and toxicity

Among the fat-soluble vitamins, vitamin K is unique, in that it functions solely as a cofactor for an enzyme (Almquist, 1971). The low concentration of this enzyme in the body explains the relatively low dietary requirement for vitamin K. The contribution of intestinal synthesis and coprophagy to the requirement further decreases the need for dietary vitamin K in many species. The vitamin K requirements established for poultry (see Table 11.2) are for birds that are prevented from access to their feces by housing in pens with raised wire floors. Factors that increase the requirement for vitamin K include infectious diseases, specific vitamin K antagonists, antibiotics, and conditions that impair absorption (e.g., high dietary levels of vitamins E or A). Dicumarols produced by molds are especially active vitamin K antagonists and may be consumed with contaminated foods. The therapeutic use of antibiotics depletes intestinal microflora and increases the need for a dietary source of vitamin K. Domestic ducks are particularly sensitive to vitamin K depletion by antibiotic therapy. In the absence of these complicating factors, vitamin K deficiencies are rarely observed in captive or wild birds (Scott and Dean, 1991).

Pathology due to a vitamin K deficiency is similar, regardless of the species or age of bird. The most predominant symptom is a reduction in blood levels of prothrombin, causing increased clotting time. Hemorrhages appear externally on areas that receive abrasions, such as the feet and wings. Internally, petechial hemorrhages in the liver and erosion of the kaolin lining of the gizzard may occur. If intestinal parasites are present, bleeding into the gut may occur. With a severe deficiency, even minor trauma may lead to internal bleeding and death. The amount of vitamin K transferred to the egg is related to the dietary supply, although yolk stores do not reach very high levels. Embryos from vitamin K-deficient chickens have high mortality, due to hemorrhaging late in incubation (Nelson and Norris, 1960; Griminger and Brubacher, 1966).

Vitamin K sources have very low toxicity. Menadione is more toxic than phylloquinones or menaquinones, but levels of over 1000 times the requirement are well tolerated in chickens.

The lack of good analytical methods has precluded extensive characterization of the vitamin K content of natural foods. Leaves of plants tend to be rich in vitamin K, while fruits and seeds are poor sources. Fat-extracted oil seeds, such as soybean meal, are particularly poor sources. Foods of animal origin are moderate in their vitamin K content. The feces of chickens contain levels of vitamin K that are more than tenfold the amount in the diet, and coprophagy is a quantitatively significant dietary source in this species. The bioavailability of phylloquinone is considered to be 100%, and the menaquinones with four, five, or six isoprene units have similar bioavailability.

Vitamin K is usually supplemented as a water-soluble form of menadione: menadione bisulfite, menadione–sodium bisulfite complex, or dimethlypyridinol bisulfite. All forms are susceptible to large losses during processing or storage; however, menadione–sodium bisulfite complex is the most resistant.

Thiamin

The thiamin in foods occurs in the free form and in phosphorylated forms, mostly as thiamin diphosphate covalently bound to proteins. Following consumption, bound thiamin is freed from proteins during digestion in the proventriculus and gizzard. Thiamin is absorbed by an active, carrier-mediated process, when present at low dietary concentrations, and by diffusion, at higher concentrations. Thiamin diphosphates are hydrolyzed in the enterocytes to free thiamin. In the blood, thiamin is transported bound to albumin, within red blood cells, and attached to a specific thiamin-binding protein. Thiamin is taken up by all cells of the body and is phosphorylated to di- and triphosphate esters. Several enzymes covalently bind thiamin diphosphate and use it as a cofactor to catalyze decarboxylation, dehydrogenase, and transketolase reactions. These enzymes are critical in the metabolism of carbohydrates, lipids, and branched-chain amino acids. Excess thiamin that is not needed by tissues is excreted in the urine and no appreciable storage pools are known. However, thiamin liberated from muscle catabolism can prevent deficiency symptoms for several weeks in adult birds that consume a deficient diet. Thiamin is transported to egg yolk in association with a specific thiamin-binding protein. Thiamin stores in the egg are sufficient to last the hatchling for only a few days of consumption of a deficient diet (Miller *et al.*, 1981).

Requirement

The thiamin requirement is relatively similar across those species that have been examined (see Table 11.2), but can be influenced by diet composition and the presence of thiamin antagonists. The prominent role of thiamin as a cofactor in enzymes involved in carbohydrate metabolism and lipogenesis translates into higher requirements when carbohydrates are the major energy source in the diet. Low-carbohydrate, high-fat diets decrease the thiamin requirement.

A variety of thiamin antagonists may be found in avian foods, including polyphenols, thiaminase, and some pharmaceuticals (Murata, 1982; Lowenstine, 1986). Polyphenols (e.g., tannic acid), found in many fruits, seeds, and nuts, react with thiamin to yield nonabsorbable thiamin disulfide. Many species of fish, crustacea, and mollusks that are commonly fed to captive birds contain thiamin-degrading enzymes and antagonists, which are collectively called thiaminases. Thiaminases causes considerable loss of thiamin upon storage of foods prior to consumption and when foods are stored in the crop or proventriculus following consumption. Captive piscivores, such as penguins, herons, egrets, gulls, and storks, often develop a thiamin deficiency if fed fish that are not supplemented with thiamin. Wild birds apparently are not affected by this problem, probably because the enzyme in very fresh fish is compartmentalized and does not gain access to thiamin. Molds also produce thiaminase, as well as several other thiamin antagonists that decrease the thiamin levels in spoiled foods. Thiaminase is heat-labile and destroyed with cooking. The anticoccidial drug Amprolium owes its activity to its antagonism of thiamin metabolism and, when used to treat birds, it increases their thiamin requirement.

Deficiency and toxicity

A thiamin deficiency is usually expressed by alterations in nervous function. The mechanism for neurological changes may be related to the critical role of thiamin-dependent enzymes in energy production via the Krebs cycle. It may also be due to undefined nonenzymatic roles of thiamin in the transmission of nerve impulses. Anorexia is usually the first symptom of a deficiency and can become very severe, causing emaciation and weakness. When given a choice, deficient chickens can discriminate against foods that do not contain adequate thiamin in order to prevent a deficiency. Other early symptoms of a thiamin deficiency may include changes in the motility of the crop, which are manifest as delayed emptying (chickens) or dilatation and regurgitation (pigeons). A continued deficiency causes a collection of symptoms related to neuromuscular defects, known as polyneuritis, including tremors, ataxia, paralysis, and convulsions. Many species have a tetanic retraction of the head, called opisthotonos or stargazing, because this unusual posture could only be useful for upward observation. Stargazing was the symptom in chickens and pigeons that led to the discovery of thiamin in the late 1800s. In deficient Peregrine Falcons, stargazing may be accompanied by convulsions and can be initiated by excitement or by turning the lights off. In Red Wattlebirds, stargazing is accompanied by frantic wing flapping, muscle spasms, and backward somersaults (Ward, 1971; Gries and Scott, 1972b; Paton and Dorward, 1983; St Claire *et al.*, 1994).

Thiamin is not very toxic. Extremely high levels (> 1000-fold the requirement) are tolerated but have an analgesic effect.

Dietary sources

Thiamin is widely distributed in foods. The richest sources are microbial (yeasts, zooplankton), foods of animal origin, and some cereal grains. In foods of plant origin, the thiamin content tends to be positively correlated with protein content. The amount of thiamin present in fresh foods is always greater than in foods that have been stored or processed. Thiamin in foods may be lost due to its instability or because its high solubility in water results in extraction during processing, thawing, etc. Much of the thiamin present in grains is in the germ and seed coat, which are removed during milling. Consequently, whole grains have several-fold more thiamin than milled grains. Supplementation of thiamin is usually in the form of thiamin-hydrogen chloride (HCl) or thiamin-NO_3.

Nectar and many fruits are very low in thiamin. It is thought that nectarivorous and frugivorous birds obtain their thiamin from insects. Severe thiamin deficiencies have been diagnosed in nectarivorous Red Wattlebirds that have been induced to overwinter in areas of Australia by the propagation of winter-blooming plants. In this case, the concurrent lack of insects, coupled with a high-carbohydrate diet, causes a deficiency (Paton and Dorward, 1983).

Riboflavin

Riboflavin in foods is complexed to proteins, usually as flavin mononucleotide (FMN) and flavin adenine dinucleotide (FAD). These complexes are liberated during protein digestion, and the subsequent hydrolytic activities of brush-border phosphatases release free riboflavin. Free riboflavin is absorbed by active transport in the small intestine and transported between tissues bound to proteins, such as albumin. In cells, most riboflavin exists as FMN and, to a lesser extent, FAD. These cofactors are associated with flavoproteins and are important in the transfer of electrons. A very large number of enzymes (> 50) involved in oxidation and reduction reactions require riboflavin as a cofactor. When flavo-proteins throughout the bird's body become saturated with these cofactors, excesses are excreted in the urine (McDowell, 1989).

The transfer of riboflavin to the egg is tightly regulated. Estrogen stimulates the hepatic synthesis of a riboflavin-binding protein, which complexes with vitellogenin and enters the yolk using the vitellogenin receptor. A similar riboflavin-binding protein is synthesized by the oviduct and it makes up about 1% of the albumen. Riboflavin in the egg is tightly bound and its unavailability may deter the growth of some bacteria. The yellow color of riboflavin gives egg white its faint yellowish color and the dietary adequacy can be inferred from the color of the albumen (Naber and Squires, 1993a,b; White, 1996).

Requirement, deficiency, and toxicity

The riboflavin requirement has been accurately determined for poultry and quail (see Table 11.2). In chickens, high levels of dietary fat or protein increase the riboflavin requirement, presumably because many of the enzymes responsible for the oxidation of these nutrients require riboflavin cofactors. High levels of dietary riboflavin do not increase tissue levels or provide an appreciable buffer against a future dietary deficit.

A riboflavin deficiency affects all tissues of the body, and symptoms are first evident in young birds as slowed growth and poor efficiency of food utilization (Gries and Scott, 1972b; Serafin, 1981; Wada *et al.*, 1996). The legs of Northern

Bobwhite Quail, domestic ducks, and domestic geese become weak and may become bowed or paralyzed. With a mild deficiency, symptoms may resemble the perosis typified by a manganese, zinc, or choline deficiency. In chickens, pigeons, and pheasants, a severe deficiency causes myelin degeneration of the peripheral nerves, affecting the sciatic nerve in particular. The sciatic nerve may swell to four times normal size and become pinched, causing continuous stimulation. The resulting muscle contractions sometimes lead to leg paralysis and the inability to extend the digits, giving a syndrome called curled-toe paralysis. Some species (turkeys, ducks, pigeons) exhibit diarrhea and dermatitis during a deficiency. The vent may become encrusted and inflamed. In adults, deficiency symptoms can be reversed with supplemental riboflavin, but with growing birds this becomes less likely as the expression of problems progresses.

The riboflavin content of the egg is proportional to the hen's intake over the range of 0–5 mg of riboflavin kg^{-1} diet, after which it plateaus. When laying hens suddenly consume riboflavin-deficient diets, the riboflavin content of the egg drops in only 2 days, indicating minimal stores. Deficient females lay eggs that have a peak in embryonic mortality around midincubation. Embryonic symptoms include slow growth, edema, shortened limbs, and improper emergence of the down, giving it a blunted appearance, known as clubbed down. With marginal deficiencies, mortality often occurs at pipping, and symptoms may include dwarfism and clubbed down (Squires and Naber, 1993; Wilson, 1997).

High dietary levels of riboflavin saturate the intestinal carrier protein and are inefficiently absorbed, rendering riboflavin virtually nontoxic. Similarly, high dietary levels fed to the hen saturate the binding protein that transfers riboflavin to the egg, preventing high levels in the egg.

Dietary sources

Riboflavin is found in most types of avian foods, but is especially rich in foods of animal origin, particularly liver. Among foods of plant origin, green leaves and fruits are good sources. Riboflavin is exceptionally stable in foods, and losses with storage are mostly due to sunlight (UV). Riboflavin is found primarily in the germ and seed coats of grains and is lost with milling. It is also very water-soluble and may be extracted if foods are soaked in water or freeze-thawed. Riboflavin in corn-based diets fed to chickens is about 60% bioavailable relative to crystalline riboflavin, which is typically used as a supplement (Chung and Baker, 1990).

Niacin

Niacin in foods occurs predominantly in bound forms. In many foods of plant origin, these complexes are not digested and are unavailable to birds. For example, much of the niacin in cereal grains is found in covalently bound complexes with small peptides and carbohydrates, collectively referred to as niacytin. Niacytin is not hydrolyzed and consequently the total niacin content of foods of plant origin is often a poor indicator of their bioavailable niacin. In animal tissues, it is mostly found in nicotinamide adenine dinucleotide (NAD) and

NAD phosphate (NADP). Dietary NAD and NADP are efficiently hydrolyzed by enzymes of the intestinal mucosa to give free nicotinamide, which is absorbed by carrier-mediated facilitated diffusion. At high dietary levels, nicotinamide is also absorbed throughout the small intestine by passive diffusion. Intestinal microflora cleave a portion of nicotinamide to give nicotinic acid, which is also absorbed by diffusion, but at a slower rate. Both nicotinamide and nicotinic acid are transported between tissues in the free form and are taken up by tissues and converted to NAD and NADP. Both NAD and NADP function as enzymatic substrates in proton (H$^+$) transfers, including more than 200 reactions involved in the metabolism of fatty acids, carbohydrates, amino acids, and nucleic acids.

Dietary tryptophan can be used to synthesize niacin, via the pathway shown in Fig. 11.6. The stoichiometry of niacin conversion into nicotinic acid is variable among species of birds. The hepatic enzyme picolinic acid carboxylase diverts 2-amino-3-acroleyl fumaric acid towards the glutaryl-coenzyme A (CoA) pathway and impairs niacin synthesis. Species (e.g., domestic ducks and turkeys) that have high picolinic acid carboxylase activity are inefficient in niacin synthesis from tryptophan. Conversely, species (e.g., pigeons and chickens) that have low picolinic acid carboxylase activity are more proficient in the conversion. To synthesize 1 mg of niacin, chickens need to consume about 50 mg of tryptophan in excess of their requirement for protein synthesis and other uses. This is equivalent to a dietary tryptophan level of about 0.1% or, expressed in another way, a diet devoid of niacin increases the tryptophan requirement of the growing chick by about 50%. For ducks, about 180 mg tryptophan are needed for each mg of niacin synthesized, reflecting a fourfold more active picolinic acid carboxylase than in chickens. A surfeit of dietary tryptophan cannot fully replace the

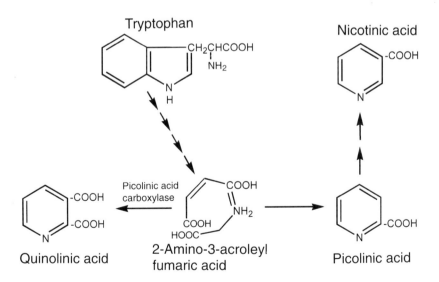

Fig. 11.6. Metabolic conversion of tryptophan to nicotinic acid. Species that have highly active picolinic acid carboxylase are inefficient at synthesizing niacin.

requirement for niacin in growing ducks (Oduho and Baker, 1993; Chen *et al.*, 1996).

Requirement, deficiency, and toxicity

The niacin requirement is extremely variable across species, reflecting the variability in each species' capacity to synthesize niacin from dietary tryptophan. The established requirement for niacin is based on diets with moderate to low levels of dietary tryptophan (see Table 11.2). Chicken hatchlings have considerable excess tryptophan in the proteins of their yolk, buffering a dietary need for niacin for several weeks posthatch. Symptoms that develop in severely deficient chicks include anorexia, slow growth, inflammation of the oral cavity, dermatitis, poor feathering, and an enlargement of the hock joints of the legs, accompanied by bowing (perosis). The perosis caused by a niacin deficiency is not as severe as that due to a manganese deficiency, and the Achilles tendon rarely slips from its condyles. The tongue of deficient chicks may turn black, due to necrosis, a condition known as black tongue. Growing ducks are particularly sensitive to a niacin deficiency and develop diarrhea and severe perosis, in addition to the above symptoms. Growing pheasants, geese, and turkeys develop perosis as their primary deficiency symptom, whereas Northern Bobwhite Quail develop stiff, short feathers and dermatitis. A severe niacin deficiency, coupled with a tryptophan deficiency, impairs egg production in hens; however, embryos can supply some of their niacin requirement through synthesis from tryptophan originating from egg-yolk proteins and may be partially protected (Serafin, 1974, 1981; Scott *et al.*, 1982; NRC, 1994).

In chickens, the toxic potential of nicotinamide is about four times that of nicotinic acid. Presumably, this is due to the faster absorption of nicotinamide by passive diffusion.

Dietary sources

Niacin is widely distributed in foods of both plant and animal origin. Niacin is typically present at higher levels and in a more bioavailable form in foods of animal origin than in those of plant origin. Because tryptophan can be used to synthesize niacin, the total niacin value of foods of animal origin is particularly great. The bioavailability of niacin in cereal grains and oil seeds is only about 10–15%. However, the niacin in grains that are in a young milky stage of development is fully available (Carpenter *et al.*, 1988).

Niacin supplementation is usually in the form of nicotinamide or nicotinic acid. These supplemental forms are very stable relative to most other vitamins and are not markedly affected by most processing or storage conditions. Relative to nicotinic acid, nicotinamide is about 125% more bioavailable to the growing chicken. The bioavailability of NAD is similar to that of nicotinamide (Oduho and Baker, 1993).

Vitamin B$_6$

Vitamin B$_6$ is the general term used to describe the vitamin activity of pyridoxine (pyridoxol), pyridoxal, pyridoxamine, and their phosphates. The free forms of these vitamers have similar vitamin activity in all species of birds and mammals that have been tested. The predominant chemical forms vary among foods of plant and animal origin: plant tissues contain mostly pyridoxine, while animal tissues contain a mixture of pyridoxal and pyridoxamine.

The various forms of vitamin B$_6$ are absorbed by passive diffusion in the small intestine and, to a limited extent, in the ceca of the chicken (Heard and Annison, 1986). Following absorption, the various forms are metabolized in the enterocyte and liver, resulting in pyridoxal and pyridoxal phosphate as the primary circulating forms. They are transported between tissues bound to albumin and other blood proteins. Cells take up free pyridoxal and phosphorylate it prior to use as a cofactor in a variety of transaminases, deaminases, trans-sulfurases, phosphorylases, and decarboxylases. Vitamin B$_6$ also serves a poorly defined regulatory role in the function of many proteins, including hemoglobin and several enzymes involved in lipid metabolism.

Most of the enzymes responsible for amino acid biosynthesis, catabolism, and interconversions require vitamin B$_6$ as a cofactor. Vitamin B$_6$ is also required for the function of glycogen phosphorylase, which mobilizes glucose from glycogen stores. The high abundance of this enzyme in muscle and liver accounts for more than half of all vitamin B$_6$ in the bird's body. Pyridoxine phosphate bound to glycogen phosphorylase can serve as a reserve to resupply other tissues during starvation, but not when a high-energy diet that is deficient in vitamin B$_6$ is consumed. Plasma aspartate aminotransferase has a low priority for B$_6$ and a diminution of its activity is one of the first signs of dietary inadequacy in chickens (Masse *et al.*, 1991).

Requirement, deficiency, and toxicity

The established minimum requirement for vitamin B$_6$ (see Table 11.2) is based on diets that have a moderate level of well-balanced protein (Daghir and Balloun, 1973). Diets high in total protein or containing unbalanced protein markedly increase the requirement for vitamin B$_6$, presumably because of the greater levels of vitamin B$_6$-dependent enzymes needed to metabolize the excess amino acids. Fortunately, most foods that have high levels of protein also have high levels of vitamin B$_6$ and deficiencies of this vitamin are rarely reported. Deficiencies can be induced by feeding diets that have a purified protein source, such as casein. Gross symptoms of a vitamin B$_6$ deficiency in young chickens and ducks are of limited diagnostic value and include weakness, poor appetite, slow growth, poor feather growth, ataxia, anemia, perosis, trembling, convulsions, and hyper-excitability. Collagen fibers in the articular cartilage and cortical bone are malformed and contribute to poor connective-tissue integrity and perosis. Deficient hens decrease egg production and their eggs have poor hatchability. The vitamin B$_6$ content of eggs reflects that in the food, and the amount needed for maximal fertility is twice that needed to maintain egg numbers. Vitamin B$_6$

is not particularly toxic when fed at high levels (Fuller *et al.*, 1961; Gries and Scott, 1972a; Masse *et al.*, 1996).

Dietary sources

Most avian foods, with the exception of some fruits and processed grains, are adequate sources of vitamin B_6. Vitamin B_6 is found predominantly in the germ and aleuronic layers of cereal grains, which are removed during milling. Whole grains and nuts are a good source of vitamin B_6 and virtually all animal tissues are an excellent source.

A large portion of the B_6 in foods of plant origin is covalently bound to proteins, as well as to glycosides (e.g., in sunflower seeds and soybeans) and other nonprotein macromolecules (Yen *et al.*, 1976). Many of the nonprotein-bound forms of vitamin B_6 are poorly digestible. Consequently, the bioavailability of vitamin B_6 in foods of plant origin is usually less than in those of animal origin. In the chicken, the bioavailability of vitamin B_6 in corn is 40% and in soybean meal 60%, relative to pyridoxine hydrochloride. Pyridoxine is far more stable than either pyridoxal or pyridoxamine and is usually supplemented as pyridoxine hydrochloride.

Pantothenic acid

Pantothenic acid occurs in most foods as a component of CoA and the acyl-carrier protein. It is released by enzymatic digestion and absorbed in the small intestine by a specific energy-dependent transport pathway. Free pantothenic acid is transported by the blood to the tissues, where it is converted to CoA. The pantothenic acid in CoA can be transferred to acyl-carrier protein. Both CoA and acyl-carrier protein function metabolically as carriers of acyl groups. Coenzyme A forms high-energy thioester bonds with carboxylic acids, especially acetic acid, and participates in the catabolic and anabolic metabolism of carbohydrates, fatty acids, and amino acids. The acyl-carrier protein is a component of the fatty acid synthase complex, where it functions to transfer intermediates during fatty acid synthesis.

Requirement, deficiency, and toxicity

The pantothenic acid requirement (see Table 11.2) is relatively unaffected by other dietary parameters, with the possible exception of a vitamin B_{12} deficiency. Although severe deficiencies are rare, due to the wide distribution of pantothenic acid, mild deficiencies may occur in birds consuming diets based on cereal grains. Mild deficiencies cause slow growth, poor feathering, and weakness in poultry, Ring-necked Pheasants, domestic ducks, Northern Bobwhite Quail and Cockatiels. More severe deficiencies in chickens and turkeys fed purified diets often include: ataxia; skin lesions at the corners of the mouth; swollen and encrusted eyelids; and cracks and, later, hemorrhages in the skin of the feet and toes. Internal pathology includes: fatty liver degeneration, thymic and bursal necrosis, and degeneration of myelin in nerves. Pantothenic acid-deficient hens do not markedly decrease egg production, but embryos from their eggs have high mortality, characterized by edema, hemorrhages, and poor feathering. Excess pantothenic acid is rapidly excreted in the urine and high dietary levels are not known to be toxic (Scott *et al.*, 1964; Gries and Scott, 1972b; Serafin, 1974; Roudybush, 1996).

Dietary sources

The distribution of pantothenic acid in foods is similar to that of vitamin B_6. Few natural foods are deficient, although domestic grains and fruits have moderate to low levels. The pantothenic acid in sorghum, wheat, and barley has a bio-availability to chickens of about 60% of crystalline pantothenic acid (Southern and Baker, 1981). Supplementation is usually as calcium pantothenate, which is stable to most conditions except heat and moisture. Only the D isoform of pantothenic acid has vitamin activity in birds and mammals.

Biotin

Most of the biotin in avian foods is covalently linked to proteins, usually at a lysine residue as ε-N-biotinyl-L-lysine (biocytin). When these proteins are digested, biocytin is liberated and then free biotin is released by intestinal biotinidase. Many sources of protein-bound biotin are resistant to digestion, and biotin is one of the least available of the vitamins. In many foods of either plant or animal origin, less than half of the biotin is available. Microflora in the ceca of chickens synthesize considerable amounts of biotin, but the extent of its digestion and absorption is not known. Biotin from the food is absorbed throughout the small intestine by facilitated diffusion and, at high concentrations, by simple diffusion. Biotin is transported in the blood bound to biotin-binding protein I and is taken up by cells throughout the bird's body. Biotin serves as a covalently bound cofactor in four carboxylase enzymes, which are central to the metabolism of lipids, glucose, some amino acids, and energy. In particular, biotin is needed for gluconeogenesis, lipogenesis, and the elongation of essential fatty acids. An appreciable amount of biotin is stored in the liver, and these stores can buffer dietary deficiencies for several weeks (White and Whitehead, 1987; Bryden, 1989; McDowell, 1989; Watkins, 1990).

Biotin is transferred to both the white and the yolk of the egg by avidin- and biotin-binding proteins, respectively. Biotin-binding protein II is synthesized by the liver of the hen and delivers biotin to the developing follicle for incorporation into the yolk. This protein binds biotin noncovalently with high affinity ($k_D \approx 1$ pmol l^{-1}) and is saturated in eggs laid by biotin-replete hens. Biotin bound to binding protein II is the primary source of biotin for the developing embryo. The yolk also contains a smaller amount of biotin bound to biotin-binding protein I. The albumen does not contain either of these proteins, but instead it contains avidin, which is a glycoprotein secreted by the oviduct during the deposition of egg white. Avidin binds biotin with extraordinarily high affinity ($k_D \approx 1$ fmol l^{-1}). The degree to which avidin is saturated with biotin is species-dependent, being low in chickens but high in turkeys and cormorants. Avidin has antibacterial properties by withholding biotin from bacteria and nutritionally depriving them. Avidin-bound biotin is poorly available to the embryo, but at the time of hatching it is swallowed by the emerging chick and can supply its biotin. More mature birds that eat eggs as food are unable to utilize avidin-bound biotin, because it is very resistant to digestion (Bush and White, 1989; White *et al.*, 1992; Subramanian and Adiga, 1995).

Requirement, deficiency, and toxicity

The established requirement for biotin (see Table 11.2) is based on experiments where husbandry conditions minimized coprophagy. Conditions that decrease intestinal synthesis (e.g., antibiotics, diarrhea) further increase the requirement. The type of dietary carbohydrate has a marked impact on the amount of biotin synthesized in chickens and may also modify the requirement (Bauer and Griminger, 1980).

The symptoms of a biotin deficiency are variable, even within a species. Part of this variability is due to the other dietary factors, especially the amount and type of fat. In chickens and turkeys, a deficiency causes slow growth, ruffled feathers, perosis, and dermatitis, characterized by lesions on the feet and, later, around the beak and eyes. The defects in bone growth that cause perosis are the result of altered eicosanoid metabolism, due to impaired elongation of dietary linoleic acid to arachidonic acid (Frigg and Brubacher, 1976; Scott *et al.*, 1982; Whitehead, 1984; Watkins et al, 1989).

In some instances, a biotin deficiency leads to sudden death without external lesions. Chicks become lethargic and develop hepatic and renal steatosis. This condition is referred to as fatty liver and kidney syndrome and can occur with a marginal deficiency, especially when dietary fat is low and lipogenesis is required. Apparently, the diversion of biotin to lipogenic enzymes diminishes the amount available for the biotin-dependent enzyme pyruvate carboxylase, causing impaired gluconeogenesis, hypoglycemia, and death. Stresses that deplete glycogen exacerbate the situation (Whitehead and Siller, 1983).

There is a positive relationship between the biotin level in the diet and that in the egg yolk (Buenrostro and Kratzer, 1984; Whitehead, 1984). High levels of biotin in the egg can provide stores in the chick that buffer a dietary deficiency for at least a week. Embryos from biotin-deficient hens display dwarfing, chondrodystrophy, and deformities of the mandibles, beak ('parrot beak'), and skeleton. If chicks hatch, they are usually ataxic as well as deformed.

Biotin deficiencies are relatively rare, except when diets are based on grains with very low biotin bioavailability or when an antagonist, such as avidin, is present in the diet. Excess biotin is readily excreted and toxicity is not known to be a problem.

Dietary sources

Biotin is present in many avian foods. Foods of animal origin (especially liver and kidney), vegetative parts of plants, legumes (peanuts, soybeans), and some fruits are particularly good sources. Many domestic grains are poor sources, especially

Table 11.6. The bioavailability (%) of biotin in feedstuffs relative to D-biotin (values are averages taken from studies on poultry summarized by Baker, 1995).

Source	Bioavailability	Source	Bioavailability
Alfalfa meal	56	Rice polishings	23
Barley	26	Rye	0
Canola meal	64	Safflower meal	32
Corn	106	Sorghum	26
Grass meal	67	Sunflower meal	95
Molasses, beet	75	Triticale	6
Oats	37	Wheat, whole	17
Peanut meal	53	Wheat, germ	55

wheat and barley, and their biotin may be almost completely unavailable (Table 11.6). The biotin content of foods of plant origin is more variable than that of most other water-soluble vitamins, being influenced by many environmental factors. Raw eggs are not a good biotin source for birds, because, following consumption, the rich levels found in the yolk are bound by avidin in the white, making this vitamin unavailable (Kratzer *et al.*, 1988).

Only the D isomer of biotin has vitamin activity and this is the form most commonly supplemented. Biotin losses may occur when dietary fats become rancid. Losses are minimized when the food has high levels of antioxidants.

Folates

Folate is the generic term used to describe a variety of compounds that are related to folic acid (pteroyl-γ-monoglutamic acid) and exhibit vitamin activity. There are more biologically active forms of folate than in any other vitamin, making their quantification in foods difficult. Folates in most of the foods consumed by birds are found primarily as mono- and polyglutamyl derivatives of tetrahydrofolic acid (FH_4). They are hydrolyzed to FH_4 by the action of folyl-γ-glutamyl hydrolases in the small intestine and absorbed by carrier-mediated active transport. The overall efficiency of folate absorption is lower than that of most of the other vitamins. Folate is transported between tissues loosely complexed with plasma proteins. Cells take up folic acid and add three to seven glutamate residues prior to utilization as a cofactor in metabolism (Eilam *et al.*, 1981; McDowell, 1989).

A wide variety of enzymes utilize folic acid derivatives to transfer single-carbon units, such as methyl, methylene, methenyl, formimino, and formyl groups (Scott *et al.*, 1982; Rennie *et al.*, 1993). Folate is needed for the synthesis of purines and pyrimidines, interconversions of glycine and serine, synthesis of choline, degradation of histidine, and conversion of homocysteine to methionine. The metabolism of folate is integrally related to the metabolism of several other single-carbon donors, including *S*-adenosylmethionine, serine, vitamin B_{12}, and choline (Fig. 11.7). These nutrients interact in ways that can markedly influence their respective requirements. The nutritional importance of these interactions is much greater in birds than in mammals, due to the bird's high rate of uric acid synthesis for nitrogen excretion, combined with a very high methionine and cysteine requirement. Uric acid synthesis consumes single-carbon units from several sources (see Fig. 6.8, p. 160). Further, methionine is often the first limiting amino acid in avian diets and this increases the need for other sources of methyl groups (e.g., choline).

The interactions between methyl donors are exemplified by the conversion of homocysteine to methionine (Fig. 11.7). Methionine, as *S*-adenosylmethionine, donates methyl groups in a large number of metabolic pathways, resulting in its conversion to homocysteine. Methionine is regenerated through the methylation of homocysteine by vitamin B_{12}. Folate (5-methyl-FH_4) is used to regenerate vitamin B_{12}. When vitamin B_{12} is deficient, folate is 'trapped' as 5-methyl-FH_4 and accumulates at the expense of other metabolically active folate pools. Thus, a vitamin B_{12} deficiency markedly increases the folate requirement. Choline can partly ameliorate the need for vitamin B_{12}, because it can be used as a source of methyl groups to regenerate methionine through a separate pathway. If homocysteine is not remethylated back to methionine because of a deficiency of

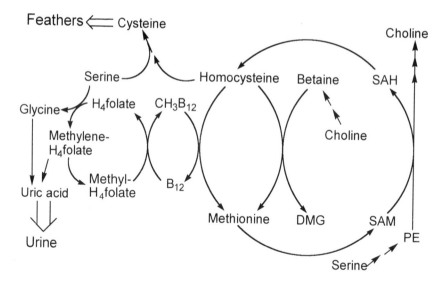

Fig. 11.7. Interactions between vitamin B_{12}, folate, choline, and methionine occur due to their interdependence in donating methyl groups. Three moles of *S*-adenosylmethionine (SAM) are consumed for each mole of choline synthesized from phosphatidylethanolamine (PE). Also, SAM is converted to *S*-adenosylhomocysteine (SAH) following methylation of proteins, nucleic acids, and the synthesis of creatine and carnitine. Choline is converted to betaine prior to donating methyl groups and producing dimethylglycine (DMG).

vitamin B_{12}, folic acid, or choline, it is oxidized and the methionine requirement is markedly increased.

Requirement, deficiency, and toxicity

High dietary levels of protein increase the folate requirement above recommended levels (see Table 11.2), due to a greater need for uric acid synthesis. Inadequate levels of other methyl donors (serine, vitamin B_{12}, methionine, betaine, and choline) also increase the requirement. Antibiotics and other medications that antagonize intestinal synthesis, as well as management practices that prevent coprophagy, add to dietary needs (Creek and Vasaitis, 1963; Stokstad and Jukes, 1987; Ryu *et al.*, 1995).

Deficiencies of folate result in impaired synthesis of deoxyribonucleic acid (DNA) and ribonucleic acid (RNA), causing reduced cell division (Maxwell *et al.*, 1988). Tissues that rely on high rates of cell division are the first to be affected and initial symptoms include macrocytic anemia and leukopenia. Asynchronous cell division causes red blood cells to have large multilobular nuclei. More severe deficiencies cause lethargy, perosis, and dermatitis. In poultry, a folic acid deficiency is more devastating to the growth, integrity, and pigmentation of feathers than deficiencies of other vitamins. Poorly developed feathers, with weak and brittle shafts, give deficient birds a very unthrifty appearance. Turkey poults display a paralysis of the neck, which is quickly fatal, but is reversed within 15 min by folate administration.

In the hen, a folate deficiency impairs the oviduct's response to estrogen and ability to form albumen (Burns and Jackson, 1979). More marginal deficiencies result in normal egg production but cause embryonic deformities, including bending of the tibiotarsus, syndactyly, and deformed mandibles. With severe deficiencies, embryonic mortality is high and usually occurs late in development or at pipping.

The relatively poor absorption of folate allows this vitamin to be tolerated at high levels. It is usually considered to be nontoxic.

Dietary sources

Folates occur in a wide variety of avian foods. Among plant sources, fast-growing leafy components, legumes (e.g., soybeans), and some fruits (e.g., citrus) are rich sources. Animal prey is also a rich source of folates, particularly the liver. In mammals, the bioavailability of folate in foods is variable, but it has not been suitably investigated in birds.

Folate is sensitive to oxidation by minerals and it may be lost from foods, because of its high water solubility. It is usually supplemented as folic acid or its sodium salt.

Vitamin B_{12}

Vitamin B_{12}, also known as cobalamin, is bound to proteins in avian foods. It is released during acidification and proteolysis in the ventriculus and gizzard. The absorption of vitamin B_{12} in birds is poorly characterized, but in mammals it is absorbed from the intestinal lumen by active transport and, at high concentrations, by simple diffusion. Active absorption is very efficient and is facilitated by intrinsic factor, which is a vitamin B_{12}-binding protein secreted by gastric parietal cells, thus protecting vitamin B_{12} from attack by digestive enzymes. Specific receptors on enterocytes in the ileum bind intrinsic factor and internalize it, together with vitamin B_{12}. Vitamin B_{12} is transported in the blood bound to another binding protein, transcobalamin, and is taken up by cells through a specific receptor. Three enzymes utilize vitamin B_{12} as a cofactor: methionine synthetase, leucine mutase, and methylmalonyl-CoA mutase. These enzymes play important roles in the metabolism of propionate, amino acids, and single-carbon units (Fig. 11.7). Considerable vitamin B_{12} is stored in liver, bound to transcobalamin. In replete birds, these pools can buffer dietary inadequacy for long periods.

Requirement, deficiency, and toxicity

Vitamin B_{12} is required at levels in the diet that are at least 15-fold less than any other nutrient (see Table 11.2). This very low requirement is probably due to a combination of low metabolic needs, low endogenous losses, and microbial synthesis in the lower gastrointestinal tract. High-protein diets increase the demand for vitamin B_{12} for use with amino acid-metabolizing enzymes, thus increasing the requirement.

A diet almost completely devoid of animal foods and management practices that prevent coprophagy are necessary to demonstrate a deficiency in poultry and Japanese Quail. Gizzard erosions and fatty liver are symptoms of a vitamin B_{12} deficiency in chickens. Signs of a folate deficiency may accompany a vitamin B_{12} deficiency, due to the trapping of folate. The vitamin B_{12} content of eggs decreases with decreasing vitamin B_{12} in the diet, but laying hens that have replete stores can consume a very deficient diet for 2–5 months before deficiency signs are evident in embryos. Embryonic symptoms are characterized by atrophy of the thigh muscles and hemorrhages. Chicks that hatch from replete hens have stores sufficient for several weeks. Vitamin B_{12} is not known to be toxic in birds (Scott *et al.*, 1982; Ward *et al.*, 1985; Squires and Naber, 1992).

Dietary sources

Vitamin B_{12} is synthesized almost exclusively by bacteria and is found only in bacteria and in foods of animal origin. The liver and, to a lesser extent, the kidneys of animal prey are a very rich source. Many insects, crustacea, and mollusks are also a good source. Grains, fruits, and leaves are devoid of vitamin B_{12} unless they have fermented. Vitamin B_{12} may be supplemented to diets as cyanocobalamin.

Choline

Most choline in avian foods is found as phosphatidylcholine (lecithin) or sphingomyelin. Choline may be released from phosphatidylcholine by the action of pancreatic phospholipases during the course of normal lipid digestion. The resulting free choline is absorbed by carrier-mediated transport into enterocytes, reacylated to phosphatidylcholine, and exported as a component of portomicrons. Intact phosphatidylcholine is also absorbed by the intestinal epithelial cells.

Choline can be synthesized to a limited extent by chickens. It is synthesized in the liver from serine and methyl groups from S-adenosylmethionine (Fig. 11.7). Three moles of methionine are required for each mole of choline synthesized, but the rate of synthesis is inadequate to meet most of the metabolic requirement in growing poultry and Northern Bobwhite Quail (Jukes, 1947; Serafin, 1974).

Choline has three primary functions in birds: structural, neurotransmitter, and methyl donor. Quantitatively, the primary role of choline is as a component of phosphatidylcholine, sphingomyelin, and very-low-density lipoproteins (VLDLs). As such, it is needed for the structure of cellular membranes, the maturation of the cartilage matrix of bone, and the export of lipids from the liver. Some choline is acetylated to give the important neurotransmitter acetylcholine. Choline is also oxidized to betaine, which is an important donor of methyl groups for transmethylation of homocysteine to methionine (Fig. 11.7). This pathway is particularly important for methionine conservation when dietary methionine is deficient (Emmert *et al.*, 1996).

Requirement, deficiency, and toxicity

Choline is usually considered a vitamin, although it does not completely fit the 'trace organic' definition, due to the large amount needed in the diet (> 1% of the dry matter). This high level is similar to the amounts of essential amino acids and fatty acids required by birds. The established requirement for choline (see Table 11.2) is increased even higher when the dietary level of protein is high, probably due to the additional need for methyl groups to synthesize and excrete uric acid. The requirement can be lowered by a surfeit of either of the methyl donors methionine or betaine. Dietary methionine in excess of that needed for protein synthesis can spare the choline requirement effectively in adult chickens, turkeys, and ducks, but its conversion to choline is limited in young chicks. Betaine effectively spares choline on a 1:1 molar basis in both young chicks and older birds; however, betaine can only replace the transmethylation functions of choline and it cannot be used to synthesize choline in order to substitute for choline's structural and neurotransmitter functions. Thus, dietary betaine can only substitute for about one-third of the choline requirement. Levels of dietary choline in excess of its requirement can ameliorate marginal deficiencies of methionine, folic acid, and vitamin B_{12} (Ketola and Nesheim, 1974; Baker, 1995).

In growing chickens, turkeys, ducks, geese, and Northern Bobwhite Quail, the predominant symptoms of a choline deficiency are slow growth and perosis (Serafin, 1974, 1981; Scott and Dean, 1991). The perosis can be very severe and involve hemorrhages at the growth plate and slippage of the Achilles tendon from its condyles. Geese may exhibit additional disorders that are symptomatic of impaired neurotransmitter levels, including excitability, curled toes, and paralysis. In egg-laying hens, the defining choline deficiency symptom is due to an inability to package hepatic triglycerides into VLDL. These triglycerides accumulate in the liver causing a condition called hepatic steatosis. For this reason, choline is often referred to as a lipotrope, because it promotes the mobilization of lipids out of the liver. Methionine, vitamin B_{12}, and folate have lipotrope activity by participating in the endogenous synthesis of choline. Duck and goose producers have occasionally utilized choline-deficient diets in the production of liver pâté, but problems with perosis have hindered this practice.

Eggs from hens fed low-choline diets may be smaller but do not contain markedly low levels of choline, and embryonic development is relatively normal. Additionally, the high levels of choline in body lipids can effectively buffer a deficiency when total dietary calories are limited.

High levels of supplemental choline, as choline chloride, depress growth. The relative contribution of choline and that of chloride need to be explored, because birds tolerate higher levels of choline as phosphatidylcholine. High levels of dietary choline may be metabolized to trimethylamine by gut microflora. This compound finds it way into eggs, where it has a fishy flavor (March and MacMillan, 1979)

Dietary sources

Almost all avian foods contain choline as a component of cellular membranes and fat stores. Generally, high-fat foods are also high in choline, including animal prey, seeds, and domestic grains, except corn. The refining (e.g., alkaline treatment and bleaching) of plant oils almost entirely removes their choline.

Choline is supplemented as choline chloride or choline bitartrate. Choline is very stable and does not suffer significant losses during storage or processing. However, choline chloride is very hygroscopic and promotes the loss of several other vitamins (e.g., thiamin, folate, pantothenic acid, vitamin K).

Vitamin C

Vitamin C is the term used to describe all compounds that exhibit the activity of L-ascorbic acid. Many species do not normally need a dietary source of vitamin C, because they can synthesize sufficient amounts from glucose. Other species require a dietary source and quickly develop deficiency symptoms otherwise. Most vitamin C activity in avian foods is in the form of free ascorbic acid. Following absorption, ascorbic acid is transported in association with plasma albumin. Although ascorbic acid is the primary form in the blood, it is usually transported into cells as the oxidized form, dehydroascorbic acid, and then reduced back to ascorbic acid in the cytoplasm. The catecholamine-synthesizing cells of the adrenal glands concentrate large amounts of vitamin C and release it during stress. Cells that synthesize reactive oxygen intermediates (e.g., hetero-phils, macrophages) also require high levels of vitamin C.

Ascorbic acid is synthesized from glucose via the glucuronic acid pathway (Fig. 11.8). Depending upon the species, ascorbic acid synthesis may occur predominantly in the liver, kidneys, or both (Table 11.7). Some species of Passeriformes completely lack the enzyme L-gulonolactone oxidase and require a dietary source of vitamin C to prevent the quick onset of deficiency symptoms (e.g., Red-vented Bulbul). All species that are unable to synthesize ascorbic acid are insectivorous or frugivorous, which gives them a reliable dietary supply. Granivores normally consume very low levels of vitamin C and all species that have been tested do not need a dietary source to prevent deficiency symptoms. Some Passeriformes that are able to synthesize ascorbic acid do so at rates two to ten times slower than species such as chickens, ducks, and Japanese Quail, which do not have a dietary requirement. Until feeding experiments are conducted, it should be assumed that species with slow rates of synthesis have a dietary requirement.

In addition to variation in vitamin C requirements among Passeriformes species, variation among Galliformes is also evident (Hanssen *et al.*, 1979; Wilson, 1989). Chickens, turkeys, and Japanese Quail do not normally have a vitamin C requirement, but growing Willow Ptarmigan and Northern Bobwhite Quail

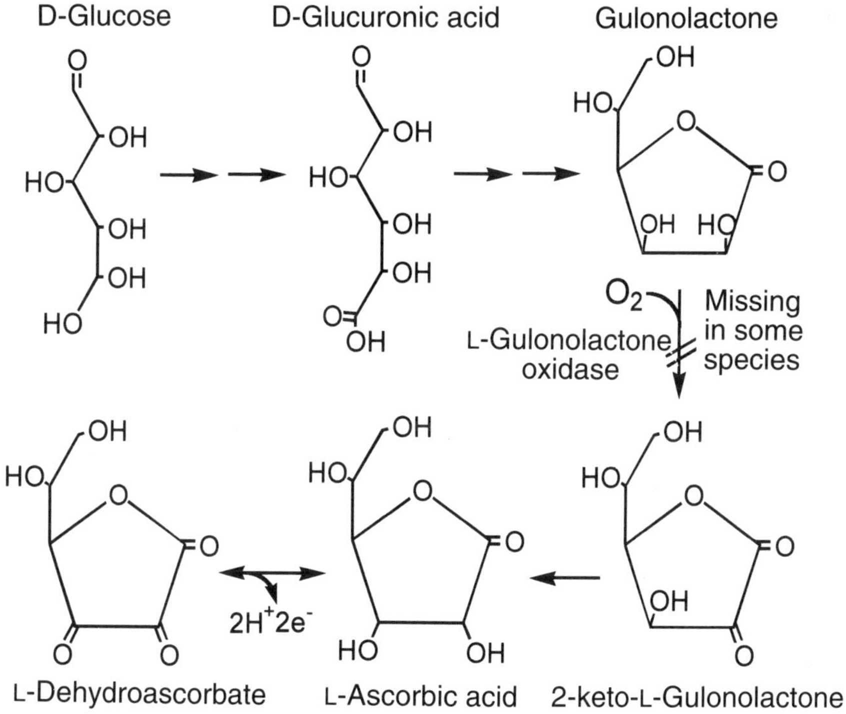

Fig. 11.8. The synthesis of L-ascorbic acid from glucose. L-Ascorbic acid can act as an antioxidant through conversion to L-dehydroascorbate.

Table 11.7. Ascorbic acid synthetic capacity and dietary requirement in different species.[*]

Order and species	Synthesis		Dietary requirement[†]	Typical diet
	Liver	Kidney		
Passeriformes				
Long-tailed Shrike	−	−	Probable	Insects, vertebrates
Bay-backed Shrike	−	−	Probable	Insects, vertebrates
Northern Shrike	−	−	Probable	Insects, vertebrates
White-browed Bulbul	−	−	Probable	Figs, fruits, insects
Red-whiskered Bulbul	−	−	Probable	Fruits, insects?
Himalayan Bulbul	−	−	Probable	Fruits, insects
Red-vented Bulbul	−	−	Yes	Insects, fruits
Common Iora	−	−	Probable	Insects
Scarlet Minivet	−	−	Probable	Insects
Clamorous Reed-Warbler	−	−	Probable	Insects
Spot-breasted Fantail-Flycatcher	−	−	Probable	Insects
Asian Paradise-Flycatcher	−	−	Probable	Insects
Pale-billed Flowerpecker	−	−	Probable	Fruits, some insects
Crimson Sunbird	−	−	Probable	Nectar, insects
Black-hooded Oriole	−	−	Probable	Insects, fruits, nectar
Barn Swallow	−	−	Probable	Insects
Blue-capped Rock-Thrush	+	−		Insect, fruits
Rufous Treepie	+	−		Omnivorous
Oriental Magpie-Robin	+	−		Insects, nectar
Asian Pied Starling	+	−		Insects, fruits
House Sparrow	++	−		Seeds, insects
Black-headed Munia	++	−		Seeds, insects
Bank Myna	+	−		Omnivorous
Hill Myna	+	−		Fruits, insects
Common Myna	+	+		Fruits, seeds, insects
House Crow	+	++		Omnivorous
House Sparrow			None[‡]	Omnivorous
White-crowned Sparrow			None[‡]	Omnivorous
Savannah Sparrow			None[‡]	Omnivorous
Piciformes				
Black-rumped Flameback	+	−		Insects, fruits, nectar
Strigiformes				
Indian Scops-Owl	−	++		Insects, vertebrates
Coraciiformes				
White-throated Kingfisher	−	++		Fish, vertebrates
Psittaciformes				
Alexandrine Parakeet	−	+++		Seeds, fruits
Cockatiel			None	Seeds, grain
Gruiformes				
Coot (Euroasian)	−	+++		Vegetation, insects

Table 11.7. Continued

Order and species	Synthesis		Dietary requirement[†]	Typical diet
	Liver	Kidney		
Cuculiformes				
Asian Koel	−	+++		Fruits, insects
Falconiformes				
Laggar Falcon		+++		Small vertebrates
Anseriformes				
Bar-headed Goose	−	+++		Vegetation
Common Pochard	−	+++		Vegetation, insects
Galliformes				
Chicken	−	+++	None	Seeds, insects
Turkey			None	Seeds, insects, fruits
Japanese Quail	−	+++	None	Seeds, insects
Willow Ptarmigan	−	+++	Yes[§]	Vegetation
Grey Partridge	−	+++		Seeds, insects
Columbiformes				
Pigeon (Domestic)	−	+++	None	Grain, seeds
Ciconiiformes				
Cattle Egret	−	++		Insects, vertebrates
Trochiliformes				
Anna's Hummingbird			None[‡]	Nectar, insects
Costa's Hummingbird			None[‡]	Nectar, insects

[*]Summarized from data from: Plimmer and Rosedale (1923); Roy and Guha (1958a,b); Chaudhuri and Chatterjee (1969); Hanssen *et al.* (1979); Murphy and King (1982); Brice and Grau (1989); Alawad *et al.* (1993); Roudybush (1996). Missing data indicate that measurements have not been made.
[†]Yes, determined by feeding deficient diets; Probable, suspected based on lack of endogenous synthesis; None, no deficiency symptoms are observed following long periods of consumption of a diet devoid of vitamin C.
[‡]Only adults have been tested.
[§]Required by growing chicks and not by adults.
−, Undetectable; +, low; ++, moderate; +++, high.

respond to dietary ascorbic acid with increased growth and livability, respectively. Although the Willow Ptarmigan is proficient at synthesizing L-ascorbic acid in its kidneys, it exhibits severe deficiency signs when its dietary supply is inadequate. Even the high endogenous synthetic rate in chickens may be inadequate during periods of severe stress such as heat, physical trauma, infection, and the consumption of some types of purified diets. Ascorbic acid supplementation of grain-based diets has been reported to improve resistance to a variety of infectious diseases and to improve wound healing. Chickens apparently have the capacity to detect the level of ascorbic acid in foods and increase their intake during periods of heat stress (Pardue and Thaxton, 1986; Kutlu and Forbes, 1993; McKee and Harrison, 1995; Kratzer *et al.*, 1996).

Metabolic functions

The functions of ascorbic acid are related to its redox properties (Gecha and Fagan, 1992). Oxidation of ascorbic acid involves two successive losses of single electrons to give dehydroascorbic acid (Fig. 11.8). Following oxidation, the enzyme ascorbic acid reductase regenerates ascorbic acid. Ascorbic acid undergoes this catalytic cycle: as a cofactor or a protector of several enzymes, most of which are involved in hydroxylation reactions; in the reduction of trace minerals (e.g., iron) during their metabolism; and as a soluble antioxidant or prooxidant. As an antioxidant, ascorbic acid prevents damage to proteins by hydrogen peroxide (H_2O_2). It is also involved in protecting cellular membranes through regenerating vitamin E by reacting with vitamin E radicals. The best-characterized function of vitamin C is as a cofactor in the hydroxylation of lysine and proline residues of procollagen and proelastin. This posttranslational modification is necessary for the proper folding and mechanical characteristics of mature collagen and elastin. In chickens, this function of ascorbic acid is especially important in the formation of collagen during wound repair (Rajkhowa *et al.*, 1996).

Dietary ascorbic acid promotes the bioavailability of dietary iron and, consequently, decreases the iron requirement. When ascorbic acid is consumed in the same meal as iron, it reduces it from the ferric (Fe^{3+}) form, which predominates in the acid environment of the proventriculus and gizzard, to the ferrous (Fe^{2+}) form. Ferrous iron forms a chelate with ascorbic acid that is soluble in the alkaline environment of the small intestine and is relatively efficiently absorbed.

Plasma levels of ascorbic acid increase with increasing dietary ascorbic acid, but these are depleted within a few days following dietary withdrawal. There are no stores of vitamin C that are able to buffer prolonged dietary deficiency. In chickens, vitamin C is not transferred to the egg, but it is synthesized by the young embryo (Pardue and Thaxton, 1986; Zwaan and Lam, 1992).

Requirement, deficiency, and toxicity

The dietary requirement for birds that are unable to synthesize ascorbic acid has not been adequately investigated. Nor is it known how much vitamin C is needed during a period of rapid growth or egg production in those species where synthetic capacity is low or moderate. In the absence of any information, the vitamin C requirements of the guinea pig (200 mg kg^{-1}), a mammal requiring vitamin C, and of fish (50 mg kg^{-1} for trout, salmon, and tilapia) are relevant.

Vitamin C may be useful to the young growing chicken during periods of stress from environmental or infectious sources (Pardue and Thaxton, 1986; Gross, 1992). However, many experiments have been equivocal. In those that have shown beneficial responses, dietary levels of between 50 and 150 mg kg^{-1} dry matter were typically most effective. Higher levels sometimes cause reductions in growth rates or efficiency of feed conversion in nonstressed poultry.

In Willow Ptarmigan chicks, a deficiency of vitamin C causes a syndrome similar to scurvy, seen in mammals (Hanssen *et al.*, 1979). The onset of symptoms occurs only 2–4 days following hatching and they begin with signs of nervousness and stress, presumably due to the role of ascorbic acid in catecholamine metabolism. Other symptoms include poor growth, diarrhea, lethargy, improper bone development, and spontaneous fractures of leg and wing bones. Deficiency signs are reversible within a week following ascorbic acid supplementation. As Willow Ptarmigan chicks mature, endogenous synthesis rates become adequate and a dietary source of vitamin C is not required. Red-vented Bulbuls are completely unable to synthesize ascorbic acid and require dietary vitamin C. When fed a deficient diet, they develop symptoms within 15 days, including weight loss, behavioral changes, lethargy, feather loss, and hemorrhages in the liver and leg joints (Roy and Guha, 1958b).

Vitamin C is relatively nontoxic. High dietary levels of vitamin C are absorbed by the intestine and then excreted by the kidneys. Very high levels of vitamin C (> 1%) cause transient diarrhea and gastroenteritis. Chronic consumption has been suspected to contribute to problems with excess iron absorption, which leads to hemochromatosis in susceptible species, but evidence is lacking. Moreover, hemochromatosis is a symptom of a vitamin C deficiency in mammals. High dietary levels of vitamin C protect against toxicity of some transition metals by promoting their excretion (e.g., selenium, lead, vanadium, and cadmium).

Dietary sources

Vitamin C is not widely distributed across avian foods. Foods of plant origin are usually the primary dietary source of vitamin C. Fruits, vegetables, and many herbs are particularly rich, but domestic grains are very low. Internal organs (e.g., adrenals, kidneys, liver) of vertebrate prey, particularly fish, are moderately good sources, but muscle is low in vitamin C.

The vitamin C content of avian foods decreases precipitously during storage. It is very susceptible to oxidation, especially in the presence of trace minerals. Vitamin C is usually supplemented as L-ascorbic acid. Because of the instability of this compound, supplements for avian feeds often are coated with ethylcellulose or fat. Such preparations last about four times longer during storage and are considerably more resistant to processing, such as pelleting. Food that has been stored for more than 4 months with unprotected vitamin C should be considered to be devoid of vitamin C activity.

Other Factors Sometimes Called Vitamins

Several other factors partially satisfy the operational definition of vitamins and might be considered required nutrients in some situations. These include: p-aminobenzoic acid, myoinositol, ubiquinones, lipoic acid, carnitine, and pyrroloquinoline quinone. Each of these substances has been shown to be required for maximal growth of chickens under some circumstances. In the case of ubiquinones, lipoic acid, and pyrroloquinoline, high levels of antibiotics or germ-free conditions are needed to demonstrate a requirement. All common diets are more than adequate in these vitamin-like compounds, and deficiency symptoms have not been reported outside of the laboratory. It is possible that avian species that have simple digestive tracts with relatively small microbial populations may be more susceptible to deficiencies.

References

Abawi, F.G. and Sullivan, T.W. (1989) Interactions of vitamin-A, vitamin-D3, vitamin-E and vitamin-K in the diet of broiler chicks. *Poultry Science* 68, 1490–1498.

AIN (1990) Nomenclature policy: generic descriptions and trivial names for vitamins and related compounds. *Journal of Nutrition* 120, 12–19.

Alawad, A., Kolb, E. and Wahren, M. (1993) The activity of the L-gulonolactone-oxidase in the kidney and the content of ascorbic acid in the kidney, in the liver and in the plasma of chickens, ducks, geese and quails. *Archiv für Geflugelkunde* 57, 185–189.

Almquist, H.J. (1971) Vitamin K. In: Sebrell, W.H. and Harris, R.S. (eds) *The Vitamins.* Academic Press, New York, pp. 418–433.

Applegate, T.J. and Sell, J.L. (1996) Effect of dietary linoleic to linolenic acid ratio and vitamin E supplementation on vitamin E status of poults. *Poultry Science* 75, 881–890.

Austic, R.E. and Scott, M.L. (1991) Nutritional diseases. In: Calnek, B.W. (ed.) *Diseases of Poultry.* Iowa State University Press, Ames, Iowa, pp. 45–71.

Baker, D.H. (1995) Vitamin bioavailability. In: Ammerman, C.B., Baker, D.H. and Lewis, A.J. (eds) *Bioavailability of Nutrients for Animals.* Academic Press, San Diego, pp. 399–431.

Bauer, K.D. and Griminger, P. (1980) Effect of dietary carbohydrates and biotin level on cecal size and biotin concentration of growing chickens. *Poultry Science* 59, 1493–1498.

Bernard, J.B., Watkins, B.E. and Ullrey, D.E. (1989) Manifestations of vitamin-D deficiency in chicks reared under different artificial lighting regimes. *Zoo Biology* 8, 349–355.

Brice, A.T. and Grau, C.R. (1989) Hummingbird nutrition – development of a purified diet for long-term maintenance. *Zoo Biology* 8, 233–237.

Brue, R.N. (1994) Nutrition. In: Ritchie, B.W., Harrison, G.J. and Harrison, L.R. (eds) *Avian Medicine: Principles and Applications.* Wingers Publishing, Lake Worth, Florida, pp. 63–95.

Bryden, W.L. (1989) Intestinal distribution and absorption of biotin in the chicken. *British Journal of Nutrition* 62, 389–398.

Buentrostro, J.L. and Kratzer, F.H. (1984) Use of plasma and egg yolk biotin of white Leghorn hens to assess biotin availability from feedstuffs. *Poultry Science* 63, 1563–1570.

Burns, R.A. and Jackson, N. (1979) The effects of folate deficiency and oestradiol administration on the plasma free amino acid concentration of the immature hen. *British Poultry Science* 20, 131–139.

Bush, L. and White, H.B. (1989) Avidin traps biotin diffusing out of chicken egg yolk. *Comparative Biochemistry and Physiology B – Comparative Biochemistry* 93, 543–547.

Calle, P.P., Dierenfeld, E.S. and Robert, M.E. (1989) Serum alpha-tocopherol in raptors fed vitamin-E-supplemented diets. *Journal of Zoo and Wildlife Medicine* 20, 62–67.

Carpenter, K.J.M., Schelstraete, V.C., Vilicich, V.C. and Wall, J.S. (1988) Immature corn as a source of niacin. *Journal of Nutrition* 118, 165–171.

Chambon, P. (1996) A decade of molecular biology of retinoic acid receptors. *FASEB Journal* 10, 940–54.

Chaudhuri, C.R. and Chatterjee, I.B. (1969) L-Ascorbic acid synthesis in birds: phylogenetic trend. *Nature* 164, 435–436.

Chen, B.J., Shen, T.F. and Austic, R.E. (1996) Efficiency of tryptophan–niacin conversion in chickens and ducks. *Nutrition Research* 16, 91–104.

Chen, P.S. and Bosmann, H.B. (1964) Effect of vitamins D_2 and D_3 on serum calcium and phosphorus in rachitic chicks. *Journal of Nutrition* 83, 133–138.

Chung, T.K. and Baker, D.H. (1990) Riboflavin requirement of chicks fed purified amino acid and conventional corn–soybean meal diets. *Poultry Science* 69, 1357–1363.

Clark, N.B., Lee, S.K. and Murphy, M.J. (1990) Vitamin D action on calcium regulation and osmoregulation in embryonic chicks. In: Wada, M., Ishii, S. and Scanes, C.G. (eds) *Endocrinology of Birds.* Springer-Verlag, Berlin, pp. 159–170.

Coates, M.E., Ford, J.E. and Harrison, G.F. (1968) Intestinal synthesis of vitamins of the B complex in chicks. *British Journal of Nutrition* 22, 493–500.

Creek, R.D. and Vasaitis, V. (1963) The effect of excess dietary protein on the need for folic acid by the chick. *Poultry Science* 42, 1136–1141.

Daghir, N.J. and Balloun, S.L. (1973) Effect of dietary protein level on vitamin B_6 requirements of chicks. *Poultry Science* 52, 1247–1251.

Dierenfeld, E.S., Sandfort, E.S. and Satterfield, W.C. (1989) Influence of diet on plasma vitamin E in captive peregrine falcons. *Journal of Wildlife Management* 53, 160–164.

Dierenfeld, E.S., Katz, N., Pearson, J., Murru, F. and Asper, E.D. (1991) Retinol and alpha-tocopherol concentrations in whole fish commonly fed in zoos and aquariums. *Zoo Biology* 10, 119–125.

Edwards, H.M. (1990) Efficacy of several vitamin-D compounds in the prevention of tibial dyschondroplasia in broiler chickens. *Journal of Nutrition* 120, 1054–1061.

Eilam, Y., Ariel, M., Jablonska, M. and Grossowicz, N. (1981) On the mechanism of folate transport in isolated intestinal epithelial cells. *American Journal of Physiology* 240, G170–G-175.

Emmert, J.L., Garrow, T.A. and Baker, D.H. (1996) Hepatic betaine-homocysteine methyl-transferase activity in the chicken is influenced by dietary intake of sulfur amino acids, choline and betaine. *Journal of Nutrition* 126, 2050–2058.

Fraser, D.R. and Emtage, J.S. (1976) Vitamin D in the avian egg. *Biochemistry Journal* 160, 671–682.

Frigg, M. and Brubacher, G. (1976) Biotin deficiency in chicks fed a wheat-based diet. *International Journal of Vitamin Research* 46, 314–321.

Fuhrmann, H. and Sallmann, H.P. (1995) The influence of dietary fatty acids and vitamin E on plasma prostanolds and liver microsomal alkane production in broiler chickens with regard to nutritional encephalomacia. *Journal of Nutritional Science and Vitaminology* 41, 553–561.

Fuller, H.L., Field, R.C., Roncalli-Amici, W. S., Dunahoo, W. S. and Ewards, H. M. (1961) The vitamin B_6 requirement of breeder hens. *Poultry Science* 40, 429–433.

Gaal, T., Mezes, M., Noble, R.C., Dixon, J. and Speake, B.K. (1995) Development of antioxidant capacity in tissues of the chick embryo. *Comparative Biochemistry and Physiology B – Biochemistry and Molecular Biology* 112, 711–716.

Gecha, O.M. and Fagan, J.M. (1992) Protective effect of ascorbic acid on the breakdown of proteins exposed to hydrogen peroxide in chicken skeletal muscle. *Journal of Nutrition* 122, 2087–2093.

Graham, D.L. and Halliwell, W.L. (1986) Malnutrition in birds of prey. In: Fowler, M.E. (ed.) *Zoo and Wild Animal Medicine.* W.B. Saunders, Philadelphia, pp. 379–385.

Gries, C.L. and Scott, M.L. (1972a) The pathology of pyridoxine deficiency in chicks. *Journal of Nutrition* 102, 1259–1268.

Gries, C.L. and Scott, M.L. (1972b) The pathology of thiamin, riboflavin, pantothenic acid and niacin deficiencies in the chick. *Journal of Nutrition* 102, 1269–1286.

Griffiths, P. and Fairney, A. (1988) Vitamin D metabolism in polar vertebrates. *Comparative Biochemistry and Physiology* 91B, 511–516.

Griminger, P. and Brubacher, G. (1966) The transfer of vitamins K$_1$ and menadione from the hen to the egg. *Poultry Science* 45, 512–519.

Gross, W.B. (1992) Effects of ascorbic acid on stress and disease in chickens. *Avian Diseases* 36, 688–692.

Halevy, O., Arazi, Y., Melamed, D., Friedman, A. and Sklan, D. (1994) Retinoic acid receptor-alpha gene expression is modulated by dietary vitamin A and by retinoic acid in chicken T lymphocytes. *Journal of Nutrition* 124, 2139–2146.

Hanssen, I., Grav, H.J., Steen, J.B. and Lysnes, H. (1979) Vitamin C deficiency in growing willow ptarmigan (*Lagopus lagopus lagopus*). *Journal of Nutrition* 109, 2260–2278.

Haq, A.U. and Bailey, C.A. (1996) Time course evaluation of carotenoid and retinol concentrations in posthatch chick tissue. *Poultry Science* 75, 1258–1260.

Harper, T.A., Boucher, R.V. and Callenbach, E.W. (1952) Influence of source and quantity of vitamin A ingested upon liver storage and survival time of bobwhite quail. *World's Poultry Science* 8, 273–283.

Hart, E.B., Steenbock, H., Lepkovsky, S. and Halpin, J.G. (1923) The nutritional requirements of baby chicks: III. The relationship of light to growth of the chicken. *Journal of Biological Chemistry* 58, 33–41.

Hassan, S. and Hakkarainen, J. (1990) Response of whole blood, erythrocyte and plasma vitamin-E content to dietary vitamin-E intake in the chick. *Acta Veterinaria Scandinavica* 31, 399–408.

Hay, A.W. and Watson, G. (1977) Vitamin D$_2$ in vertebrate evolution. *Comparative Biochemistry and Physiology* 56B, 375–380.

Heard, G.S. and Annison, E.F. (1986) Gastrointestinal absorption of vitamin B-6 in the chicken (*Gallus domesticus*). *Journal of Nutrition* 116, 107–120.

Heine, U.I., Roberts, A.B., Munoz, N.S., Roche, N.S. and Sporn, M.B. (1985) Effects of retinoid deficiency on the development of the heart and vascular system of the quail embryo. *Virchows Archiv* 50, 135–143.

Hencken, H. (1992) Chemical and physiological behavior of feed carotenoids and their effects on pigmentation. *Poultry Science* 71, 711–715.

Honour, S.M., Trudeau, S., Kennedy, S. and Wobeser, G. (1995a) Experimental vitamin A deficiency in mallards (*Anas platyrhynchos*): lesions and tissue vitamin A levels. *Journal of Wildlife Diseases* 31, 277–288.

Honour, S., Kennedy, S., Trudeau, S. and Wobeser, G. (1995b) Vitamin A status of wild mallards (*Anas platyrhynchos*) wintering in Saskatchewan. *Journal of Wildlife Diseases* 31, 289–298.

Jacob, J. and Ziswiler, V. (1982) The uropygial gland. In: Farner, D.S., King, J.R. and Parkes, K.C. (eds) *Avian Biology*, Vol. III. Academic Press, New York, pp. 199–224.

Jakobsen, K., Engberg, R.M., Andersen, J.O., Jensen, S.K., Lauridsen, C., Sorensen, P., Henckel, P., Bertelsen, G., Skibsted, L.H. and Jensen, C. (1995) Supplementation of broiler diets with all-rac-alpha- or a mixture of natural source rrr-alpha-, gamma-, delta-tocopheryl acetate. 1. Effect on vitamin E status of broilers *in vivo* and at slaughter. *Poultry Science* 74, 1984–1994.

Jensen, L.S. (1968) Vitamin E and essential fatty acids in avian reproduction. *Federation Proceedings* 27, 914–921.

Jiang, Y.H., McGeachin, R.B. and Bailey, C.A. (1994) Alpha-tocopherol, beta-carotene, and retinol enrichment of chicken eggs. *Poultry Science* 73, 1137–1143.

Jukes, T.H. (1947) Choline. *Annual Review of Biochemistry* 1947, 194–211.

Kear, J. (1986) Ducks, geese, swans, and screamers (Anseriformes): nutrition. In: Fowler, M.E. (ed.) *Zoo and Wild Animal Medicine.* W.B. Saunders, Philadelphia, pp. 335–341.

Ketola, H.G. and Nesheim, M.C. (1974) Influence of dietary protein and methionine levels on the requirement for choline by chickens. *Journal of Nutrition* 104, 1484–1489.

Kim, Y.S. and Combs, G.F. (1993) Effects of dietary selenium and vitamin-E on glutathione concentrations and glutathione S-transferase activities in chick liver and plasma. *Nutrition Research* 13, 455–463.

Kling, L.J. and Soares, J.H. (1980) Vitamin E deficiency in the Japanese quail. *Poultry Science* 59, 2352–2354.

Kratzer, F.H., Knollman, K., Earl, L. and Buenrostro, J.L. (1988) Availability to chicks of biotin from dried egg products. *Journal of Nutrition* 118, 604–608.

Kutlu, H.R. and Forbes, J.M. (1993) Self-selection of ascorbic acid in coloured foods by heat stressed broiler chicks. *Physiology and Behavior* 53, 103–110.

Lawrence, L.M., Mathias, M.M., Nockels, C.F. and Tengerdy, R.P. (1985) The effect of vitamin E on prostaglandin levels in the immune organs of chicks during the course of an *E. coli* infection. *Nutrition Research* 5, 497–502.

LeVan, L.W., Schnoes, H.K. and DeLuca, H.F. (1981) Isolation and identification of 25-hydroxy-vitamin D2 25-glucuronide: a biliary metabolite of vitamin D2 in the chick. *Biochemistry* 14, 1250–1256.

Lowenstine, L.J. (1986) Nutritional disorders of birds. In: Fowler, M.E. (ed.) *Zoo and Wild Animal Medicine.* W.B. Saunders, Philadelphia, pp. 201–212.

McDowell, L.R. (1989) *Vitamins in Animal and Human Nutrition.* Academic Press, San Diego.

McIlroy, S.G., Goodall, E.A., Rice, D.A., McNulty, M.S. and Kennedy, D.G. (1993) Improved performance in commercial broiler flocks with subclinical infectious bursal disease when fed diets containing increased concentrations of vitamin E. *Avian Pathology* 22, 81–94.

McKee, J.S. and Harrison, P.C. (1995) Effects of supplemental ascorbic acid on the performance of broiler chickens exposed to multiple concurrent stressors. *Poultry Science* 74, 1772–1785.

Mainka, S.A., Cooper, R.M., Black, S.R. and Dierenfeld, E.S. (1992) Serum alpha-tocopherol in captive and free-ranging raptors. *Journal of Zoo and Wildlife Medicine* 23, 72–76.

Mainka, S.A., Dierenfeld, E.S., Cooper, R.M. and Black, S.R. (1994) Circulating alpha-tocopherol following intramuscular or oral vitamin-E administration in Swainsons hawks (*Buteo swainsonii*). *Journal of Zoo and Wildlife Medicine* 25, 229–232.

March, B.E. and MacMillan, C. (1979) Trimethylamine in the caeca and small intestine as a cause of fishy taints in eggs. *Poultry Science* 58, 93–97.

Masse, P.G., Vuilleumier, J.P. and Weiser, H. (1991) Aspartate aminotransferase activity in experimentally induced asymptomatic vitamin-B6 deficiency in chicks. *Annals of Nutrition and Metabolism* 35, 25–33.

Masse, P.G., Rimnac, C.M., Yamauchi, M., Coburn, S.P., Rucker, R.B., Howell, D.S. and Boskey, A.L. (1996) Pyridoxine deficiency affects biomechanical properties of chick tibial bone. *Bone* 18, 567–574.

Maxwell, M.H., Whitehead, C.C. and Armstrong, J. (1988) Haematological and tissue abnormalities in chicks caused by acute and subclinical folate deficiency. *British Journal of Nutrition* 59, 73–80.

Millar, R.I., Smith, L.T. and Wood, J.H. (1977) The study of the dietary vitamins D requirements of ringnecked pheasant chicks. *Poultry Science* 56, 1739 (abstract).

Miller , M.S., Buss, E.G. and White, H.B. (1981) Thiamin deposition in eggs is not dependent on riboflavin-binding protein. *Biochemistry Journal* 198, 225–226.

Murata, K. (1982) Actions of two types of thiaminase on thiamin and its analogues. *Annals New York Academy of Science* 378, 146–157.

Murphy, M.E. and King, J.R. (1982) Semi-synthetic diets as a tool for nutritional ecology. *Auk* 99, 165–167.

Naber, E.C. (1993) Modifying vitamin composition of eggs: a review. *Journal of Applied Poultry Research* 2, 385–393.

Naber, E.C. and Squires, M.W. (1993a) Vitamin profiles of eggs as indicators of nutritional status in the laying hen – diet to egg transfer and commercial flock survey. *Poultry Science* 72, 1046–1053.

Naber, E.C. and Squires, M.W. (1993b) Early detection of the absence of a vitamin premix in layer diets by egg albumen riboflavin analysis. *Poultry Science* 72, 1989–1993.

Narbaitz, R. (1987) Tole vitamin D in the development of the chick embryo. *Journal of Experimental Zoology* 1, 15–22.

Nelson, T.S. and Norris, L.C. (1960) Studies on the vitamin K requirement of the chick. *Journal of Nutrition* 73, 135–141.

Nestler, R.B., Derby, J.V. and DeWitt, J.B. (1948) Storage by bobwhite quail of vitamin A fed in various forms. *Journal of Nutrition* 36, 323–329.

Nestler, R.B., DeWitt, J.B. and Derby, J.V. (1949) Vitamin A storage in wild quail and its possible significance. *Journal of Wildlife Management* 13, 265–271.

Nichols, D.K. and Montali, R.J. (1987) Vitamin E deficiency in captive and wild piscivorous birds. *Proceedings Annual Meeting American Association of Zoo Veterinarians*, 186–190.

Nichols, D.K., Wolff, M.J., Phillips, L.G. and Montali, R.J. (1989) Coagulopathy in pink-backed pelicans (*Pelecanus rufescens*) associated with hypervitaminosis E. *Journal of Zoo and Wildlife Medicine* 20, 57–61.

Norman, A.W. (1990) The avian as an animal model for the study of the vitamin-D endocrine system. *Journal of Experimental Zoology*, Suppl. 4, 37–45.

Norman, A.W. and Hurwitz, S. (1993) The role of the vitamin D endocrine system in avian bone biology. *Journal of Nutrition* 123, 310–316.

NRC (1994) *Nutrient Requirements of Poultry.* National Academy Press, Washington, DC.

Oduho, G.W. and Baker, D.H. (1993) Quantitative efficacy of niacin sources for chicks: nicotinic acid, nicotinamide, NAD and tryptophan. *Journal of Nutrition* 123, 2201–2206.

Panda, S.K. and Rao, A.T. (1994) Effect of a vitamin E selenium combination on chickens infected with infectious bursal disease virus. *Veterinary Record* 134, 242–243.

Pardue, S.L. and Thaxton, J.P. (1986) Ascorbic acid in poultry: a review. *World's Poultry Science* 42, 107–123.

Paton, D.C. and Dorward, D.F. (1983) Thiamin deficiency and winter mortality in red wattlebirds, *Anthochaera carunculata* (Aves: Melphagidae) in suburban Melbourne. *Australian Journal of Zoology* 31, 147–154.

Pellett, L.J., Andersen, H.J., Chen, H. and Tappel, A.L. (1994) Beta-carotene alters vitamin E protection against heme protein oxidation and lipid peroxidation in chicken liver slices. *Journal of Nutritional Biochemistry* 5, 479–484.

Plimmer, R.H. and Rosedale, J.L. (1923) The rearing of chickens on the intensive system. Part IV. Vitamin C requirements of chickens and other birds. *Biochemistry Journal* 17, 787–797.

Rajkhowa, T.K., Katiyar, A.K. and Vegad, J.L. (1996) Effect of ascorbic acid on the inflammatory–reparative response in the punched wounds of the chicken skin. *Indian Journal of Animal Sciences* 66, 120–125.

Rennie, J.S., Whitehead, C.C. and Armstrong, J. (1993) Biochemical responses of broiler chicks to folate deficiency. *British Journal of Nutrition* 69, 801–808.

Romach, E.H., Kidao, S., Sanders, B.G. and Kline, K. (1993) Effects of RRR-alpha-tocopheryl succinate on IL-1 and PGE2 production by macrophages. *Nutrition and Cancer* 20, 205–214.

Rombout, J., Vanrens, B., Sijtsma, S.R., Vanderweide, M.C. and West, C.E. (1992) Effects of vitamin-A deficiency and Newcastle disease virus infection on lymphocyte sub-populations in chicken blood. *Veterinary Immunology and Immunopathology* 31, 155–166.

Roudybush, T. (1996) Nutrition. In: Rosskopf, W. and Woerpel, R. (eds) *Diseases of Cage and Aviary Birds*. Williams & Wilkins, Baltimore, pp. 218–234.

Roy, R.N. and Guha, B.C. (1958a) Species difference in regard to the biosynthesis of ascorbic acid. *Nature* 182, 319–320.

Roy, R.N. and Guha, B.C. (1958b) Production of experimental scurvy in a bird species. *Nature* 182, 1689–1690.

Ryu, K.S., Pesti, G.M., Roberson, K.D., Edwards, H.M. and Eitenmiller, R.R. (1995) The folic acid requirements of starting broiler chicks fed diets based on practical ingredients. 2. Interrelationships with dietary methionine. *Poultry Science* 74, 1456–1462.

St Claire, M.C., Lidl, G.M., Bermudez, A.J., Franklin, C.L. and Beschwilliford, C. (1994) Vitamin B-responsive cystic enteritis in a pigeon colony. *Laboratory Animal Science* 44, 79–81.

Schneider, J. (1986) Vitamin stability and activity of fat-soluble vitamins as influenced by manufacturing processes. *Proceedings of the NFIA Nutrition Institute* 1–6.

Scott, M.L. and Dean, W.F. (1991) *Nutrition and Management of Ducks*. M.L. Scott of Ithaca, Ithaca, New York.

Scott, M.L., Holm, E.R. and Reynolds, R.E. (1964) Studies on the pantothenic acid require-ments of young ring-necked pheasants and bobwhite quail. *Poultry Science* 43, 676–682.

Scott, M.L., Nesheim, M.C. and Young, R.J. (1982) *Nutrition of the Chicken*, 3rd edn. M.L. Scott & Associates, Ithaca.

Serafin, J.A. (1974) Studies on the riboflavin, niacin, pantothenic acid and choline requirements of young bobwhite quail. *Poultry Science* 53, 1522–1532.

Serafin, J.A. (1981) Studies on the riboflavin, pantothenic acid, nicotinic acid, and choline requirements of young Embden geese. *Poultry Science* 60, 1910–1915.

Sheehy, P.J.A., Morrissey, P.A. and Flynn, A. (1991) Influence of dietary alpha-tocopherol on tocopherol concentrations in chick tissues. *British Poultry Science* 32, 391–397.

Sheehy, P.J.A., Morrissey, P.A. and Flynn, A. (1994) Consumption of thermally-oxidized sunflower oil by chicks reduces alpha-tocopherol status and increases susceptibility of tissues to lipid oxidation. *British Journal of Nutrition* 71, 53–65.

Shen, H., Summers, J.D. and Leeson, S. (1981) Egg production and shell quality of layers fed various levels of vitamin D. *Poultry Science* 60, 1485–1490.

Sklan, D., Yosefov, T. and Friedman, A. (1989) The effects of vitamin-A, beta-carotene and canthaxanthin on vitamin-A metabolism and immune responses in the chick. *International Journal for Vitamin and Nutrition Research* 59, 245–250.

Sklan, D., Melamed, D. and Friedman, A. (1994) The effect of varying levels of dietary vitamin-A on immune response in the chick. *Poultry Science* 73, 843–847.

Soares, J.H. (1995) Calcium bioavailability. In: Ammerman, C.B., Baker, D.H. and Lewis, A.J. (eds) *Bioavailability of Nutrients for Animals*. Academic Press, San Diego, pp. 95–118.

Southern, L.L. and Baker, D.H. (1981) Bioavailable pantothenic acid in cereal grains and soybean meal. *Journal of Animal Science* 53, 403–408.

Squires, M.W. and Naber, E.C. (1992) Vitamin profiles of eggs as indicators of nutritional status in the laying hen – vitamin-B12 study. *Poultry Science* 71, 2075–2082.

Squires, M.W. and Naber, E.C. (1993) Vitamin profiles of eggs as indicators of nutritional status in the laying hen – vitamin-A study. *Poultry Science* 72, 154–164.

Stokstad, E.L.R. and Jukes, T.H. (1987) Sulfonamides and folic acid antagonists: a historical review. *Journal of Nutrition* 117, 1335–1342.

Subramanian, N. and Adiga, P.R. (1995) Simultaneous purification of biotin-binding proteins-I and -II from chicken egg yolk and their characterization. *Biochemical Journal* 308, 573–577.

Suttie, J.W. (1980) The metabolic role of vitamin K. *Federation Proceedings* 39, 2730–2739.

Tengerdy, R.P., Lacetera, N.G. and Nockels, C.F. (1990) Effect of beta-carotene on disease protection and humoral immunity in chickens. *Avian Diseases* 34, 848–854.

Thompson, J.N., McHowell, G.A., Pitt, A.J. and Houghton, C.I. (1965) Biological activity of retinoic acid ester in the domestic fowl: production of vitamin A deficiency in the early chick embryo. *Nature* 205, 1006–1010.

Vericel, E., Budowski, P. and Crawford, M.A. (1991) Chick nutritional encephalomalacia and prostanoid formation. *Journal of Nutrition* 121, 966–969.

Vieira, A.V., Kuchler, K. and Schneider, W.J. (1995) Retinol in avian oogenesis – molecular properties of the carrier protein. *DNA and Cell Biology* 14, 403–410.

Wada, Y., Kondo, H. and Itakura, C. (1996) Peripheral neuropathy of dietary riboflavin deficiency in racing pigeons. *Journal of Veterinary Medicine and Science* 58, 161–163.

Walters, M.R. (1992) Newly identified actions of the vitamin D endocrine system. *Endocrine Reviews* 13, 719–764.

Ward, F.P. (1971) Thiamin deficiency in a peregrine falcon. *Journal of American Veterinary Medicine Association* 159, 599–601.

Ward, N.E., Jones, J. and Maurice, D.V. (1985) Influence of propionic acid for depleting laying hens and their progeny of vitamin B_{12}. *Nutrition Reports International* 32, 1325–1330.

Watkins, B.A. (1990) Dietary biotin effects on desaturation and elongation of C-14-linoleic acid in the chicken. *Nutrition Research* 10, 325–334.

Watkins, B.A., Bain, S.D. and Newbrey, J.W. (1989) Eicosanoic fatty acid reduction in the tibiotarsus of biotin-deficient chick. *Calcified Tissue International* 45, 41–46.

White, H.B. (1996) Sudden death of chicken embryos with hereditary riboflavin deficiency. *Journal of Nutrition* 126, S1303–S1307.

White, H.B. and Whitehead, C.C. (1987) Role of avidin and other biotin-binding proteins in the deposition of biotin in chicken eggs. *Biochemistry Journal* 241, 677–684.

White, H.B., Orth, W.H., Schreiber, R.W. and Whitehead, C.C. (1992) Availability of avidin-bound biotin to the chicken embryo. *Archives of Biochemistry and Biophysics* 298, 80–83.

Whitehead, C.C. (1984) Biotin intake and transfer to the egg and chick in broiler breeder hens on litter or in cages. *British Poultry Science* 25, 287–292.

Whitehead, C.C. and Siller, W.G. (1983) Experimentally induced fatty liver and kidney syndrome in the young turkey. *Research in Veterinary Science* 34, 73–80.

Will, B.H., Usui, Y. and Suttie, J.W. (1992) Comparative metabolism and requirement of vitamin-K in chicks and rats. *Journal of Nutrition* 122, 2354–2360.

Wilson, H.R. (1989) Chick mortality in bobwhite quail as affected by supplemental ascorbic acid. *Poultry Science* 68, 1418–1420.

Wilson, H.R. (1997) Effects of maternal nutrition on hatchability. *Poultry Science* 76, 134–143.

Wilson, S. and Duff, S.R.I. (1991) Effects of vitamin or mineral deficiency on the morphology of medullary bone in laying hens. *Research in Veterinary Science* 50, 216–221.

Wobeser, G. and Kost, W. (1992) Starvation, staphylococcosis, and vitamin A deficiency among mallards overwintering in Saskatchewan. *Journal of Wildlife Diseases* 28, 215–222.

Xu, T., Leach, R.M., Hollis, B. and Soares, J.H. (1997) Evidence of increased cholecalciferol requirement in chicks with tibial dyschondroplasia. *Poultry Science* 75, 47–53.

Yen, J.T., Jensen, A.H. and Baker, D.H. (1976) Assessment of the concentration of biologically available vitamin B6 in corn and soybean meal. *Journal of Animal Science* 45, 269–274.

Zwaan, J. and Lam, K.W. (1992) Comparison of ascorbic acid levels in the eye and remainder of the chicken embryo during development. *Experimental Eye Research* 54, 411–413.

APPENDIX

Common English names and corresponding genus and species names of birds referred to in the text, as listed in Clements (1991).

English name	Genus and species
Adelie Penguin	*Pygoscelis adeliae*
Alexandrine Parakeet	*Psittacula eupatria*
American Bittern	*Botaurus lentiginosus*
American Black Duck	*Anas rubripes*
American Kestrel	*Falco sparverius*
American Robin	*Turdus migratorius*
Anna's Hummingbird	*Calypte anna*
Asian Koel	*Eudynamys scolopacea*
Asian Paradise-Flycatcher	*Terpsiphone paradisi*
Asian Pied Starling	*Sturnus contra*
Atlantic Puffin	*Fratercula artica*
Bald Eagle	*Haliaeetus leucocephalus*
Bank Myna	*Acridotheres ginginianus*
Bar-headed Goose	*Anser indicus*
Barnacle Goose	*Branta leucopsis*
Barn Owl	*Tyto alba*
Barn Swallow	*Hirundo rustica*
Bar-tailed Godwit	*Limosa lapponica*
Bay-backed Shrike	*Lanius vittatus*
Belted Kingfisher	*Magaceryl alcyon*
Black Brant	*Branta bernicla nigricans*
Black-crowned Night Heron	*Nycticorax nycticorax*
Black-headed Munia	*Lonchura malacca*
Black-hooded Oriole	*Oriolus xanthornus*
Black-rumped Flameback	*Dinopium benghalense*
Black Vulture	*Coragyps atratus*
Blue-capped Rock-Thrush	*Monticola cinclorhynchus*
Blue Grouse	*Dendragapus obscurus*
Blue-throated Hummingbird	*Lampornis clemenciae*

English name	Genus and species
Blue Tit	*Parus caeruleus*
Boat-tailed Grackle	*Quiscalus major*
Brant Goose (Atlantic)	*Branta bernicla*
Brolga (Australian Crane)	*Grus rubicunda*
Budgerigar	*Melopsittacus undulatus*
Buffy Fish Owl	*Ketupa ketupa*
Bullfinch (Eurasian)	*Pyrrhula pyrrhula*
Canada Goose	*Branta canadensis*
Canary	*Serinus canarius*
Canvasback	*Aythya valisineria*
Cape Gannet	*Morus capensis*
Cape Griffon	*Gyps coprotheres*
Cardinal (Northern)	*Cardinalis cardinalis*
Caribbean Flamingo	*Phoenicopterus ruber ruber*
Carmin Bee-eater	*Merops nubicus*
Cattle Egret	*Bubulcus ibis*
Cedar Waxwing	*Bombycilla cedrorum*
Chestnut-mandibled Toucan	*Ramphastos ambiguus*
Chicken, domestic	*Gallus gallus*
Chuckar Partridge	*Alectoris graeca*
Clamorous Reed-Warbler	*Acrocephalus stentoreus*
Cockatiel	*Nymphicus hollandicus*
Common Buzzard	*Buteo buteo*
Common Eider	*Somateria mollissima*
Common Iora	*Aegithina tiphia*
Common Murre	*Uria aalge*
Common Myna	*Acridotheres tristis*
Common Pheasant	*Phasianus colchicus*
Common Pochard	*Aythya ferina*
Common Quail	*Coturnix coturnix*
Common Redpoll	*Carduelis flammea*
Common Starling	*Sturnus vulgaris*
Coot (Euroasian)	*Fulica atra*
Copper Pheasant	*Syrmaticus soemmerringii*
Costa's Hummingbird	*Calypte costae*
Crimson Sunbird	*Aethopyga siparaja*
Crow (American)	*Corvus brachyrhynchos*
Dark-eyed Junco	*Junco hyemalis*
Darwin's Rhea	*Pterocmia pennata*
Double-crested Cormorant	*Phalacrocorax auritus*
Eared Dove	*Zenaida auriculata*

continued overleaf

English name	Genus and species
Elegant Crested Tinamou	*Eudroma elegans*
Emperor Penguin	*Aptenodytes forsteri*
Emu	*Dromaius novaehollandiae*
Eurasian Oystercatcher	*Haematopus ostralegus*
European Bee-eater	*Merops apiaster*
European Goldfinch	*Carduelis carduelis*
European Robin	*Erithacus rubecula*
Evening Grosbeak	*Hesperiphona vespertina*
Gambel's Quail	*Callipepla gambelii*
Glaucous-winged Gull	*Larus glaucescens*
Golden-crowned Kinglet	*Regulus satrapa*
Golden-winged Warbler	*Vermivora chysoptera*
Gouldian Finch	*Poephlia gouldii*
Grasshopper Sparrow	*Ammodramus savannarum*
Great Blue Heron	*Ardea herodias*
Greater Flamingo	*Phoenicopterus ruber roseus*
Greater Rhea	*Rhea americana*
Great Horned Owl	*Bubo virginianus*
Great Tit	*Parus major*
Greenfinch (European)	*Carduelis chloris*
Grey-lag Goose	*Anser anser*
Grey Parrot (African)	*Psittacus erithacus*
Grey Partridge	*Perdix perdix*
Herring Gull	*Larus argentatus*
Hill Myna	*Gracula religiosa*
Himalayan Bulbul	*Pycnonotus leucogenys*
Hoatzin	*Opisthocomus hoatzin*
Horned Lark	*Eremophila alpestris*
House Crow	*Corvus splendens*
House Finch	*Carpodacus mexicanus*
House Martin	*Delichon urbica*
House Sparrow	*Passer domesticus*
Imperial Eagle	*Aquila heliaca*
Indian Scops-Owl	*Otus bakkamoena*
Indigo Bunting	*Passerina cyanea*
Jackass Penguin	*Spheniscus demersus*
Japanese Quail	*Coturnix japonica*
Kakapo	*Strigops habroptilus*
Keel-billed Toucan	*Ramphstos sulfuratus*
Kestrel (Eurasian)	*Falco tinnunculus*
King Penguin	*Aptenodytes patagonicus*
Kiwi (Brown)	*Apteryx australis*

English name	Genus and species
Laggar Falcon	*Falco jugger*
Lapwing (Northern)	*Vanellus vanellus*
Laughing Kookaburra	*Dacelo novaeguineae*
Leach's Storm-Petrel	*Oceanodroma leucorhoa*
Lesser Flamingo	*Phoenicopterus minor*
Lesser Snow Goose	*Ansercaerulescens caerulescens*
Limpkin	*Aramus guarauna*
Little Egret	*Egretta garzetta*
Little Owl	*Athene noctua*
Little Penguin	*Eudyptula minor*
Loggerhead Shrike	*Lanius ludovicianus*
Long-eared Owl	*Asio otus*
Long-tailed Shrike	*Lanius schach tricolor*
Mallard	*Anas platyrhynchos*
Maned Duck	*Chenonetta jubata*
Masked Finch	*Poephila personata*
Mistle Thrush	*Turdus viscivorus*
Mistletoebird	*Dicaeum hirundinaceum*
Muscovy Duck	*Cairina moschata*
Nduk Eagle Owl	*Bubo vasseleri*
New Holland Honeyeater	*Phylidonyris novaehollandiae*
Northern Bobwhite Quail	*Colinus virginianus*
Northern Shrike	*Lanius excubitor*
Oilbird	*Steatornis caripensis*
Orange Dove	*Ptilinopus victor*
Oriental Magpie-Robin	*Copsychus saularis*
Osprey	*Pandion haliaetus*
Ostrich	*Struthio camelus*
Pale-billed Flowerpecker	*Dicaeum erythrorhynchos*
Pelican, Brown	*Pelecanus occidentalis*
Pelican, Dalmatian	*Pelecanus crispus*
Pelican, Pink-backed	*Pelecanus rufoccens*
Peregrine Falcon	*Falco peregrinus*
Phainopepla	*Phainopepla nitens*
Pigeon, domestic	*Columba livia*
Pileated Woodpecker	*Dryocopus pileatus*
Pintail Duck	*Anas acuta*
Pygmy Parrot	*Micropsitta bruijnii*
Rainbow Lorikeet	*Trichoglossus haematodus*
Red-billed Quelea	*Quelea quelea*
Red Crossbill	*Luxia curvivostra*

continued overleaf

English name	Genus and species
Red Jungle Fowl	*Gallus gallus*
Red Knot	*Calidris canutus*
Red-necked Stint	*Calidris ruficollis*
Redpoll (Common)	*Carduelis flammea*
Redshank (Common)	*Tringa totanus*
Red-tailed Hawk	*Buteo jamaicensis*
Red-vented Bulbul	*Pycnonotus cafer*
Red Wattlebird	*Anthochaera carunculata*
Red-whiskered Bulbul	*Pycnonotus jocosus*
Resplendent Quetzel	*Pharomacrus mocino*
Ring-necked Pheasant (Common)	*Phasianus colchicus*
Rock Dove	*Columba livia*
Rock Ptarmigan	*Lagopus mutus*
Roseate Spoonbill	*Ajaia ajaja*
Ruddy Duck	*Oxyura jamaicensis*
Ruddy Turnstone	*Arenaria interpres*
Ruffed Grouse	*Bonasa umbellus*
Rufous Hummingbird	*Selasphorus rufus*
Rufous Treepie	*Dendrocitta vagabunda*
Saddle-billed Stork	*Ephippiorhynchus senegalensis*
Sage Grouse	*Centrocercus urophasianus*
Salmon-crested Cockatoo	*Cocatua moluccensis*
Sanderling	*Calidris alba*
Savannah Sparrow	*Passerculus sandwichensis*
Scarlet Ibis	*Eudocimas rubra*
Scarlet Minivet	*Pericrocotus flammeus*
Scarlet Tanager	*Piranga olivacea*
Screech Owl (Western)	*Otus kennicottii*
Shag	*Phalacrocorax aristotelis*
Sharp-beaked Ground Finch	*Geospiza difficilis*
Snail Kite	*Rostrhamus sociabilis*
Snow Goose	*Anser caerulescens*
Song Sparrow	*Melospiza melodia*
Spot-breasted Fantail-Flycatcher	*Rhipidura albogularis*
Spur-winged Goose	*Plectropterus gambensis*
Sword-billed Hummingbird	*Ensifera ensefera*
Tawny Owl	*Strix aluco*
Toco Toucan	*Ramphastas toco*
Tooth-billed Bowerbird	*Scenopoeetes dentirostris*
Tree Sparrow	*Spizella arborea*
Tree Swallow	*Tachycineta bicolor*
Turkey, Wild or domestic	*Meleagris gallopavo*

English name	Genus and species
Turkey Vulture	*Cathartes aura*
Turtle Dove	*Streptopelia turtur*
Western Gull	*Larus occidentalis*
Whimbrel	*Numenius phaeopus*
White-browed Bulbul	*Pycnonotus luteolus*
White-chinned Petrel	*Procellaria aequinoctialis*
White-crowned Sparrow	*Zonotrichia leucophrys gambelii*
White-headed Munia	*Lonchura maja*
White Ibis	*Eudocimus albus*
White-naped Crane	*Crus vipio*
White-throated Kingfisher	*Halcyon smyrnensis*
White throated Sparrow	*Zonotrichia albicollis*
White Wagtail	*Motacilla alba*
Willow Ptarmigan (Alaskan)	*Lagopus lagopus scoticus*
Wilson's Storm-Petrel	*Oceanites oceanicus*
Wood Duck	*Aix sponsa*
Wood Pigeon	*Columba palumbus*
Yellow-eyed Junco	*Junco phaeonotus*
Yellow-headed Blackbird	*Xanthocephalus xanthocephalus*
Yellow-rumped Warbler	*Dendroica coronata*
Zebra Finch	*Taeniopygia guttata*

Reference

Clements, J.F. (1991) *Birds of the World: a Check List.* Ibis Publishing Company, Vista, California.

Index

Note: page numbers in *italics* refer to figures and tables